Volume 33

Ceramic Transactions

MATERIALS IN MICROELECTRONIC AND OPTOELECTRONIC PACKAGING

Edited by Hung C. Ling
AT&T Bell Laboratories
Koichi Niwa
Fujitsu Laboratories
Vishwa N. Shukla
Texas Instruments, Inc.

The American Ceramic Society
Westerville, Ohio

Proceedings of the International Symposium on Materials for Optoelectronic and Microelectronic Packaging, presented at the Third International Ceramic Science and Technology Congress, held in San Francisco, CA, November 1-4, 1992.

Library of Congress Cataloging-in-Publication Data

Materials in microelectronic and optoelectronic packaging / edited by
Hung C. Ling, Koichi Niwa, Vishwa N. Shukla
p. cm. -- (Ceramic transactions, ISSN 1042-1122 ; v. 33)
"Proceedings of the International Symposium on Materials for Optoelectronic and Microelectronic Packaging, presented at the Third International Ceramic Science and Technology Congress, held in San Francisco, CA, November 1-4, 1992"--CIP t.p. verso.
Includes index.
ISBN 0-944904-63-7
1. Microelectronic packaging--Materials. 2. Optoelectronic devices--Materials. I. Ling, Hung C. II. Niwa, Koichi. III. Shukla, Vishwa N. IV. International Symposium on Materials for Optoelectronic and Microelectronic Packaging (1992 : San Francisco, Calif.) VI. Series.
TK7874.M3437 1993
621.381'046--dc20 93-11050
CIP

Printed in the United States of America.

1 2 3 4–97 96 95 94 93

ISSN 1042-1122
ISBN 0-944904-63-7

Preface

The area of microelectronics has seen tremendous growth in the degree of integration, resulting in ever more functionality being compressed onto smaller and smaller real estate on both the silicon chips and on circuit boards. While VLSI and ULSI technologies continue to march forward, the bottleneck is now increasingly focused on the chip-to-chip communication. Microelectronic packaging has migrated from packaging of individual IC chips to multilayer multichip modules, incorporating many IC chips and other passive devices (capacitors, resistors, etc.) in a single package. As optical fiber is being increasingly deployed in the office and home environment, low-cost and robust packaging of optoelectronic devices becomes equally important. Frequently, different classes of materials (ceramics, metals, polymers) are used together to solve unique packaging problems. Many materials issues arise in these kinds of endeavors, resulting from differing systems requirements, materials compatibility concerns, and long-term reliability issues of new materials being developed. Increasingly, scientists and engineers of various backgrounds are working together to develop interdisciplinary systems solutions.

This volume contains a collection of papers presented at the International Symposium on Materials for Microelectronic and Optoelectronic Packaging, held at the Third Ceramic Science and Technology Congress, San Francisco, CA, November 1-4, 1992. This symposium was cosponsored by the Electronics Division of the American Ceramic Society and the Ceramic Society of Japan. Forty-three papers were presented at the conference, addressing major areas of optoelectronic packaging, microelectronic packaging, processing/structure/properties, characterization techniques, and thick film/metallization technology. Thirty-six papers were peer reviewed and are included in this volume.

We acknowledge and appreciate the contributions from the speakers, session chairs, manuscipt reviewers, and Society officials to make both the symposium and the proceedings successes.

Hung C. Ling
Koichi Niwa
Vishwa N. Shukla

Contents

Section I. Optoelectronic Packaging

Section II. Microelectronic Packaging

Section III. Processing, Structure, and Properties

Section IV. Characterization Techniques

Section V. Thick Film/Metallizaton Technology

Section I. Optoelectronic Packaging

OPTICAL COMPONENTS AND PACKAGING TECHNIQUES FOR FIBER OPTIC COMMUNICATION SYSTEMS

Kuniharu Kato
NTT Opto-electronics Laboratories, Tokai-Mura, Ibaraki-Ken, 319-11 Japan

ABSTRACT

Optical components and packaging techniques for fiber optic communication systems are reviewed. The former can be roughly classified into three types; optical fiber type, discrete type and planar- lightwave-circuit (PLC) type components.

I. INTRODUCTION

Fiber optic communication systems are being widely used in communication networks such as long-haul transmission networks, subscriber networks in order to realize fiber-to-the-home (FTTH) or fiber-to-the-curb (FTTC) and local area networks (LANs) [1]-[3]. Optical components for the FTTH/FTTC network must be inexpensive, highly stable and exhibit low-loss. One of the most important problems in designing and fabricating these components is to achieve efficient, easy and reliable coupling with single-mode fibers since they need highly accurate adustment less than a few microns. In order to realize these objectives, a number of studies and developments have been made.

In this paper, Section II describes the main functions of the optical components required in FTTH/FTTC networks. Section III presents key technologies and a structural classification of the optical components. Sections IV through VI review three types of optical component; the optical-fiber type, the discrete type and the planar-lightwave-circuit (PLC) type and packaging techniques. PLC-type components generally incorporate on all planar lightwave circuits fabricated from such materials as InP, GaAs (semiconductor), Silica glass, or Lithium Niobate. In this paper, the focus will be on silica-based waveguide devices.

II. OPTICAL FUNCTIONS IN FTTH NETWORK

Figure 1 shows the main optical functions required in FTTH/FTTC networks. These functions include optical connections between optical fibers, optical branching for distributing and monitoring optical signals, optical switching for flexible network utilization, optical amplification for optical signals and optical multiplexing for increasing transmission capacity. Optical multiplexing is generally classified into frequency division multiplexing (FDM) with wavelength spacings of less than about 1 nm and wavelength division multiplexing (WDM) with wider wavelength spacings.

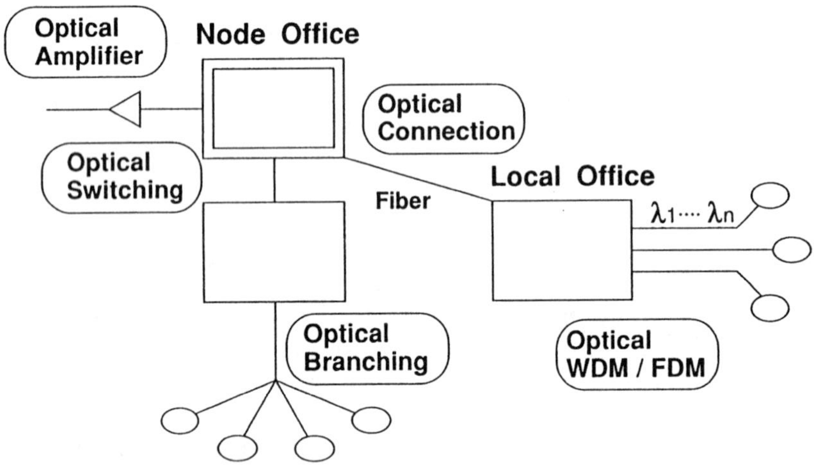

Figure 1 Optical functions in FTTH/FTTC network.

III. KEY TECHNOLOGIES AND STRUCTURAL CLASSIFICATION

In order to realize the various optical components, three key technologies are needed, that is, fabrication technology for fabricating materials and optical devices, design and packaging technologies. Packaging technology includes several techniques which include optical coupling, fiber attachment, the mechanical mounting of optical devices, cooling, sealing, and electrical connection.

Optical components can be classified into three types; optical fiber type, discrete type and PLC type. **Figure 2** shows an optical coupler as an example. Optical fiber type couplers are generally fabricated by fusing and heating fibers and they exhibit low loss. Discrete-type couplers are composed of lenses and prisms and require many parts. PLC type couplers are fabricated by glass deposition and patternning techniques and have the potential for large-scale integration. PLC technology offers ease of optical integration without the need for alignment between optical device elements on a PLC. Practical optical components used in fiber optic communication systems have at least one or two optical fibers for input or output optical signals and their structure must usually be complex in order to achieve highly functional performance.

Type	Structure (Ex. 1x2 coupler)	Features
Optical Fiber	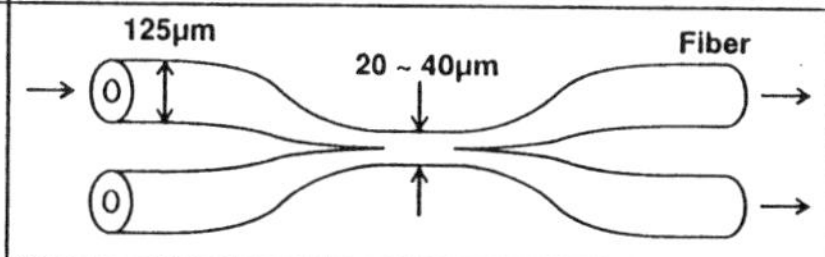	• Fusing , Heating • Low Loss
Discrete	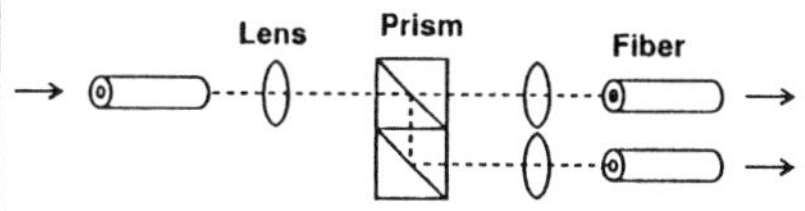	• Many Discrete Components • Many Parts
P L C*	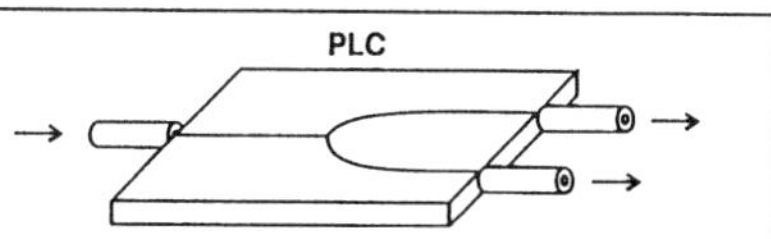	• Patterning • Integration

* Planar Lightwave Circuit

Figure 2 Structural classification of optical components.

IV. OPTICAL FIBER TYPE COMPONENTS

(1) Optical Connector

Optical connectors are the most important components for making connections between optical fibers in optical fiber networks. To date, several types[4]-[7] of optical connectors have been developed for use with single-mode fiber.

Figure 3 shows the SC connector series which features (a) low loss and high return loss through the use of ceramic ferrule and physical contact techniques and (b) compactness and durability through the use of a push-pull coupling mechanism. COSMO connectors have been newly developed to realize extremely high packing density[8]. At present, optical connectors are manually assembled. It is necessary to simplify and automate the assembly processes in order to reduce the assembly cost which currently accounts for half the total cost of the connector.

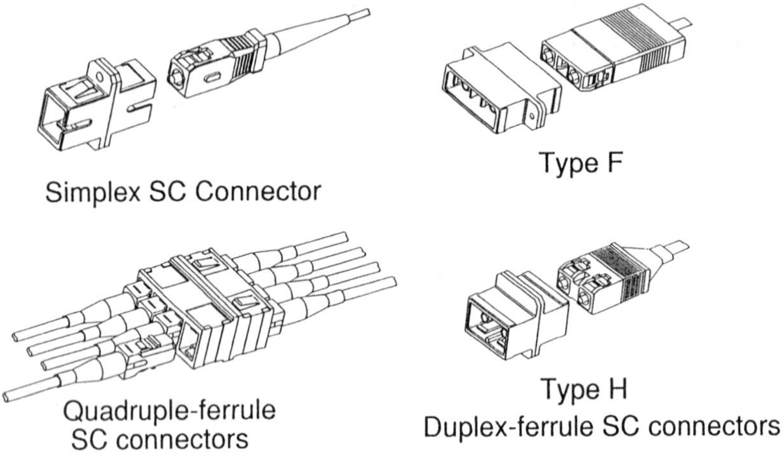

Figure 3 SC connector series.

(2) Optical Fiber Coupler and NxN Star Coupler

Optical fiber couplers and NxN star couplers are important components in the distribution and monitoring of optical signals.

Optical fiber couplers (2x2) are generally fabricated by one of two methods. One is a mechanical-polishing method[9], in which the fiber coupler is fabricated by polishing each fiber and placing the two polishing faces in contact. The other is a fusion-elongation method[10], in which the coupler is fabricated by fusing and elongating two optical fibers to form an optical coupling region. At present, the latter method is currently the most popular because of its relative ease.

Fiber-type NxN star couplers are also fabricated by one of two methods. One involves the combination of 2x2 optical fiber couplers. With the other the coupler is fabricated by simultaneously fusing many optical fibers[11]. In order to obtain a large-scale NxN star coupler, a very large number of 2x2 fiber couplers must be used in the former configuration. The latter can be compact, however, it is difficult to use it to fabricate large-scale NxN optical star couplers. When these fiber couplers are packaged in a case, careful attention is required in order to avoid the occurrence of stress in the fused and tapered fiber region because of its mechanical weakness and sensitivity.

(3)Optical Fiber Switch

Moving optical fiber switches[12]-[14] are inherently suitable for practical use in terms of their loss, size, driving power and cost. A 1x2 latching-type optical fiber switch[14] is shown in **Fig.4**. This switch is composed of a movable fiber covered with a magnetic alloy tube and two stationary fibers fixed in V-grooves engraved on half-cylindrical ceramic ferrules. The movable fiber can be switched by feeding a positive or negative pulse current to the solenoid coil and maintained without a holding current. This 1x2 switch exibits a low loss of 0.34 dB, a high return loss of 49 dB and stable operation[15]. The movable and stationary fibers are optically aligned automatically in the V-grooves. Therefore, precise alignment is not required in the switch assembly process. All parts of the switch can be economically mass-produced.

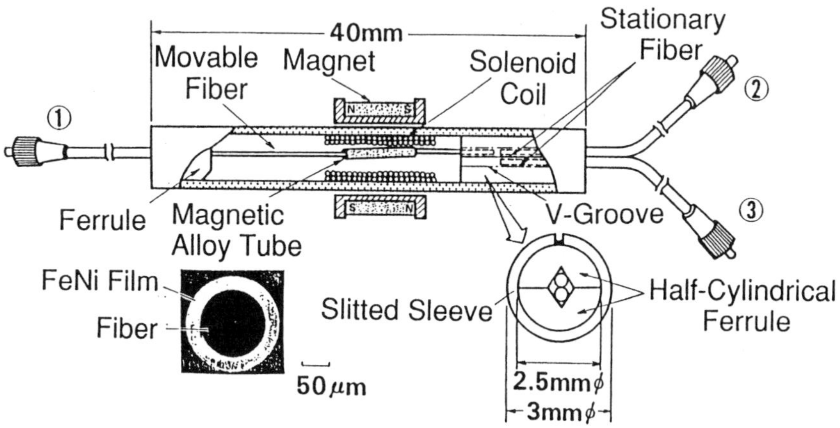

Figure 4 1x2 latching-type optical fiber switch.

(4) Optical Fiber Amplifier Module

An Erbium-doped optical fiber amplifier (EDFA) module[16] consists of an LD sub-module as a pumping light source, a fiber coupler, an Er-doped fiber and a polarization insensitive optical isolator. Each component is connected by a fusion splice technique. The Er-doped fiber which is 37 m long is coiled around a small bobbin.

This module is a good example of a module composed of optical fiber type and discrete type components. In terms of ease manufacture, however, there is a disadvantage in that it is difficult to automate the assembly process. This is because the assembly requires interconnections to be made between optical fibers which involves handling the components and their fiber ends.

V. DISCRETE TYPE COMPONENTS

(1) LD Module

Many types of LD module have been developed[17]-[19]. Among them, one which incorporates a molded aspheric glass lens holds the promise of low loss and low cost. **Figure 5** shows a cross sectional view of this module[18]. The one lens coupling system reduces the number of components and simplifies the assembly process in comparison with two-lens or three-lens systems. The assembly process is as follows. First, the LD package is adjusted in the two directions perpendicular to the optical axis. After the adjustment, the package is bonded to the case by resistance welding. Second, the single-mode (SM) fiber is optically adjusted in three planes and then the metal ferrule is bonded to the metal holder with a YAG laser welding. Finally, the SM fiber attached to

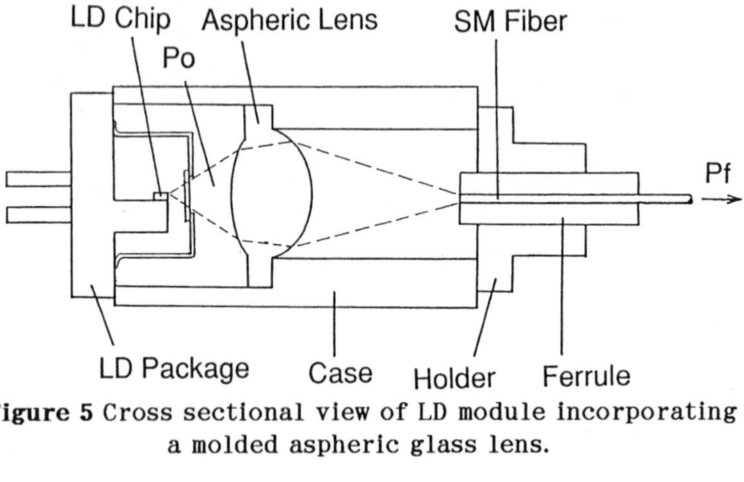

Figure 5 Cross sectional view of LD module incorporating a molded aspheric glass lens.

the holder is adjusted and the holder is bonded to the case with YAG laser welding. An average coupling loss of 2.2 dB is attained by using 1.3-μm LD packages. Automatic alignment apparatus is available for assembling these LD modules.

(2) WDM Transceiver Module

Wavelength division multiplexing (WDM) transceiver modules are important components of WDM systems, and these modules have been intensively studied[20]-[22]. **Figure 6** shows the configuration of a WDM transceiver module. LD and APD packages are integrated on an alumina ceramic substrate together with a fiber package and a glass block[22]. These packages are adjusted beforehand to bring the optical axes of the ball lens and LD (or APD) chip into line with the center of the package holder. WDM transceiver modules assembly techniques[22][23] have been proposed for multi and single mode fibers. The assembly for a single-mode fiber is carried out by utilizing two jigs[22] with precisely formed V-grooves which accurately position the module components. The LD package is set in a V-groove on one jig and the other packages and the glass block are set in V-grooves on the other jig. This technique is adjustment-free with the exception of the optical alignment between the LD and the single-mode fiber packages. These components are soldered to the ceramic substrate while their positional accuracies are maintained by the jigs. Importantly, the

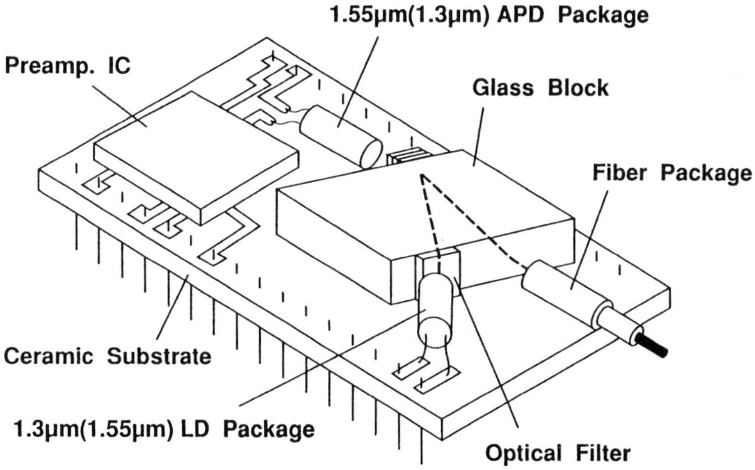

Figure 6 Configuration of WDM transceiver module.

cost of the jigs does not affect the module cost since a large number of the modules can be fabricated using the two jigs.

VI. PLC TYPE COMPONENTS

Silica-based PLC has many attractive features such as low propagation loss, compatibility with optical fiber, potential for large-scale integration and durability in the natural environment. Therefore, several studies have been made on fabricating functional PLC components[24]-[33].

(1) Optical Splitter and Star Coupler Modules

Figure 7 shows an 1x8 splitter module based on Y-branch elements[30]. The PLC chip and fibers are fixed in glass frames[34]. In an output-fiber frame, eight single-mode fibers are arrayed in V-grooves formed on a glass plate with spacings of 250 μm. In an input-fiber frame, one fiber is fixed in a V-groove. Both glass frames are adjusted and connected to the PLC sub-module with a newly developed UV curable epoxy adhesive. The humidity durability is improved by blending the adhesive with silane coupling agents[35]. The connection loss due to misalignment is less than 0.1 dB/point. In heat-cycle test from -40 to 85℃ and a high humidity test at 70℃/90%(RH), the loss increase is less than 0.2 dB[36].

To date, the fabrication and packaging of integrated-optic 19x19[37], 64x64 and 144x144[38] star couplers, first proposed by AT&T[39], have been reported. It consists of two arrays of channel waveguides separated by a slab waveguide region. **Figure 8** shows a packaged 144x144 star coupler[40]. Both ends of the star coupler chip are fixed in metal frames and their endfaces are optically polished. Eight single-mode fibers are arrayed on glass V-grooves in a fiber holder. The fiber alignment and attachment process is as follows. First, eight pairs of fiber-to-waveguide alignments are made simultaneously at the uppermost array. Second, the glass regions of the star coupler chip and the fiber holders are bonded together with UV curable adhesive to obtain a high return loss, and their metal frames are immediately welded together with a YAG laser. Each subsequent eight pairs of fiber-to-waveguide alignments are then made and then attached at the input and output its sides. Once all the interconnections have been made, the star coupler is fastened in its case with screws. This case is housed in a bookshelf-type unit. The 144x144 star coupler module exhibits an average insertion loss of 26.6 dB.

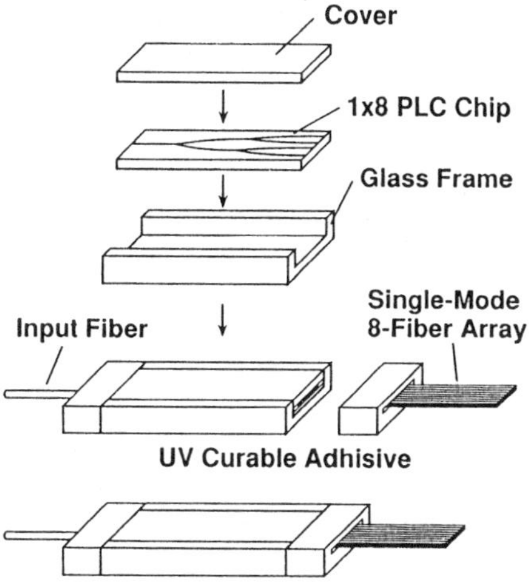

Figure 7 1x8 PLC splitter module.

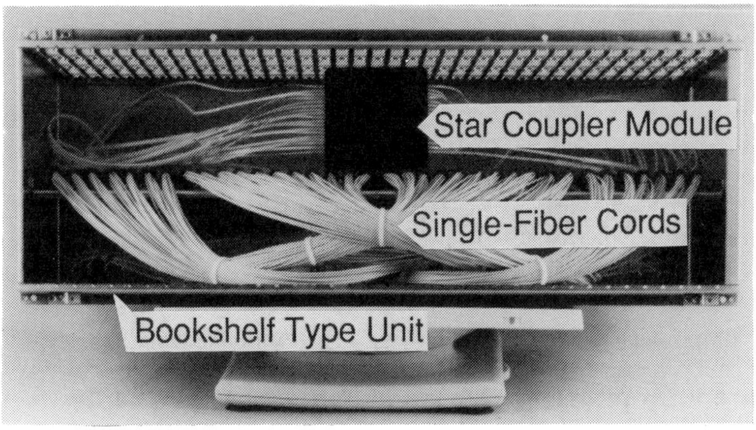

Figure 8 Photograph of packaged 144x144 star coupler.

(2) Thermo-Optic 8x8 Matrix Switch

Multiple switch elements can be integrated on a single substrate to form a large-scale optical matrix switch. A thermo-optic 8x8 matrix switch[42] consists of 64 active and 48 inactive switch elements. Fifteen switch element stages are arranged along the 35 cm long "serpentine" waveguide array on a

52x53 mm^2 Si substrate. Printed boards which supply electric power are attached to a heat-sink substrate. The average insertion loss of the fabricated switch is 3.8 dB[41]. The required switching power is 0.5 W and the switching response time is 1.3 msec.

(3) PLC-Type WDM Transceiver

Figure 9 shows the structure of a PLC-type WDM transceiver module[43]. It consists of three functional sub-modules, an LD sub-module, a PLC sub-module for multi/demultiplexing and a PD sub-module with an optical fiber for the input and output signals. A fiber with a hemispherical end is used to couple the LD sub-module to the LD chip. The PD sub-module is butt-coupled to the PLC sub-module. The sub-modules are packaged in metal frames respectively and optically aligned and bonded by YAG laser welding. A prototype module exhibited LD-to-fiber and fiber-to-PD coupling losses of 6.2 dB and 2.5 dB, respectively. The module size is 40x20x5 mm^3.

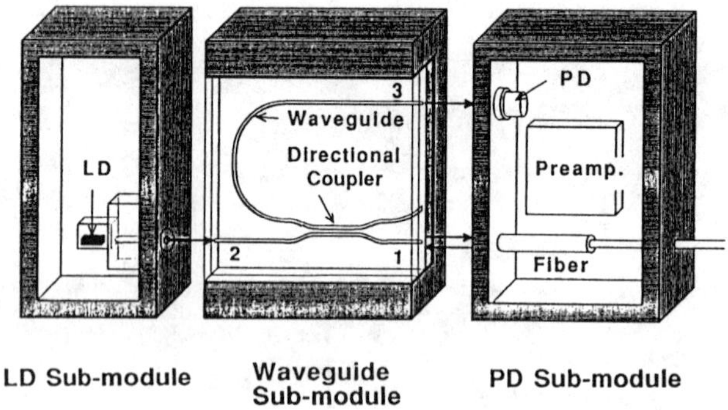

Figure 9 Structure of PLC-type WDM transceiver module.

VII. SUMMARY AND FUTURE SUBJECT

This paper discussed several optical components and packaging techniques, focusing particularly on optical coupling and fiber attachment techniques.

Optical components for single mode fibers require highly accurate adjustment of within a few microns when they are assembled. There have been many detailed studies on adjustment-free techniques which use precise V-grooves and reliable bonding techniques which use UV-curable adhesive and YAG laser welding, as reviewed in this paper. Optical integration using PLCs reduces the cost and size of components and adjustment points between optical device elements. In terms of integrated PLC packaging, simple, low loss, and highly reliable fiber attachment will continue to be the most important problems. There are two promising techniques for overcoming these problems. One is a mode-field conversion technique[44]-[46] which can reduce the coupling loss between different mode-field waveguides and which can expand the misalignment tolerance between optical waveguides. The other is a self-alignment technique[47]-[50] which can simplify the coupling process between fibers and waveguides. Further advances in these techniques will contribute greatly to the successful development of the three types of optical components described.

Acknowledgement

The author would like to thank all related members for discussion, and M.Nakahara and M.Kawachi for their encouragement.

REFERENCES

[1]P.Kaiser,"Single mode fiber technology for the subscriber loop,"in Proc. IOOC-ECOC'85, pp.125-131 (1985).
[2]T.Miki,"Fiber-optic subscriber networks and system development," IEICE Trans., E74, 1, pp.93-100 (1991).
[3]J.Minowa, N.Tokura, and K.Nosu,"Development of fiber-optic local area networks in Japan," J.Lightwave Technol., LT-3, 3, pp.438-448 (1985).
[4]W.C.Young,L.Curtis,T.L.Williford and P.Kaiser,"Design and performance of the biconic connector used in the FT3 lightwave system,"in Proc. 30th IWCS, pp.411-417 (1981).
[5]N.Suzuki, Y.Iwahara, M.Saruwatari, and K.Nawata,"Ceramic capillary connector for 1.3-μm single-mode fibers,"Electron Lett., 15, 25, pp.809-811 (1979)
[6]E.Sugita, R.Nagase, K.Kanayama, and T.Shintaku, "SC-type single-mode optical fiber connectors," J.Lightwave Technol., 7, 11, pp.1689-1696 (1989)
[7]A.W.Carlisle,"Small-size high-performance lightguide connector for LAN's," in Proc. OFC'85, TUQ18, pp.74-75 (1985).
[8]S.Iwano, R.Nagase, K.Kanayama, E.Sugita and Y.Ando,"Multi-port

optical backpanel connector featuring narrow diameter ferrules and self-retentive mechanism," in Proc. LEOS'91, E03.7, p.17 (1991)
[9]R.A.Berch, G.Kotler, and H.J.Shaw,"Single-mode fiber-optic directional coupler,"Electron. Lett., 16, pp.260-261 (1980).
[10]I.Yokohama, J.Noda, and K.Okamoto,"Fiber-coupler fabrication with automatic fusion-elongation process for low excess loss and high coupling-ratio accuracy," J.Lightwave Technol., LT-5, 7, pp.910-915 (1987).
[11]A.Eisenmann and E.Weidel,"Analysis of a fused biconical single-mode star coupler,"Optics Lett.,14,15,pp.814-816 (1989).
[12]C.M.Miller, R.B.Kummer, S.C.Mettler, and D.N.Ridgway,"Single-mode optical fiber switch,"Electronics Lett.,16,20,pp.783-784 (1980).
[13]K.Hogari and T.Matsumoto, "Electrically driven fiber-optic micromechanical on/off switch and its application to subscriber transmission systems,"J.Lightwave Technol., 8, 5, pp.722-727 (1990).
[14]S.Nagaoka,"Micro-magnetic alloy tubes for switching and splicing single-mode fibers," 4th IEEE Micro Mechanical Systems, pp.86-91 (1991).
[15]S.Nagaoka,"Low insertion loss and high return loss latching type 1x8 single-mode fiber switch,"1992 Fall Natl. Conv. IEICE Jpn., B-690 (1992). (in Japanese)
[16]M.Shimizu, M.Horiguchi, M.Yamada, I.Nishi, J.Noda, T.Takeshita, M.Okayasu, S.Uehara, and E.Sugita,"Compact and highly efficient fiber amplifier modules pumped by a 0.98-μm laser diode,"J.Lightwave Technol., 9, 2, pp.291-296 (1992).
[17]M.Saruwatari and T.Sugie,"Efficient laser diode to single-mode fiber coupling using a combination of two lenses in confocal condition,"J.Quantum Electron., QE-17, pp.1021-1027 (1981).
[18]K.Kawano,"Coupling characteristics of lens systems for laser diode modules using single-mode fiber,"Appl. Opt., 25, pp.2600-2605 (1986).
[19]K.Kato and I.Nishi,"Low-loss laser diode module using a molded aspheric glass lens,"IEEE Photon. Technol. Lett., 2, 7, pp.473-474 (1990).
[20]Y.Tamura and E.Nishimura,"Single mode fiber WDM module with transmitter and receiver,"in Proc. OEC'86, C2-3, pp.154-155 (1986).
[21]B.Hillerich,M.Rode and H.Gottsmann,"Single-mode fiber WDM unit for duplex subscriber link using a substrate with embossed alignment grooves,"J. Lightwave Technol.,7,11,pp.1654-1660 (1989).
[22]Y.Tachikawa, K.Kato, and T.Oguchi,"Two channel wavelength division multiplexing module using LD for single-mode fiber-optic

transmission system,"Paper Tech.,Group on Opt. Quan. Elec. IECE Japan, OQE86-194, pp.69-75 (1986). (in Japanese)
[23]K.Kato, Y.Tachikawa, T.Oguchi, Y.Fujii, and I.Nishi, "Three-channel WDM transceiver module assembled by adjustment-free process using aligned LED and PD ball-lens packages,"J. Lightwave Technol., 6, 4, pp.492-499 (1988).
[24]N.Takato, M.Yasu, and M.Kawachi,"Low loss high silica single mode channel waveguides,"Electron. Lett., 22, pp.321-322 (1986).
[25]B.H.Verbeek, C.H.Henry, N.A.Olsson, K.J.Orlowsky, R.F.Kazarinov, and B.H.Johnson,"Integrated four-channel Mach-Zehnder multi/demultiplexer fabricated with phosphorous doped SiO2 waveguides on Si,"J. Lightwave Technol.,6,6,pp.1011-1015 (1988).
[26]S.Valette, S.Renard, H.Denis, J.P.Jadot, A.Founier, P.Phlippe, P.Gidon, A.M.Grouillet, and E.Desgranges, "Si-based integrated optics technologies," Solid State Technol.,Feb.,pp.69-74 (1989).
[27]H.Yanagawa, T.Shimizu, S.Nakamura, and I.Ohyama, "Single-chip 32x32 star coupler with tapered fiber-coupling regions," in Proc., OFC'91, TuE3, p.15 (1991).
[28]K.Imoto,A.Takahiro, and H.Sano,"Silica glass waveguide structure and its application to a multi/demultiplexer," in Proc. ECOC'88,pp.577-580 (1988).
[29]J.W.Parker,"ESPRIT II OLVES - A European collaborative programme in optical interconnection," in Proc. PS'90,12B2 (1990).
[30]R.Kashyap, B.J.Ainslie, and G.D.Maxwell, "Second harmonic generation in GeO2 ridge waveguide,"Electron. Lett.,25,3,pp.208-209 (1989).
[31]G.Barbarossa and E.Mc.Goldrich,"Silica-based integrated optical hybrid on silicon for phase diversity coherent receiver," in Proc. LEOS'90, OE12.3/ThWW3 (1990).
[32]D.Welbourn,C.Beaumont and M.Nield,"Directional couplers in a high-index silica waveguide," in Proc. IPR'91, ThA3 (1991).
[33]T.Miyashita,"Fiber optic component technology in PIRI,"NTT Review, 1, pp.105-109 (1990).
[34]H.Suda, S.Kobayashi, and M.Yamaguchi,"Silica waveguide single-mode splitters and switches," IEEE Workshop on Passive Networks for the local loop, 5.2, London (1990).
[35]N.Murata and K.Nakamura,"UV-curable adhesive for optical communications," J. Adhesion, 35, pp.251-267 (1991).
[36]Y.Yamada, F.Hanawa, T.Kitoh, and T.Maruno,"Low-loss and stable fiber-to-waveguide connection utilizing UV curable adhesive,"IEEE Photon. Technol. Lett.,4,8,pp.906-908 (1992).

[37]H.M.Presby and C.A.Edwads,"Packaging of glass waveguide silicon devices,"Optical Engineering, 31, 1 pp.141-143 (1992).
[38]K.Okamoto, H.Okazaki, Y.Ohmori, and K.Kato,"Fabrication of integrated-optic NxN star coupler,"IEEE Photon. Technol. Lett., 4, 9, pp.1032-1035 (1992).
[39]C.Dragone, C.H.Henry, I.P.Kaminow, and R.C.Kistler, "Efficient multichannel integrated optics star coupler on Silicon," IEEE Photon. Technol. Lett., 1, 8, pp.241-243 (1989).
[40]K.Kato, K.Okamoto, H.Okazaki, Y.Ohmori, and I.Nishi,"Packaging of large scale integrated-optic NxN star coupler,"in Proc. OEC'92, PD-1, pp.2-3 (1992).
[41]A.Sugita, M.Okuno, T.Matsunaga and M.Kawachi,"Strictly nonblocking 8x8 integrated optical matrix switch with silica-based waveguides on silicon substrate," in Proc. ECOC'90, pp.545-548 (1990).
[42]M.Okuno, K.Kato, Y.Ohmori and T.Matsunaga,"Improved 8x8 integrated optic matrix switch using silica-based planar lightwave circuit,"in Proc. OEC'92, 17A-2, pp.264-265 (1992).
[43]I.Nishi, S.Suzuki, Y.Tachikawa, S.Kobayashi, H.Suda and N.Takato, "WDM transceiver module using a silica-based waveguide multi/demultiplexer,"in Proc. GRIN'91, MD1, pp.70-73 (1991).
[44]R.C.Alferness,"Efficient coupling of a semiconductor laser to an optical fiber by means of an adiabatic tapered waveguide on silicon,"in Proc. OFC'90, WM11, p.108 (1990).
[45]K.Shiraishi, Y.Aizawa, and S.Kawakami,"Beam expanded fiber using thermal diffusion of the dopant,"J. Lightwave Technol., 8, 8, pp.1151-1161 (1990).
[46]H.Hanafusa and M. Horiguchi,"Thermally-diffused expanded core fibers for low-loss and inexpensive photonic components," Electron. Lett.,27, 21, pp.796-797 (1992).
[47]Y.Yamada and M.Kobayashi,"Single-mode optical fiber connection to high-silica waveguide with fiber guiding groove," J. Lightwave Technol., 15, 12, pp.1716-1720 (1987).
[48]E.J.Murphy,"Fiber attachment for guided wave devices,"J. Lightwave Technol., 6, 6, pp.862-871 (1988).
[49]K.P.Jackson, E.B.Flint, D.Lacey, J.M.Trewhella, T.Caulfield and S.Sibley, "Compact planar-processed package for multichannel, fiber-optic computer interconnects," in Proc. OFC'92, p.121 (1992).
[50]S.Day, R.Bellerby, G.Cannell and M.Grant,"Silicon-based fiber pigtailed 1x16 and 2x16 power splitters," in Proc. ECOC'92, WeP2.15 (1992).

LOW-COST, LOW-PROFILE PACKAGING TRENDS FOR INCREASINGLY COMPLEX PHOTONIC DEVICES

R. R. Genova, AT&T-BL, SolidState Technology Center, Breinigsville, PA, USA 18031

ABSTRACT

Market demands for increasingly more complex photonic devices housed in robust, market penetration enabling low cost packages with low profiles (tight board spacings) are being addressed by combining AT&T's silicon based waveguide technologies with ceramic packaging.

Regardless of market segment (Long-haul transmission, SONET, Fiber-in-the-Loop (FITL), Fiber Optic CATV, Optical Interconnect, or Optical Processing/Computing) designers are constantly being challenged to enhance device functionality while significantly reducing their market price.

This presentation focuses upon key factors for on-going market success in this environment including: component integration, rapid "design to market" response, flexible manufacturing capabilities, innovative waveguide devices and packaging techniques and Total Quality Management.

DRIVING FORCES

You have probably heard the marketing idiom "Smaller is Better". In many industries there is a lot of truth in this adage. A clear example of this is SONY's approach to consumer products. Their Walkman™ is an excellent testament. SONY targeted the personal stereo market segment because it had matured into what the marketeers refer to as the "growth" portion of the product life cycle. This is the point where standards have emerged and significant volumes are occurring. SONY recognized that reducing size, weight, and costs while adding features and functionality all in a "user friendly" package were driving forces in the personal stereo marketplace. By challenging their designers with objectives dictated by these driving forces, SONY now enjoys market leadership and high profits in personal stereos.

The silicon based waveguide team at AT&T believes emerging system standards such as SONET, FDDI, OSS, ATM, together with dramatic component price reductions is making fiber deployment preferable to copper in ever increasing numbers of applications. It therefore becomes plausible to argue that the photonic components marketplace is crossing the threshold into the "growth" portion of the product life cycle. This has significant ramifications for component designers.

The photonic components market is fragmenting into a wide variety of rapidly growing segments. As component designers, our challenge is to provide our customers with what they want, when they want it, and at a price their markets will support. Driving forces include: flexibility, rapid product introduction timeframes, increased functionality, reduced size, low cost production, high reliability, and superior quality.

We are addressing these forces through the disciplines of: Concurrent Engineering, Platform Based Technology, Flexible Manufacturing, and Total Quality Management. Our approach to photonic component design and production based on platform technologies, specifically silicon based waveguides, had its genesis in the mid-1980's. Anticipating the future direction and needs of our marketplace, a strategic decision was made to embrace a revolutionary philosophy in photonics technology.

Reasoning that if AT&T were to sustain its position as a market leader, we needed a technology platform capable of addressing the driving forces mentioned above. Silicon waveguide technology takes advantage of decades of AT&T's experience and capital investment in large volume silicon wafer processing. Silicon is more precise and flexible than fusion-based devices ... especially for complex and/or highly integrated structures. Finally, our choice of waveguide processing has been demonstrated to enjoy lower optical loss per centimeter than does silica-based (e.g. flame hydrolisis) deposition technology.

"TO BOLDLY GO WHERE NO PHOTON HAS GONE BEFORE"

Conventional wisdom holds that one need learn to walk before learning to run. Our silicon waveguide technology was no exception. To be a truly cost effective project, we chose an "orderly migration" strategy. This strategy held that we would develop a series of increasingly more complex design functions guided by the following constraints: 1) Functions must be compatible with one another facilitating combination and inter-operability, 2) Key designs would be "spun-off" to gain experience with packaging and production, and 3) All design functions had to posses the ability to be manufactured in low cost (high yield) fashion using volume silicon wafer processing methodologies.

Consistent with "walking first", we began with precision V-groove etching and metalization of silicon piece parts. These devices are used as submounts for our laser module devices. (Ref. Figures 1 & 2). Silicon submounts are robust, inexpensive, good thermal dissipators, and lend themselves to wafer level volume production using the imbedded silicon processing infrastructure. Value added functionality such as micro-strip lines, solder dams, and terminating resistors are easily added.

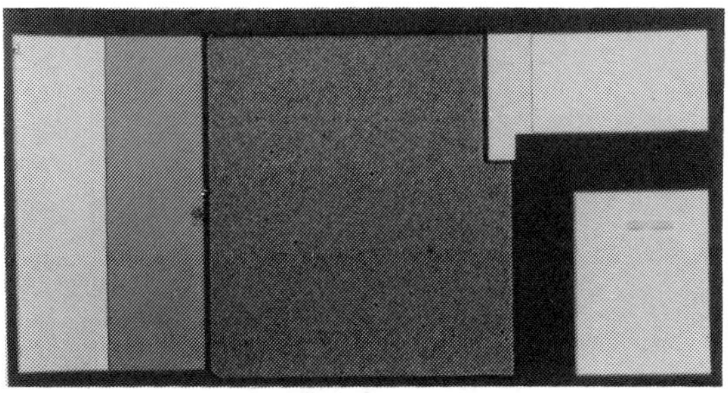

Figure 1 Laser Submount Si Chip

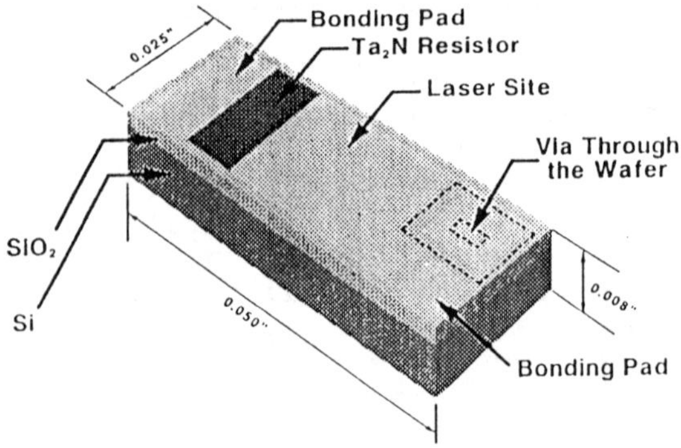

Figure 2 Above Submount's Drawing

Taking advantage of the Si <111> crystalline structure allowed us to fabricate a turning mirror in the V-groove that could subsequently be metalized. Having successfully developed a robust mirror metalization process we were able to think in terms of bending photons instead of electrons. This miniature silicon based optical bench (SiOB) has many benefits[1]. It requires minimal fiber alignment, provides excellent mechanical stability, permits chip level component mounting in a planar fashion (thus multiple devices may be populated while still in the silicon wafer stage) and allows much lower profile devices since the fiber is at 90° to the component's active area (Ref. Figure 3).

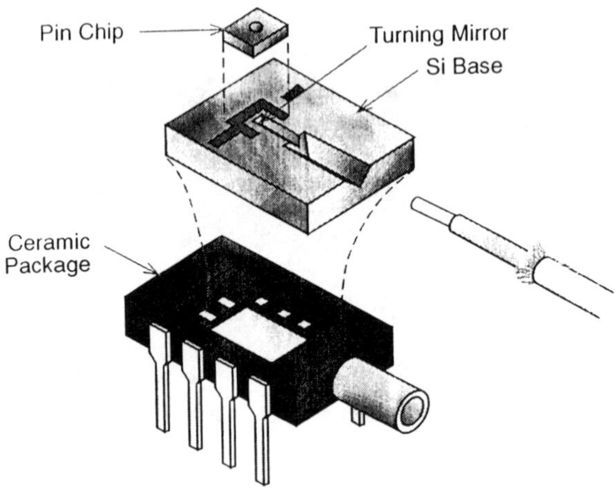

Figure 3 PIN Si OSA Drawing

As a baseline for component cost reductions we chose the very aggressive system cost objectives presented by Fiber-In-The-Loop (FITL). Figure 4 demonstrates our orderly migration plan and how we expected to employ our silicon waveguide technology platform to achieve these aggressive targets. Along the way, key design milestones were selected. These milestones did not necessarily mean that product(s) would be "spun off" for commercial market introduction. A separate economic evaluation occurred at each milestone point and dictated our course of action relative to commercial product offerings.

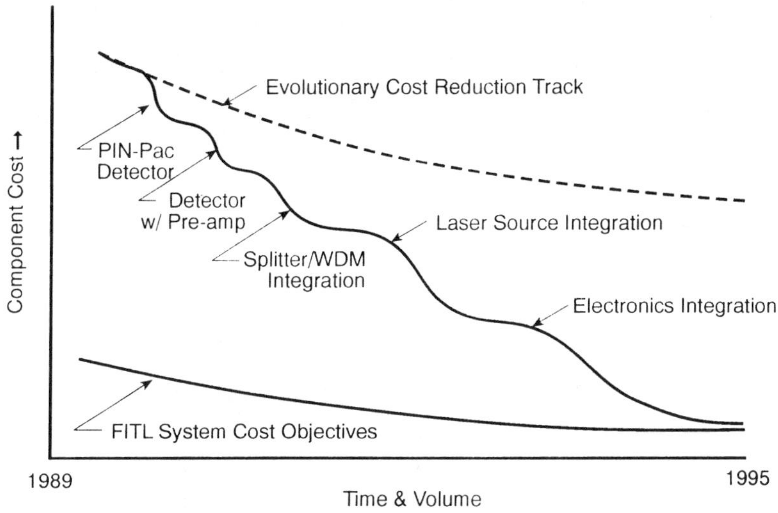

Figure 4 Silicon Technology Migration Plan

The first product we designed for the component market-place was a low profile photodetector based on AT&T's high performance InGaAsP PIN. We named this device PIN-Pac (Ref. Figures 5 & 6). It met with gratifying success in the marketplace. Since its introduction, PIN-Pac has enjoyed a number of design enhancements including integration of pre-amp ICs for both analog and digital applications.

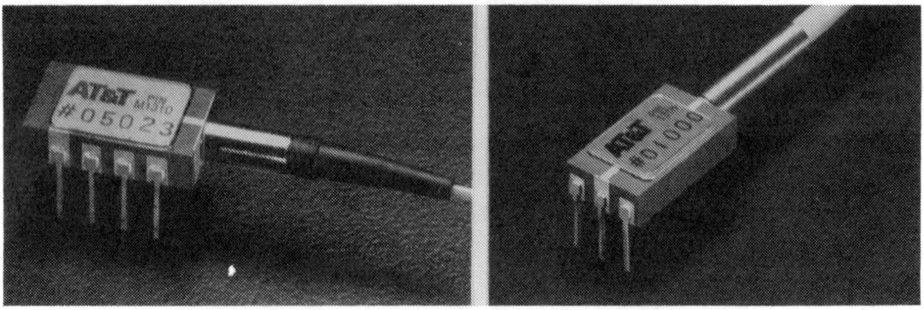

Figure 5 PIN-Pac Photodiodes

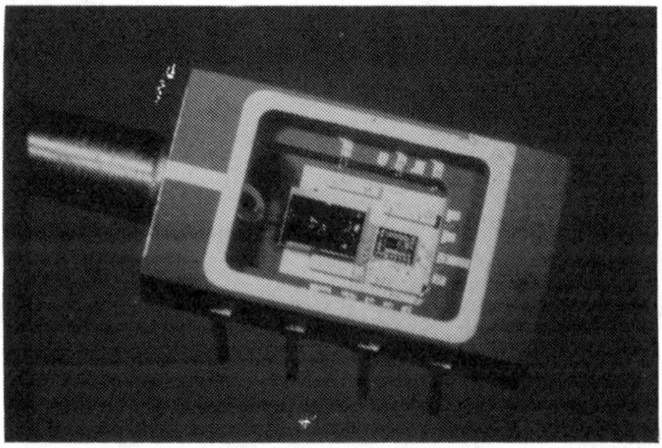

Figure 6 PIN-Pac w/amp: Internal View

We then established capabilities to do splitters and WDM structures (Figure 7). Stand alone versions of these devices were not intended for market-place introduction because we knew their cost exceeded fused fiber technology, but it was a necessary step in our developmental evolution (Figure 8). Concatenated waveguide structures such as a 1.55um/1.3um WDM together with a 3dB splitter were developed for target markets and to which they are currently being provided (Figure 9).

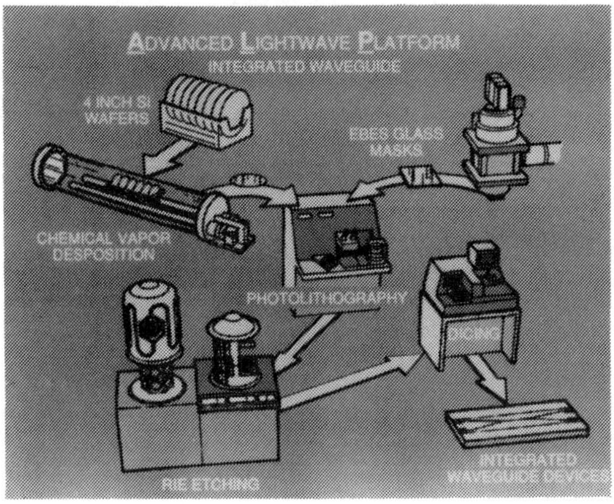

Figure 7 Si Waveguide Wafer Processing

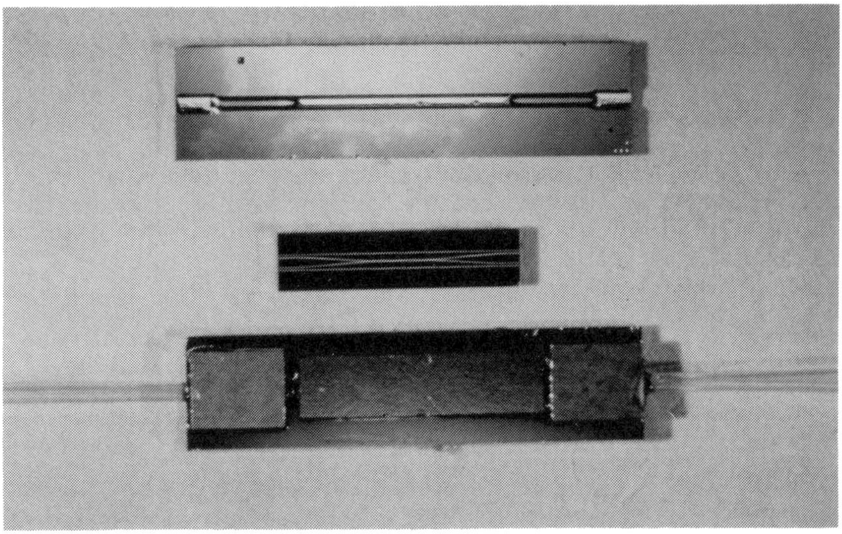

Figure 8 Individual Splitter

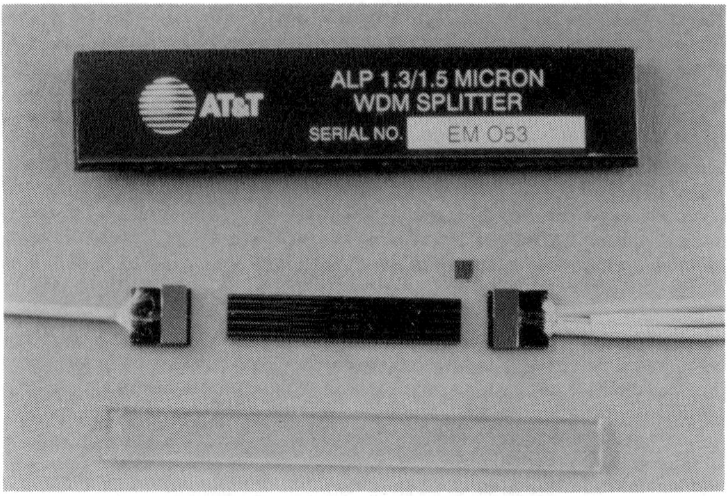

Figure 9 WDM/Splitter

Though the WDM/Splitter structure offers some unique advantages, our silicon waveguide technology is targeted at higher levels of functional integration. The merit of our approach is the significant cost reductions enjoyed when combining, for example, a splitter, WDM, detector PIN(s), laser, monitor PIN and electronics in a single package. Feasibility models of such complex devices have been produced and developmental prototypes will be generated throughout 1993. Figure 10 shows a conceptual view of a complex waveguide structure targeted at bi-directional SONET speed transmission over a single fiber. The same fundamental SiOB approach has achieved microwave operational speeds.[2]

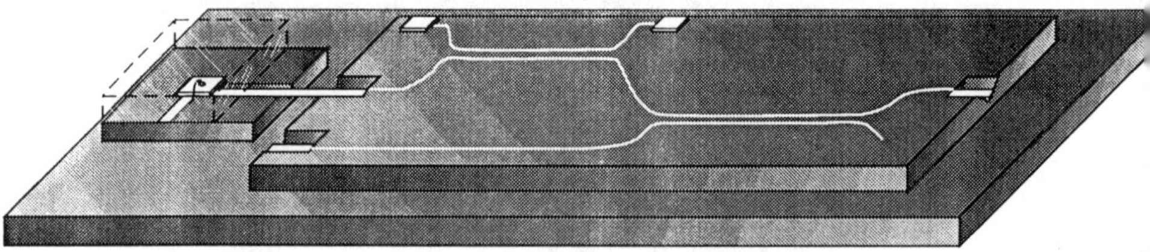

Figure 10 Bi-directional Transceiver Drawing

PACKAGING CONSIDERATIONS

As you have seen from the examples shown above, we have selected low profile (<140 mil) alumina based ceramic packages as our primary package platform. Alumina based (AlO_3) ceramic packages provide a number of benefits. They accomodate thermal management techniques with a thermal expansion coefficient (~7.0 PPM/^{o}C) similar to popular heat sink materials copper tungsten (CuW ~7.0 PPM/^{o}C) and Beryllium Oxide (BeO ~7.0 PPM/^{o}C). This rate is reasonably well matched to silicon (Si ~3.4 PPM/^{o}C), Indium Phosphide (InP ~4.5 PPM/^{o}C) and Indium Gallium Arsenide Phoshpide (InP ~5.2 PPM/^{o}C) over the -40°C to +85°C characteristic operating range. These packages are flexible with regard to: size (length, width, height, cavity size), patterns for internal metalization, multiple internal layers, number and types (Butterfly, DIP, SIP, Pad) of pinouts and location of those pinouts.

A major advantage derived from this packaging platform when teamed with our silicon bench technology is the amount of package space available for value-added component integration. This is a significant departure from conventional low-cost coaxial packaging technology is where the majority of the package design deals with holding and aligning the fiber. In our low-cost packaging approach, just the opposite is the case. We've elected to minimize fiber handling issues and instead, concentrate on integrating the components our customers want.

What about hermeticity? Hermeticity, at this point, is one of those "great unknown" customer driven issues. If it were "free", every customer would expect hermeticity as standard. But in any packaging scenario, hermeticity is not "free". The combination of design considerations and resultant yield impacts contributes to product costs. We are addressing the hermeticity questions in a three phased manner: Epoxy, Hermetic Packages, and Hermetic Sub-assemblies.

In our standard product, like the PIN-Pac detector, we seal the package with a specially formulated, mil-spec qualified epoxy developed internally at AT&T Bell Labs. Products made with this epoxy are environmentally "leakproof" to such testing as 85°C @ 85% RH (Relative Humidity) have a "hermeticity" rating in the $\geq 10^{-4}$ range.

AT&T has used this epoxy seal for years in thousands of our alumina based ceramic production devices. Some of them have been out in the field for more than 8 years and to my knowledge, we have not experienced a single epoxy related field failure. Unfortunately, other component vendors did not spend as much effort on epoxy R&D and used formulations that failed in the field. These have given epoxy sealing a negative perception.

In certain instances where $\geq 10^{-4}$ is not acceptable (aerospace, ruggedized military applications, certain Bellcore specified applications) our packages' standard fiber feedthrough ferrule and its metal lid may be solder sealed. We have demonstrated the ability to achieve $\geq 10^{-}$ hermetic sealing of these alumina based ceramic packages using standard solder sealing processes.

Another method of dealing with hermeticity requirements, and one that offers particular attractiveness, is to make individual sub-assemblies hermetic. In the silicon based waveguide sub-assembly environment where metallization and glass layers are an integral part of device fabrication, the addition of hermetic seal metallization patterns are just another part of the metallization mask(s). If we carefully design both the piece parts and the fabrication processes, we have the opportunity to achieve hermeticity at virtually no incremental cost.

PARTNERING FOR SUCCESS

As system objectives evolve, devices with increasing complexity and application specific customization are being required. System OEMs are realizing that economics will prohibit commodity device driven designs. Low cost goals together with abbreviated "time to market" needs will only be met by creative partnerships with component vendors. The key to success in these relationships will be the selection of a component vendors with established technology platforms. AT&T Microelectronics - Lightwave Business Unit is such a component vendor and welcomes opportunities to discuss "strategic partnering".

SUMMARY

The combination of silicon waveguide platform technology with alumina based ceramic packages hold great promise for a sustainable, mutually beneficial future. Devices based on this teaming of technologies will facilitate operation at speeds from dc to multi-gigabits, function well over a wide range of environmental conditions and enjoy manufacturing economics that make sense well into the 21st century.

References:

1. G. Blonder, IEEE LEOS '90 Conference Digest, LEOS 1990 annual meeting, Boston, MA, USA, page 173.

2. E. Ackerman, et.al. "A 3 to 6GHz Lightwave/Microwave Transceiver Module for Microwave Fiber-Optic Communications" 1991 MTT-S Symposium Digest, Vol. 2, pp. 577-579.

SILICA ON SILICON AND ALUMINA CERAMICS FOR INTEGRATED OPTICS AND OPTICAL INTERCONNECTIONS

C. Jacob Sun and T. Miyashita, Photonic Integration Research, Inc., 2727 Scioto Parkway, Columbus, OH 43221

ABSTRACT

In the past 10 years, great progress has been made in the field of integrated optics based on the silica on silicon technology. The flame hydrolysis deposition(FHD) together with photolithography & dry etching techniques is utilized to fabricate various types of low loss integrated optical circuits. In this paper, the technology is reviewed and the performance of these miniature optical circuits for optical communication networks and computer chip level interconnections is also discussed. Recent progress regarding forming silica waveguides on alumina ceramic substrates with the same technology for potential applications in optoelctronic integration is also presented.

1. INTRODUCTION

Since the early 1980's, the optical fiber preform manufacturing process, flame hydrolysis deposition(FHD) has been adapted to make silica glass deposition on silicon and quartz wafers[1,2]. With the utilization of microelectronic patterning techniques, photolithography and reactive ion etching[RIE], various types of channel waveguide circuits, such as 1xN, 2xN, NxN, wavelength division multi/demultiplexer(WDM), optical frequency division multi/demultiplexer(FDM), and thermo-optic switches have been made for applications in telecommunication, data communication, or sensing devices. The performance of these circuits has been progressively improved over the years, e.g., the waveguide loss has dropped from 0.2 dB/cm to 0.01 dB/cm [3] and more complex circuits are being made, e.g., wavelength insensitive coupler(WINC) [4] made of two directional couplers provides wavelength flat coupling and 1x3 WDM made of a Y-splitter and a directional coupler faciliates wavelength demultiplexing and splitting by the same circuit. Easy integration of multiple circuit elements on a single chip to perform various functions is one of the major advantages of this technology. The excellent process controlability, good material compatibility with optical fibers(high silica to high silica), and the mass productivity of the processes are the other factors making this technology a practical route to realize low cost and high performance components for expanding point-to-point fiber optic communications to more sophisticated subscriber networks.

In addition to the silicon and quartz as waveguide substrates, alumina ceramics is also attractive because of its popularity as microelectronic packaging substrates. With slight modification of the current technology, relatively low loss silica waveguides have been formed on several different types of ceramic substrates for potential hybrid optoelectronic integration.

In the following sections, each step of the technology is first reviewed. Next, the performance of various circuits are discussed. Finally, some recent results about silica waveguides on alumina ceramics are described.

2. SILICA WAVEGUIDE DEVICE FABRICATION

Figure 1 describes the manufacturing process for regular silica waveguide devices. It starts with FHD & consolidation to make planar waveguides, photolithography & RIE to form channel waveguide circuits, FHD & consolidation to clad the channel waveguide circuits, circuit testing to evaluate the circuit performance, fiber pigtailing & packaging to make the devices, and finally the reliability testing to assure the device performance. This section reviews each step of the process and related issues.

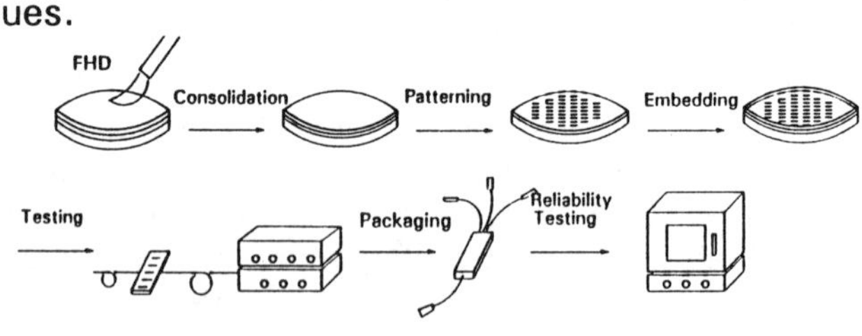

Figure 1. Silica waveguide device fabrication.

2.1. FHD & Consolidation

Flame hydrolysis deposition(FHD) is one of the common methods for making high purity glass. Optical fiber preform fabrication processes vapor-phase axial deposition(VAD) and outside vapor deposition(OVD) are the typical examples. The FHD for making planar deposition on wafers is conducted at a much lower rate, because the target deposition thickness is generally a few μms for single-mode devices and tens to 100 μms for multimode devices. The vapor of raw materials such as $SiCl_4$, $TiCl_4$, $GeCl_4$, PCl_3, and BCl_3 is fed through an oxy-hydrogen flame to be hydrolyzed into oxides and deposited onto planar substrates. GeO_2 and TiO_2 are used to raise the refractive index of the silica glass for core layer while the addition of small amounts of B_2O_3 and P_2O_5 lowers the sintering temperature of the glass. After deposition, wafers with the as-deposited porous glass layers are transfered into an electrical furnace for sintering into compact glass(consolidation). The sintering temperature is around 1200 to 1400°C depending upon the glass composition.

The finished planar waveguides are tested for their thickness and refractive index. The thickness is normally measured with microscope or with prism coupler method and the refractive index is determined with interference microscope or prism coupler method. The typical thickness variation across a wafer and between wafers is ± 2%. The index variation is typical ± 0.01% for 0.30%Δ single mode and ± 0.1% for 2.0%Δ multimode waveguides. The loss of the planar waveguides is normally too low to be detected with available methods. It is believed to be < 0.01dB/cm.

2.2. Photolithography & RIE

In the second step of the process, planar waveguides are patterned into channel waveguide circuits by photolithography & RIE. A layer of pattern transfering material such as amorphous silicon film is deposited onto the planar waveguide before the photoresist patterning because of the thick silica layer etching. RIE with fluorinated hydrocarbon gases is used to etch through the waveguide core portion. The etching rate is 5 - 10 μm per hour and the lateral dimension control is ± 1% for shallow etching(10-20 μm) and ± 2% for deep etching(50 - 100 μm). RIE produces etched channel waveguide with smooth sidewalls and good perpendicularity. Microstructure details such as guiding blocks, grooves, and windows are all precisely patterned onto the wafers simultaneously with the waveguide circuits. This is another advantage of this technology over other approaches such as ion- exchange in glass or fused fiber technology.

2.3. Waveguide Circuit Cladding

The etched single-mode circuits are normally embedded in thick cladding(20 - 40 μm) by FHD & consolidation. Multimode circuits are cladded either by the same method or by thin film deposition methods(CVD or sputtering) to provide a 3 - 5 μm thick SiO_2 cladding layer. Figure 2 shows the cross section of a single-mode 1x8 splitter near the last Y-junctions. Figure 3 demonstrated a multimode channel waveguide with thin cladding. This waveguide was rounded to provide better fiber to waveguide coupling. The waveguide and fiber geometry mismatch is negligible for single-mode devices but quite significant for multimode devices [5,6].

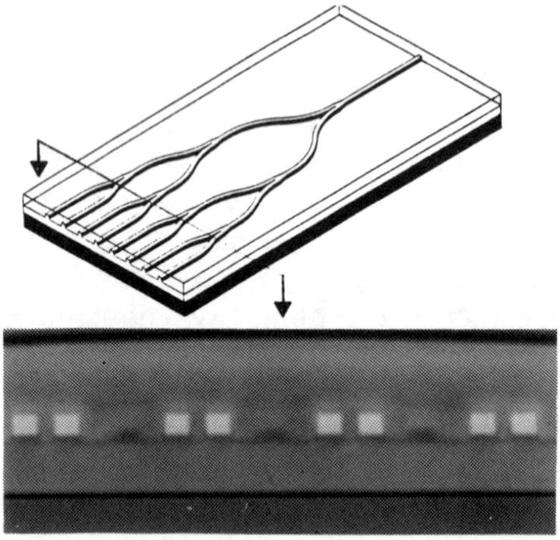

Figure 2. A single-mode 1x8 splitter.

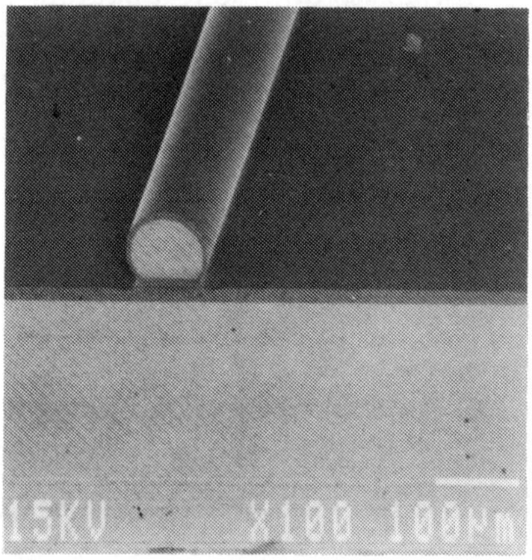

Figure 3. A multimode waveguide. (© 1991 IEEE)

2.4. Waveguide Characterization

After the circuits are made on the wafers, they are diced and sent to test for their loss, splitting uniformity, wavelength dependent properties, polarization dependence, and other characteristics. Fully automated test stations normally test 20 - 30 circuits of standard products per hour. This is the critical step for the product quality and process control. It provides the feedback data for the new product design and manufacturing process control.

2.5. Fiber Pigtailing & Packaging

An effective and low cost method for waveguide-to-fiber connection is critical to the practical use of the waveguide devices in fiber optic networks. Recently, a great deal of efforts have been devoted into this part of the process, especially for single-mode devices because of the small dimension of its waveguides and therefore tight requirement for waveguide-to-fiber alignment. Various types of fiber bonding and packaging techniques have been reported to realize low loss fiber-to-waveguide coupling [7,8,9]. The UV cure adhesive technique is used by PIRI for standard products to connect fibers to waveguide circuits. Input and output fibers are first bonded to V-grooved blocks with epoxy. These blocks are then butt-jointed to the waveguide circuit end faces with UV cure adhesive(Figure 4). The waveguide-fiber connection loss by this method is normally less than 0.1 - 0.2 dB per point.

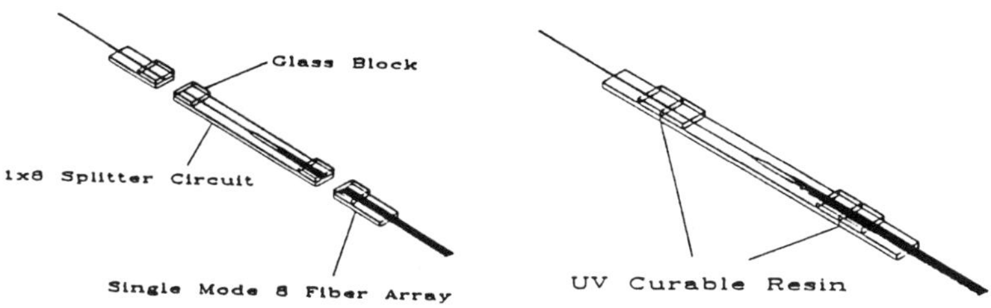

Figure 4. Fiber to waveguide connection

2.6. Reliability Testing

The final step for a waveguide device fabrication is the reliability testing. Packaged devices are tested through both environmental and mechanical tests. Test results for some devices will be presented later when the circuit performance is discussed.

3. SILICA WAVEGUIDE CIRCUITS ON SILICON

This technology virtually can make waveguides with thickness from a few μms up to 200μms and refractive index difference(Δ) up to 2 ~ 3%. Table 1 lists the characteristics of the standard single mode and multimode waveguides. Two types of single mode waveguides are normally used. One has the cross-sectional dimension of 8μm x 8μm and refractive index difference(Δ) of 0.30%. The other type is 6μm x 6μm with Δ of 0.75%. The former is used for regular single mode devices because its parameters match exactly with standard single mode fibers. The latter type is used for circuits with small radius of curvature to reduce the bending loss. The typical straight waveguide loss of both types of single-mode waveguides is 0.05 - 0.1 dB/cm (Figure 5). Waveguides of 40μm x 40μm with 1%Δ and 80μm x 80μm with 2%Δ are used for our standard multimode circuits because of their minimized coupling loss with regular 50μm and 100μm optical fibers, respectively [5]. Various types of integrated optical circuits made of these types of silica waveguides have been fabricated in recent years. This section first reviews the basic elements for these integrated optical circuits and then discusses the charateristics and performance of some of the circuits.

Table 1. Standard waveguides

Type	Refractive Index (Δ)	Core Size (μm)	Loss (dB/cm)
Single-Mode Low Δ	0.30%	8 X 8	< 0.1
Single-Mode High Δ	0.75%	6 X 6	< 0.1
Multimode	1%	40 X 40	~ 0.1
Multimode	2%	80 X 80	~ 0.1

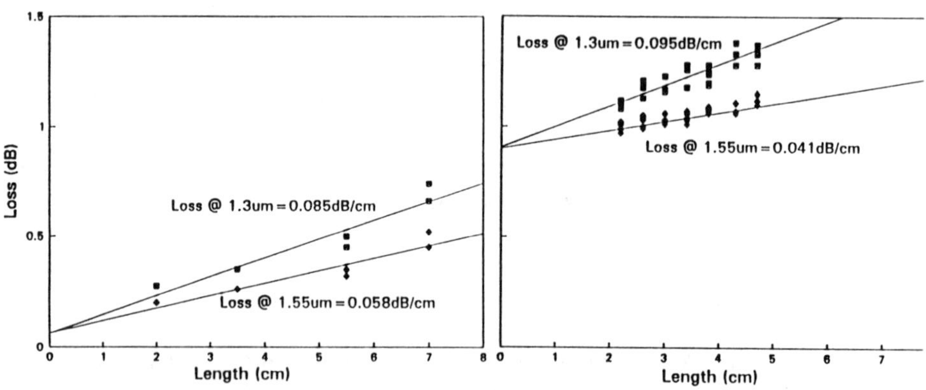

(a) (b)

Figure 5. Single-mode straight waveguide loss for (a) low Δ (b) high Δ.

3.1. Basic Circuit Elements

Table 2 lists the basic circuit elements for the integrated optical circuits.

Curved waveguides are necessary elements for making compact circuits. Curves made of single mode waveguides with 0.75%Δ exhibit negligible loss for bend radii as small as 5mm.

A small angle Y-branch provides low loss(<0.1dB) and wavelength-flat splitting. It is a simple but important element for circuits used in subscriber systems.

A directional coupler is made of two narrowly spaced waveguides. With different spacings and waveguide lengths, light can be coupled between these two waveguides with various coupling ratios. Therefore, it can be used to form splitters with any splitting ratios. It is also used to form wavelength division multi/demultiplexer(WDM) because of its wavelength-dependent coupling nature. Precise process control is necessary to achieve as-designed coupling properties. Figure 6 shows the cross section of an embedded directional coupler.

Table 2. Basic circuit elements for integrated optical circuits

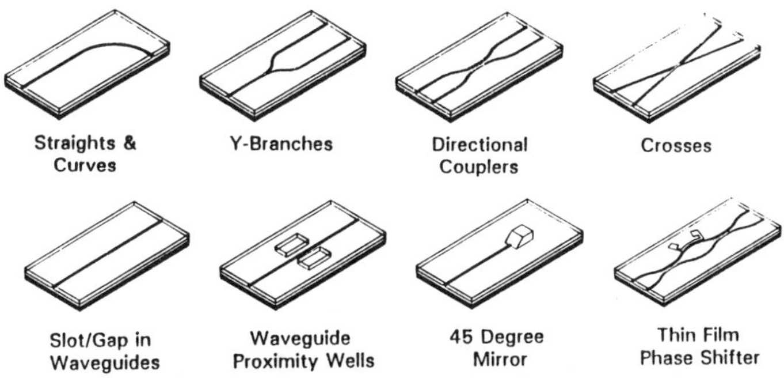

Figure 6. The cross section of a directional coupler.

Cross elements are useful for making more compact circuits. The loss and crosstalk are negligible if the angle is greater than 30 degree.

Slots in waveguides and proximity wells near circuits can be formed simultaneously while the circuits are etched. They provide the convenience of hybridizing waveguide circuits with optoelectronic elements, filters, polarizers, mirrors, and other micro-optic elements.

Thermal-optic phase shifters can be formed by depositing thin-film heaters on top of waveguides. They are used to make T-O swithes and tune the FDM circuits.

3.2. Examples of Integrated Optical Circuits

In the following, the charateristics and performance for several integrated optical circuits made of the above basic elements are described:

1XN splitters-----1xN splitters are formed by cascading Y-branches. A single mode 1x8 chip is about 25mm x 5mm. Figure 7 shows the typical insertion loss of the 1X8 circuits for wavelengths at 1.3 and 1.55μm for both TE and TM modes. The average insertion loss is 9.5 dB and the uniformity(maximum loss - minimum loss) is 0.5 dB. The polarization dependence is less than $\pm$0.2dB. Normally, the packaged devices have 0.3 dB higher loss due to the fiber pigtailing & packaging step. Environmental testing results for such 1x8 and 1x4packaged devices are shown in Figure 8. Single-mode splitters up to 1x128 have been made with an excess loss of 3dB and uniformity of $\pm$1.0dB [10].

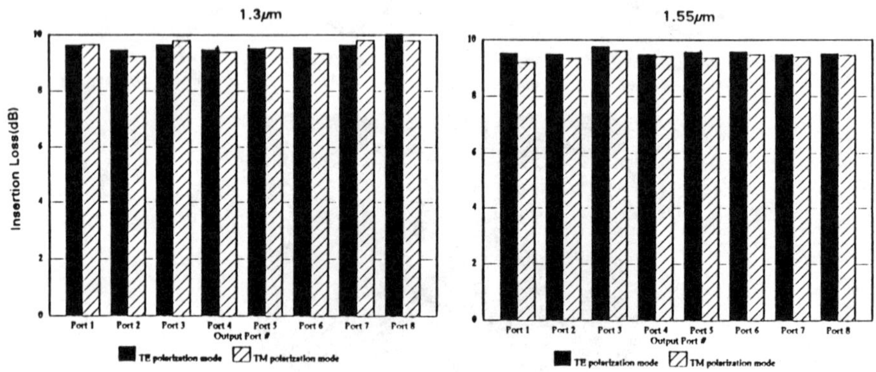

Figure 7. Insertion loss of 1x8 splitters.

WDM -----As mentioned above, a directional coupler can be used to form a WDM with properly designed waveguide parameters. However, this type of WDM normally has higher cross-talk and is only applicable for large wavelength spacings. Mach-Zehnder interferometer type circuits (Figure 9(a)) have also been fabricated to perform as WDM's [11]. They consist of two 3dB directional couplers for λ_2 and the connecting arms have a path difference ΔL, which causes zero phase difference for λ_2 and π for λ_1. This type of WDM can virtually multi/demultiplex any combinations of wavelengths. Figure 9(b) illustrates the configration for a narrowly spaced WDM or optical frequency-division multiplexer(FDM). FDM's with frequency spacings from 100 GHz to 1 GHz have been successfully fabricated [11]. By using the interference filters, a third type of WDM has also been fabricated. The thin film filter can be precisely inserted in the guiding groove, which is etched during the waveguide patterning.

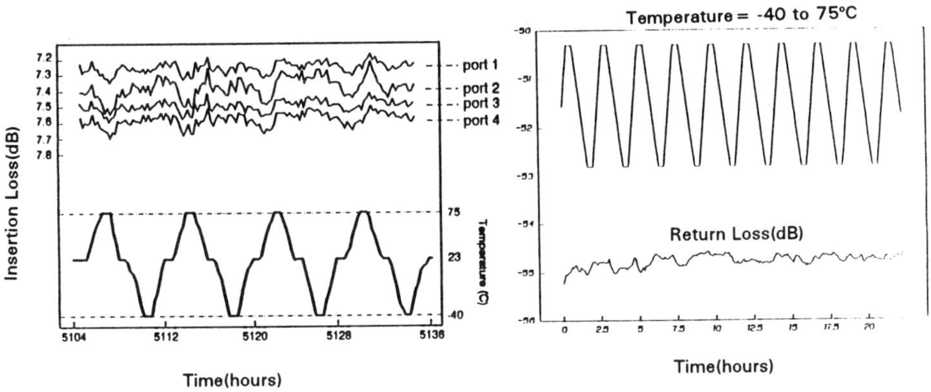

(a) (b)

Figure 8.(a) Temperature cycling tests for insertion loss of 1x4 splitters.
(b) Temperature cycling tests for return loss of 1x8 splitters.

Wavelength insensitive couplers(WINC)-----This is also a Mach-Zehnder interferometer type device made of two directional couplers and two connecting waveguide arms with a small path difference[4]. The typical variation of the coupling ratio for a 50% WINC is ± 2% from wavelength of 1.2 to 1.6 μm.

T-O switches-----A thin film heater placed on one of the arm of a Mach-Zehnder interferometer is able to generate phase difference between the two arms because of the thermal optic effect(dn/dT =

10^{-5}/°C) in silica glass. The switch time is in msecs and the crosstalk is better than 18dB over extended temperature ranges(Figure 10)

Other integrated optical circuits such as 2XN splitters, wavelength insensitive switches[12], and multichannel FDM's have also been fabricated by combining the above circuits together.

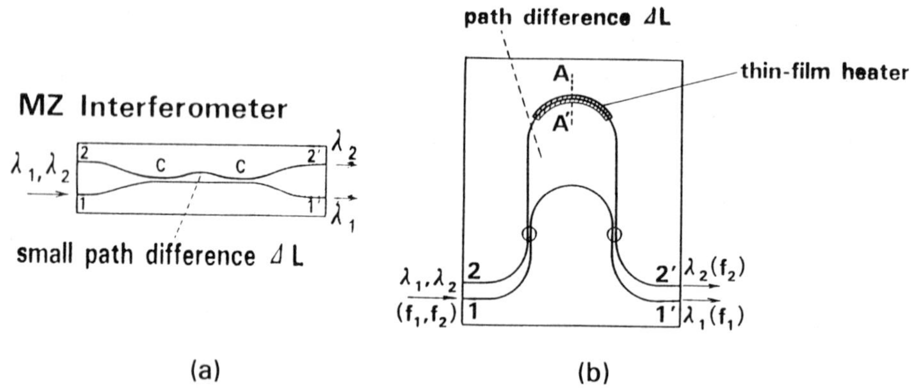

Figure 9. (a) Single-mode WDM. (b) Single-mode FDM.

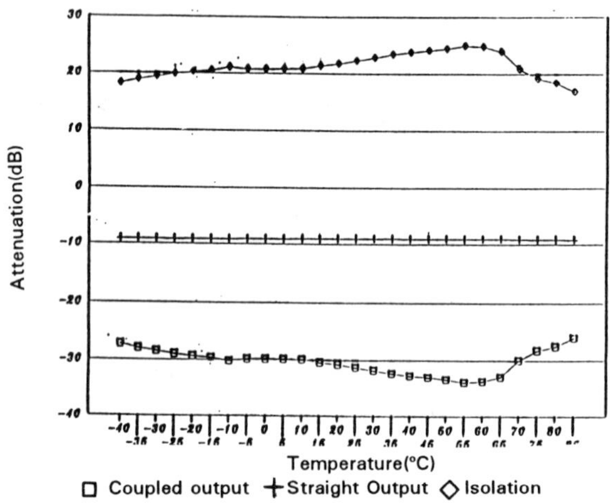

Figure 10. T-O switch performance over extended temperature range.

3.3. Silica Waveguides as Optical Interconnects

Besides the application for integrated optical circuits, silica waveguides are also useful for the hybrid optoelectronic integration. This was demonstrated in a recent work by fabricating a compact multi-channel transceiver module(size of 9 x 11 mm) having metal wirings, flip-chip bonded optoelectronic components, and high silica waveguides connecting laser diode arrays and receivers to fibers [13]. Figures 11(a) and (b) are SEM pictures for the transmitter and receiver parts of the transceiver module. In addition to acting as interfaces between the O-E components and optical fibers to allow higher packing density and greater flexibility for interconnections, silica waveguides also offer the potential of incorporating the above mentioned integrated optical circuits into O-E modules for optical signal processing.

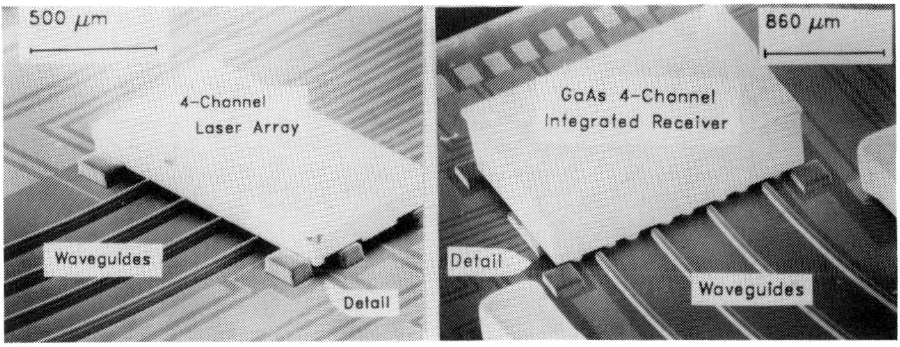

(a) (b)

Figure 11. SEM photos for (a) a 4-channel laser array bonded to the waveguide substrate (b) a 4-channel receiver bonded to the same substrate.(© 1992 IEEE)

4. SILICA WAVEGUIDES ON ALUMINA CERAMIC SUBSTRATES

As described in the last section, it is promising to utilize the current silica waveguide technology for hybrid O-E integration. In a recent work, the same technology has been applied to form silica waveguides on alumina ceramics because of its wide use as electronic chip carriers [14]. Good quality and low loss silica waveguides were formed on various types(99%, 96%, and 92% Al_2O_3) of alumina ceramic substrates. This section briefly reviews the results of this work.

The main difficulty of growing reliable silica glass on alumina ceramics lies in the large difference of thermal expansion coefficients between silica and alumina. It is around $0.5 \times 10^{-6}/°C$ for pure silica and $7 \times 10^{-6}/°C$ for alumina. Figure 12 is the estimated film stress for doped silica film on silicon and alumina substrates. Apparently, in order to reduce the film stress, higher content of dopants in the silica glass is necessary when the substrate is alumina instead of silicon. Two layers of glass films with 70 - 85 mol% SiO_2 were formed on the ceramics by FHD and consolidation. Both layers contained B_2O_3, P_2O_5, and GeO_2. The top layer had more GeO_2 for higher refractive index. Figure 13 is a photograph for the cross section of a two-layer planar waveguide on 99% alumina substrate. The refractive index difference between the core and cladding is 1% for this sample. Photolithography & RIE followed the FHD & consolidation to pattern the planar waveguides into channel waveguides. A thin SiO_2 cladding was sputtered onto the waveguides afterwards. Figure 14 shows the SEM photographs of etched channel waveguides on the ceramic substrates.

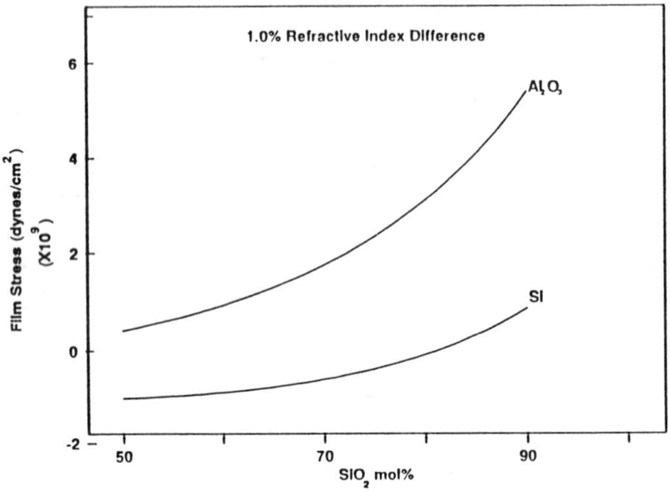

Figure 12. Estimated thermal stress of high silica film deposited on silicon and alumina substrates.(©1992 IEEE)

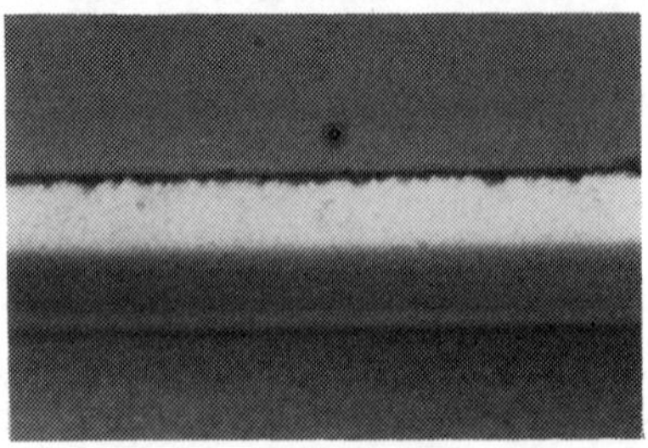

Figure 13. The cross section of a planar waveguide on alumina. (©1992 IEEE)

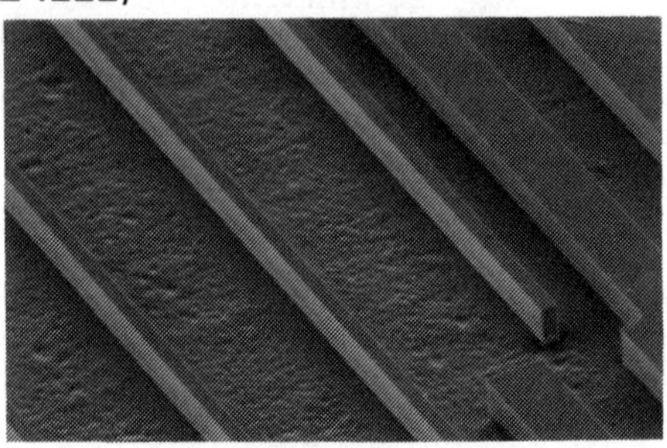

Figure 14. SEM photo of etched channel waveguides.(©1992 IEEE)

The waveguide loss was measureed to be 0.3 - 0.4dB/cm at 0.85 μm and around 0.1dB/cm at 1.3μm. These results show that the FHD silica waveguide based technology is equally applicable to alumina ceramic substrates. Even though more work needs to be done before a common ceramic substrate having both electrical metal wiring and optical waveguides is realized, this new progress proves the possibility of combining silica waveguide technology with alumina ceramic multi-chip module technology to provide dense and compact O-E modules for computer interconnections.

5. CONCLUSION

After about a decade of development, the silica waveguide based integrated optical devices are being tested in field trials for practical use. Results of the reliability and ruggedness testing are being monitored to further advance the circuit design and fabrication techniques as well as the device packaging methods. With its nature of good process controllibility and mass productivity, the technology proves to be a practical route for spreading the use of integrated optical devices in future fiber optic networks. Furthermore, some recent results demonstrate its potential for working with ceramic multi-chip module technology to faciliate advanced O-E integration on ceramics.

ACKNOWLEDGEMENTS

The authors would like to thank Dr. S. Sumida, M. Myers, K. Schmidt and Dr. S. Kobayashi for their helpful discussions and C. Hughey, A. Davis, F. Kiger, and M. Schumacher for the testing results. We also wish to thank Dr. K. Jackson of IBM T.J. Watson Research Center for his generosity of providing SEM photographs of his work.

REFERENCES

1. M. Kawachi, M. Yasu, and T. Edahiro, "Fabrication of SiO_2-TiO_2 glass planar optical waveguides by flame hydrolysis deposition," Electron. Lett., Vol 19, p. 583-584, 1983.
2. M. Kawachi, M. Yasu, and M. Kobayashi, "Flame hydrolysis deposition of SiO_2-TiO_2 glass planar optical waveguides on silicon," Jpn. J. Appl. Phys., Vol. 22, No. 12, pp. 1932, 1983.
3. T. Kominato, Y. Ohmori, H. Okazaki, and M. Yasu, "Very low loss GeO_2 doped silica waveguides fabricated by flame hydrolysis deposition method," Electron. Lett., Vol. 26, No. 5, pp. 327-328, 1990.

4. K. Jinguji, N. Takato, A. Sugita, and M. Kawachi, "Mach-Zehnder interferometer-type optical waveguide coupler with wavelength-insensitive coupling ratio," Integrated Photonics Research '90, Hilton Head, SC, 1990.
5. Y. Yamada, M. Kawachi, M. Yasu, and M. Kobayashi, "High silica multimode channel waveguide structure for minimizing fiber-waveguide-fiber coupling loss," J. Lightwave technology, Vol. LT-4, pp. 277-282, 1986.
6. C.J. Sun, W.M. Myers, K.M. Schmidt, and S. Sumida, "Silica-based circular cross-sectioned channel waveguides," IEEE Photonics Technol. Lett., Vol. 3, No. 3, 1991.
7. Y. Yamada, F. Hanawa, T. Kitoh, and T. Maruno, "Low-loss and stable fiber-to-waveguide connection utilizing UV curable adhesive," IEEE Photonics Technol. Lett., Vol. 4, No. 8, pp. 906-908, 1992.
8. H.M. Presby, S. Yang, A.E. Willer, and C.A. Edwards, "Connectorized integrated star couplers on silicon," Optical Engineering, Vol. 31, No. 6, pp. 1323-1327, 1992.
9. S. Kobayashi, T. Kitoh, Y. Hida, S. Suzuki, and M. Yamaguchi, "Y branching silica waveguide 1x8 splitter module," Electron. Lett., Vol. 26, No. 11, pp. 707-708, 1990.
10. H. Takahashi, Y. Ohmori, and M. Kawachi, "Design and fabrication of silica-based integrated-optic 1x128 power splitter," Electron. Lett., Vol. 27, No. 23, pp. 2131-2133, 1991.
11. N. Takato, K. Jinguji, M. Yasu, H. Toba, and M. Kawachi,"Silica-based single-mode waveguides on silicon and their application to guided-wave optical interferometers," J. Lightwave Technol., Vol. 6, pp. 1003-1010, 1988.
12. T. Kitoh, n. Takato, K. Jinguji, M. Yasu, and M. Kawachi, "Novel broad-band optical switch using silica-based planar lightwave circuit," IEEE Photonics Technol. Lett., Vol. 4, No. 7, pp. 735-737, 1992.
13. K.P. Jackson, E.B. Flint, M.F. Cina, D. Lacey, J.M. Trewhella, T. Caulfield, and S. Sibley, "A compact multi-channel transceiver module using planar-processed optical waveguides and flip-chip optoelectronic components," ECTC '92, San Diego, CA, 1992.
14. C.J. Sun, W.M. Myers, K.M. Schmidt, and K.P. Jackson, "High silica waveguides on alumina substrates for hybrid optoelectronic integration," IEEE Photonics Technol. Lett., Vol 4, No. 6, 1992.

Silica Guided-Wave Devices

KATSUYUKI IMOTO

Advanced Research Center, Hitachi Cable, Ltd.,
Hitachi, Ibaraki-ken, 319-14, JAPAN
+81-294-423151

Abstract

Core glass film for optical waveguide is fabricated using new technology and material. Fusion technology by CO_2 laser beam irradiation is developed for splicing the waveguide and single-mode fiber. The waveguide structure is used for many passive devices and hybrid integration of optical active components.

Introduction

To enhance the quality of optical fiber communication systems and expand application fields, it is advantageous to use optical passive components with optical processing and controlling function such as directional couplers, star couplers, multi/demultiplexers, and switches in addition to optical active devices and fibers.

It is preferable that these optical passive components have such features as small size, high reliability, adjustment-free optical coupling, automatic assembly. For application in optical fiber communication systems, various type components have been proposed and developed. In the stage of early development, combined use of discrete components such as lenses, prisms have been developed and some of these components are practically still used. However, these components are not necessarily efficient in building economical and reliable systems. To solve these problems, fiber type components have been developed and are widely used in practice. Guided-wave type components have attracted great interest recently. As shown in Table 1, these guided-wave type components have great possibility in meeting the above mentioned requirements. Particularly, silica glass guided-wave components can be expected to have more advantageous properties than fiber type components[1,2]. Because these components can be realized with a loss as low as 0.1 dB/cm and a fiber coupling loss of less than 0.05 dB.

Table 1. Comparison of Optical Passive Components

Item	Fiber Type	Waveguide Type
Size	Large	Medium *
Loss	Very Small	Large *
Isolation	Poor	Good
Reliability	Good	Good
Productivity	Good	Excellent
Cost Reduction	Difficult	Possible *

* subject to be solved

This paper summarizes a proposed fabrication method of a silica glass waveguide, and fusion splicing technology of waveguide and fiber by CO_2 laser beam irradiation[3,4]. The waveguide structure is proposed for realizing devices such as multi/demultiplexers, star couplers, and hybrid integration of optical active components[5-9]. To attain low cost and small size, this paper also describes the waveguide of high refractive index difference with $SiO_xN_yH_z$ core glass film[10]. Furthermore, this paper reports that the waveguide with small warp in substrate and small roughness of core side-etching properties can be obtained.

Fabrication Method of Waveguide

Silica embedded channel waveguide is fabricated on a quartz glass substrate by the processes of core glass film formation, metal film deposition, photolithography, dry-etching, cladding glass formation and its sintering as shown in Fig. 1. In these processes, it is very important to form core glass film with uniform refractive index and thickness. For this film formation method, we have investigated electron beam vapor deposition[3,4,9], atmospheric pressure chemical vapor deposition[5,6], and plasma CVD method[10]. And as one of the other methods, flame hydrolysis deposition method has been studied[1]. Figure 2 shows the schematic diagram of core glass film formation with the electron beam vapor deposition method. Core glass film of 8 μm thickness is deposited on a substrate at a temperature of 350 ℃. The growth rate of film can be increased to 18 Å/sec. Two kinds of SiO_2-TiO_2 and SiO_2-GeO_2 glass films are fabricated. By this process, the deviation of refractive index and thickness of these glass films can be controlled within ±0.02 % and ±0.3 %, respectively.

This low deviation rate is one of the features of this fabrication method.

Next, WSi_x film of about 1 μm thickness is deposited on core glass film by a sputtering process. The core ridges are formed by processing film waveguide through photolithography and reactive ion etching processes. The etching of WSi_x film masked with photoresist is carried out by NF_3 gas. The etching rate of core glass film masked with WSi_x film is about 0.14 μm/min by CHF_3 gas. The etching selectivity ratio between glass film and WSi_x mask is more than 20. Moreover, the decrease ratio of the width for etching depth of core glass is about 0.1. The core ridge side angle of about 86° is obtained about by increasing WSi_x film thickness and optimizing etching gases.

Finally, cladding glass of $SiO_2-B_2O_3-P_2O_5$ is formed by flame hydrolysis deposition. This glass is obtained by sintering after the deposition of soot glass particles by vapor phase reaction of raw materials.

The propagation loss of SiO_2-GeO_2 waveguide is less than 0.1 dB/cm and lower than that of SiO_2-TiO_2 waveguide with a higher melting

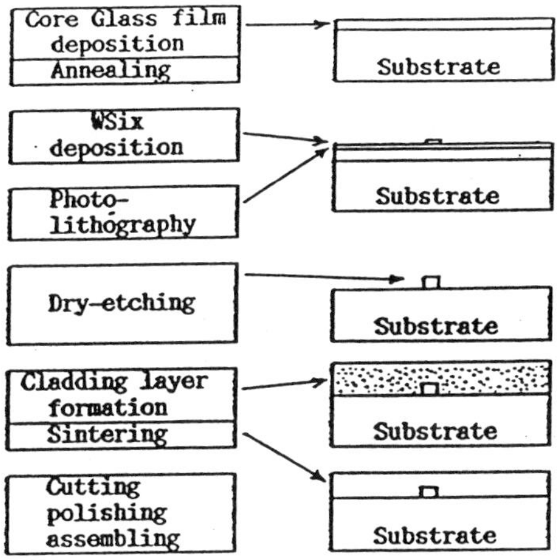

Fig. 1 High-silica glass embedded channel waveguide procedure.

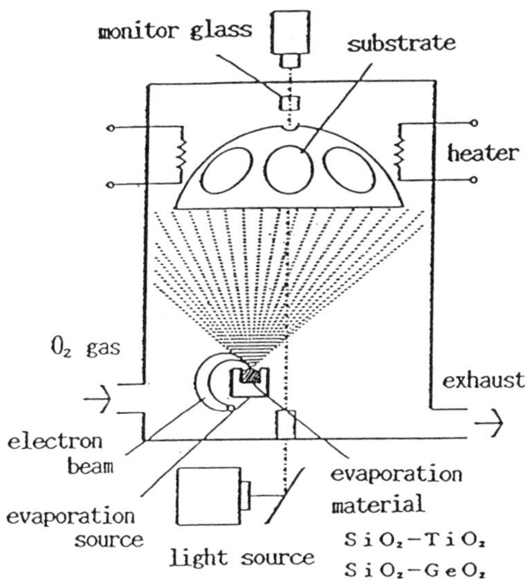

Fig. 2. Core glass formation with electron beam vapor deposition method

temperature. The refractive index difference Δ between core and cladding is 0.25 %.

The relationships between loss and GeO_2 concentration, and those between loss and Δ is shown in Fig. 3. Loss tends to the increase with increasing GeO_2 concentration. This is due to the increase of scattering loss that depend on the inhomogeneous dispersion of dopant in SiO_2 and the roughness of core side-etching caused by etching rate difference between SiO_2 and dopants, as well as the increase of absorption loss depending on the dopant.

Splicing of Waveguide and Fiber

The splicing between optical elements of silica glass waveguide and fiber pigtails are made by fusion technology with CO_2 laser as shown in Fig. 4. This fusion splicing is effective technology to satisfy high strength and reliability at splicing portion.

Histogram of the splicing loss is shown in Fig. 5. Although the average value of splicing loss is less than 0.5 dB, the deviation shows fairly scattered values in comparison with conventional fiber splicing. This is due to the poor perpendicularity on both edges of the waveguide and fiber. But, it was confirmed that loss value did not change after acceleration test of 1500G.

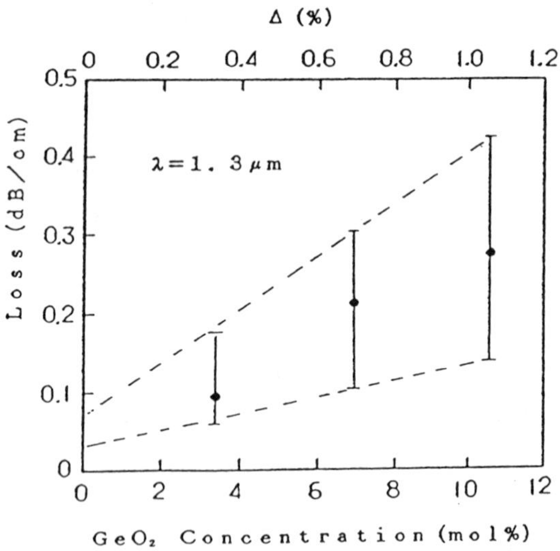

Fig. 3 Relationship between loss and GeO_2 concentration, as well as Δ.

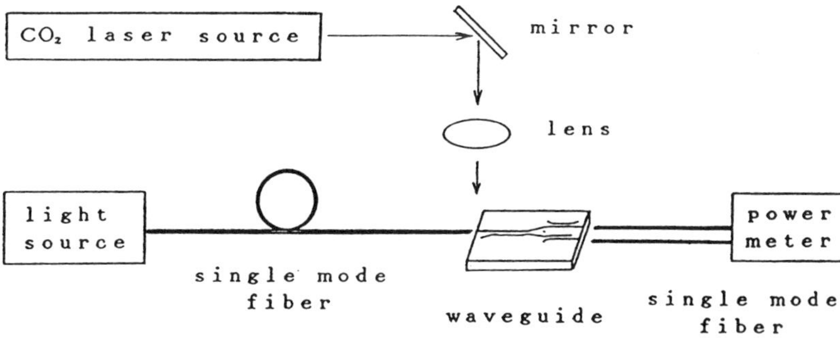

Fig. 4 Fusion splicing method with CO_2 laser.

<u>Guided-Wave Optical Devices</u>

<u>A. Wide Wavelength Spacing Type Multi/Demultiplexers</u>

The multi/demultiplexer of wide wavelength spacing with high stopband rejection are developed by using two types of new monolithic integration method.

The first one consists of three directional couplers of the same structure, and two output ports are arranged in parallel as shown in Fig. 6(a). The bandwidth which can guarantee a stopband rejection loss in excess of 20 dB is about 100 nm, which is 2.8 times as broad as the one by conventional technology. For example, The measured demultiplexing characteristics of 2-channel multi/demultiplexer for 1.3/1.55 μm wavelengths is shown in Fig. 7. These measured results closely agree with calculations. The insertion loss is about 2 dB.

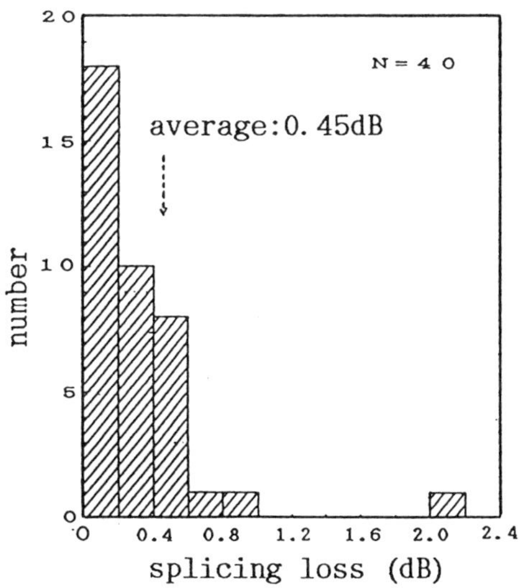

Fig. 5 Histograms of splicing loss.

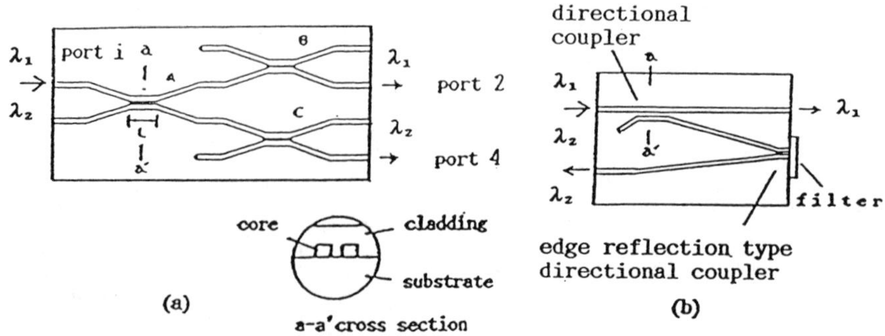

Fig. 6 Multi/demultiplexers: (a) are arranged two output ports in pararell, and (b) are arranged on opposite sides at waveguide edges each other.

The second type of multi/demultiplexer in Fig. 6(b) incorporates two output ports arranged on opposite sides at waveguide edges to lower crosstalk between channels for bidirectional optical transmission. This multi/demultiplexer consists of conventional and reflection type directional coupler, and interference filter. The loss and the crosstalk are less than 2 dB and -40dB, respectively.

B. Narrow Wavelength Spacing Type Multi/Demultiplexer

Center wavelength control of Mach-Zehnder type multi/demultiplexer with narrow wavelength spacing is achieved by irradiating CO_2 laser beam onto one of the two waveguides as shown in Fig. 8(a). Optical path length between two waveguides can be changed by evaporation and diffusion of doping materials in cladding glass (SiO_2-P_2O_5-B_2O_3). It is possible to tune center wavelength in range of more than 20 nm, and the precision of controllable center wavelength is about 0.5 nm. Fig. 8(b) shows an

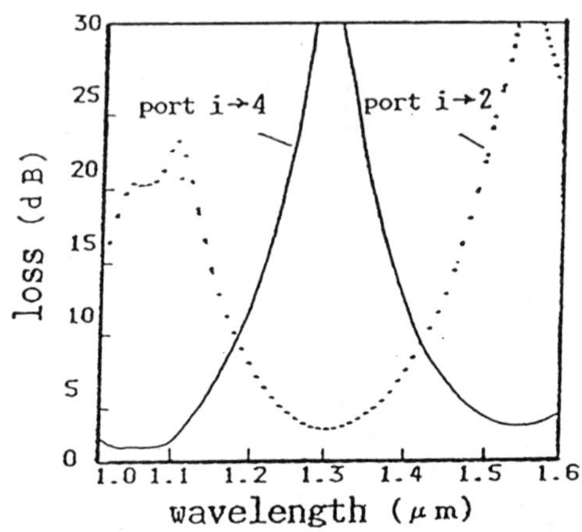

Fig. 7 Demultiplexing characteristics of 1.3/1.55μm multi/demultiplexer.

example of the measured demultiplexing characteristics of multi/demultiplexer for 1.29/1.31 μm wavelengths. The insertion loss of 3 dB and crosstalk of lower than -20 dB are obtained.

C. Cascade Type Mach-Zehnder Multi/Demultiplexer

To obtain low crosstalk at narrow-band multi/demultiplexer, a cascade type Mach-Zehnder multi/demultiplexer as shown in Fig. 9(a) is developed. Center wavelength control technology is also used with this device. The loss and the crosstalk are less than 3.5 dB and -40 dB, respectively as shown in Fig. 9(b). The cascaded structure is effective to get low crosstalk in guided-wave device.

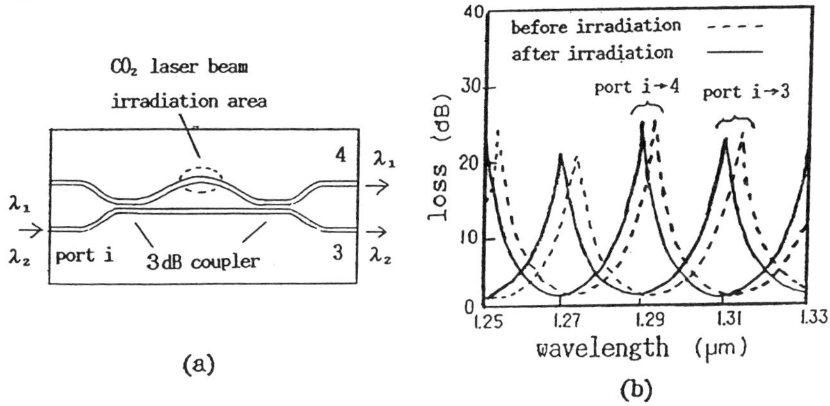

Fig. 8 Mach-Zehnder type multi/demultiplexer: (a) schematic diagram, and (b) demultiplexing characteristics for before and after CO_2 laser beam irradiation.

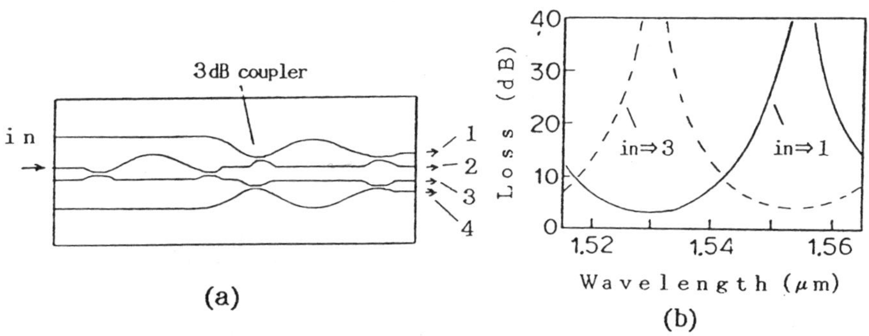

Fig. 9 Cascade type Mach-Zehnder multi/demultipler: (a) schematic diagram, and (b) demultiplexing characteristics.

D. Star Coupler

The star couplers of Nxn and 1xn types are developed. The nxn type star coupler consists of n directional couplers of the same structure. The 1xn type star couplers are constructed Y branch waveguides. The properties of these couplers are shown in Table 2.

E. Bidirectional Optical Device

A 1.3/1.55 μm bidirectional optical device is developed. This device consists of the above mentioned guided-wave type multi/demultiplexer, FP-LD, InGaAs-Pin-PD, and fiber pigtails. An exterior view of the device is shown in Fig. 10. These components are assembled in an air-tight structure package by using hybrid integration technology. This device can be expected to realize smaller size, and also to have reasonable fiber output

Table 2. Properties of Star Couplers

Item	1×N				N×N	
Dividing Number N	2	4	8	16	4	8
Insertion Loss (dB)	≦4.0	≦8.5	≦12.0	≦15.0	≦8.5	≦12.0
Power Deviation (dB)	≦1.0	≦1.0	≦1.5	≦2.0	≦1.0	≦3.0
Operating Wavelength (μm)	1.3 or 1.55					
Return Loss (dB)	≧35					
Directivity (dB)	≧50					

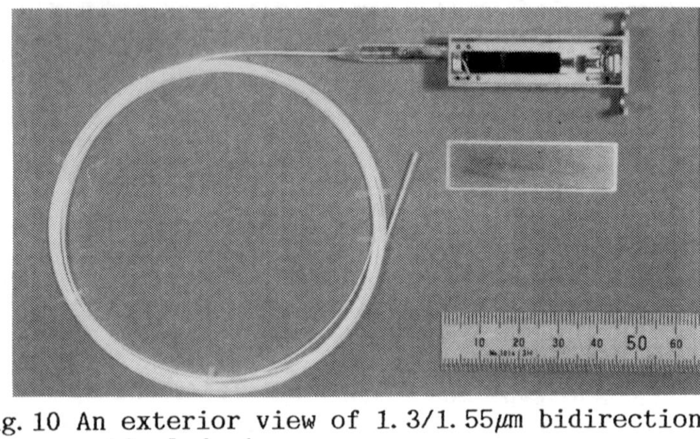

Fig. 10 An exterior view of 1.3/1.55μm bidirectional optical device.

Table 3. Properties of 1.3/1.55μm Bidirectional Optical Device

Item	Propertiy
Threshold Current Ith (mA)	≦35
Optical Output Power (mW)	≧0.25 (at Ith+20mA)
Rise and Fall Time (ns)	≦1.0
Responsivity (mA/mW)	≧0.25
Isolation between 1.3/1.55 μm (dB)	≧40
Dimensions (mm)	50L×14W×10H

power and photodetector responsivity for optical fiber communication as shown in Table 3.

Approach of Low Cost and Small Size Waveguide

Silica guided-wave components have many advantageous properties as compared with fiber type components as shown in Table 1. The size of these guided-wave components are smaller than that of fiber type components. However, the loss of guided-wave components is inferior to fiber type components. To get low loss, it is very important to make the guided-wave components far smaller in size by increasing Δ. The reduction in size can be expected to lead to low cost. That is, to achieve the high Δ of waveguide is an effective approach to get low losses, very small size, and lower cost. Particularly, this high Δ is the best approach to get smaller size of components such as Mach-Zehnder type filter and ring resonator for optical communication systems. For example, the chip size of the Mach-Zehnder type filter can be expected to be reduced by a factor of about 100 by increasing Δ ten times as shown in Fig. 11[10]. Directional coupler can also be obtained to be smaller in size with increasing Δ as shown in Fig. 12[11].

To obtain high Δ waveguide in conventional method, GeO_2, P_2O_5, TiO_2 and so on are used as dopants for the refractive index control of the core or the cladding. However, this method has two problems in getting high Δ. The first problem is that the loss of waveguide tends to increase with increasing the dopant concentration as shown in Fig. 3.

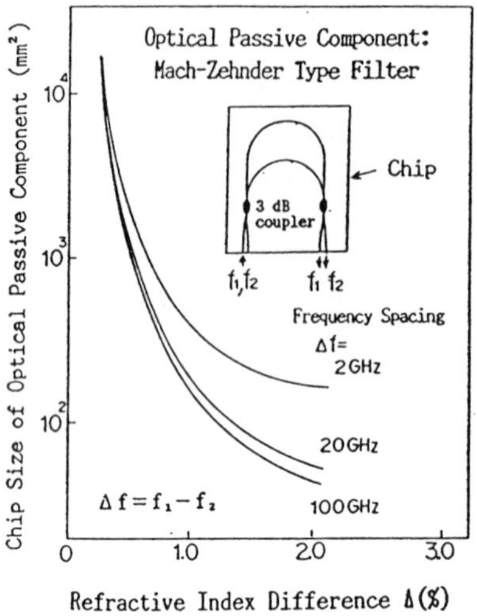

Fig. 11 Relationship between chip size of optical filter and refractive index difference.

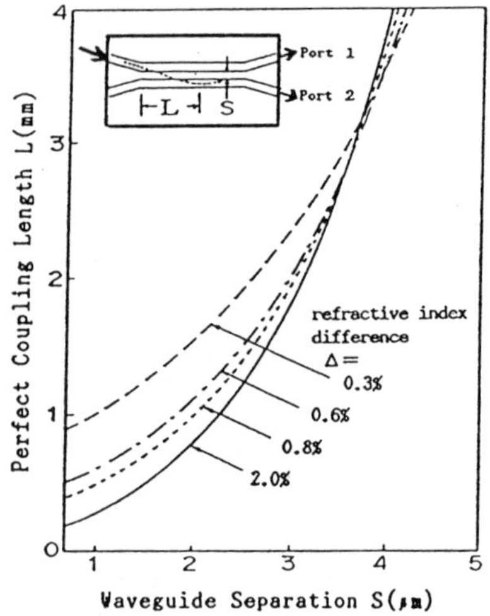

Fig. 12 Relationship between perfect coupling length and waveguide separation of optical coupler.

The second problem is poor patterning accuracy on the photolithography and dry-etching due to the increase of substrate warp by the difference of the thermal expansion coefficient of the core, cladding, and substrate. The upper limit of Δ in the conventional method is considered to be about 1.5 % in practice use because of above mentioned problems.

To solve these problems, high Δ waveguide with SiOxNyHz core glass film formed on a quartz glass substrate have been developed.

The fabrication method of this waveguide is the same as shown in Fig. 1 except for the formation method of the core glass film. This core glass film is deposited with face down type of plasma CVD method as shown in Fig. 13. SiOxNyHz film is deposited on the quartz substrate through chemical vapor reaction at low temperature (250℃) in SiH_4, N_2O, and N_2 gas under plasma

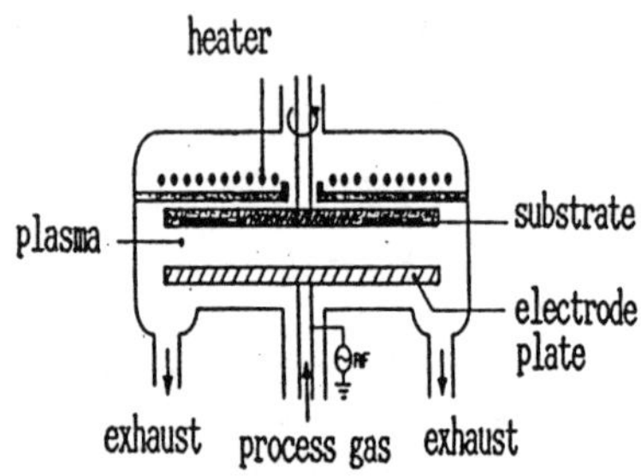

Fig. 13 Schematic diagram of core glass film by plasma CVD method.

atmosphere. As a result, Δ of this SiOxNyHz film can be increased by about 7 % by controlling nitrogen concentration as shown in Fig. 14. This high Δ is one of the features of this waveguide with SiOxNyHz core glass. The second feature is that the chip size of guided-wave type component can be reduced in size by a factor of more than 10. The third feature is low loss property which can be expected by reducing the size. The forth feature is that the warp of quartz glass substrate formed SiOxNyHz core glass film is very small compared with the conventional core glass film as shown in Fig. 15. The fifth feature is that the etching roughness of core side on patterning of core glass film in dry-etching process is very small as shown in Fig. 16. This is an impotant feature, since the scattering loss that depend on the inhomogenious structure of the waveguide can be considerably reduced. Furthermore, fusion splicing between fiber and waveguide can be applied by CO_2 laser beam irradiation, since core, cladding, and substrate of this waveguide consists of the same material that belong to silica glass systems. This application of the fusion splicing is an important condition to obtain high reliability at splicing of fiber and waveguide in practice use.

In addition, the proposed plasma CVD method with face down type can suppress the growth of particles on deposited film. Figure 17(a) shows the photograph of microscope dark view pattern on the glass film surface

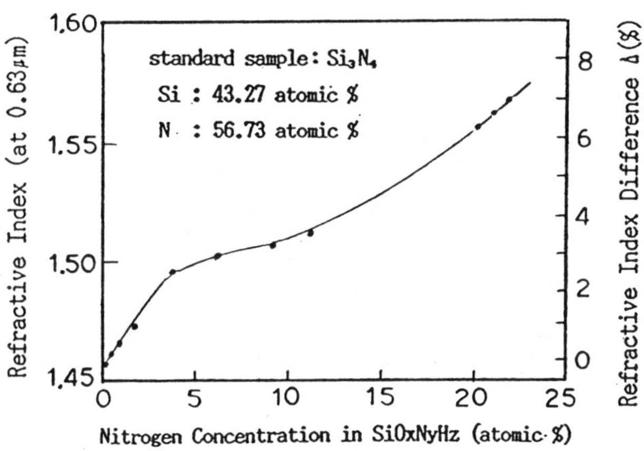

Fig. 14 Relationship between refractive index difference and nitrogen concentration in SiOxNyHz film.

deposited by this proposed method. Distribution of particles on deposited film made by face up type plasma CVD method are shown in Fig. 17(b). Particles on deposited film by the proposed method is very few. This also contributes to low loss of the waveguide.

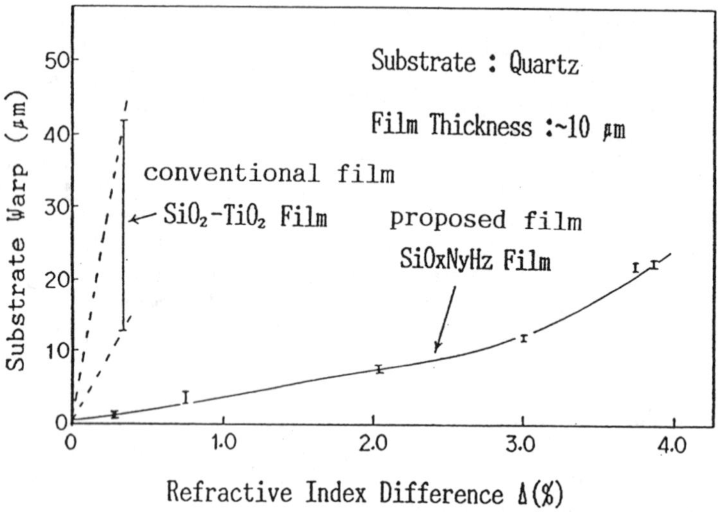

Fig. 15 Relationship between refractive index difference and warp of substrate formed glass film.

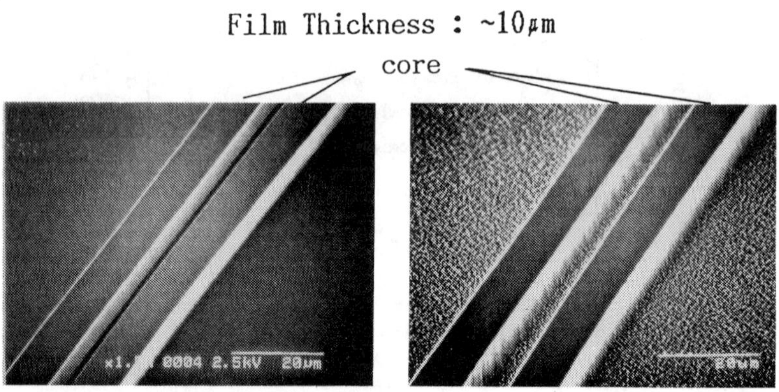

(a) (b)

Fig. 16 SEM photograph of core ridge pattern formed by dry-etching process: (a) proposed core pattern, and (b) conventional core pattern.

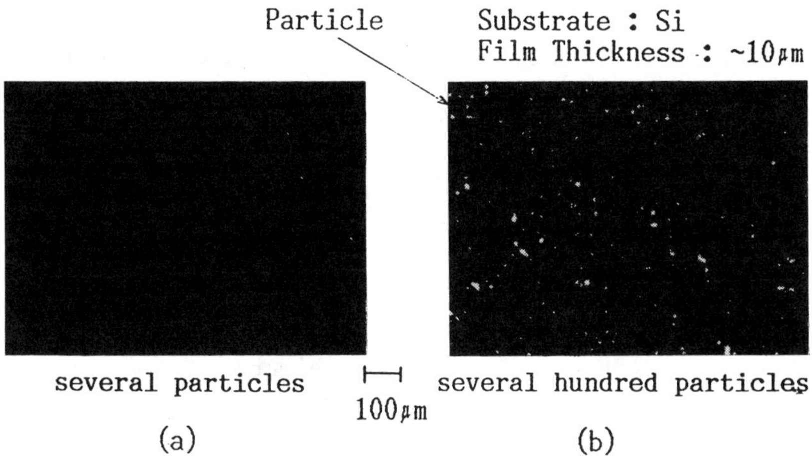

Fig. 17 Photograph of microscope dark view pattern on glass film surface: (a) proposed plasma CVD method, and (b) conventional plasma CVD method.

Conclusions

Many guide-wave devices such as multi/demultiplexers, star couplers, and hybrid integration of optical active components have been developed by using silica optical waveguide structure. This waveguide with low loss and excellent structure parameters has been fabricated on a quartz glass substrate using low temperature glass formation, photolithography, reactive ion etching, and flame hydrolysis glass deposition processes. Fusion technology has also been developed for splicing of waveguide and single-mode fiber by with CO_2 laser beam irradiation. To realize low cost and the small size of the device, the waveguide of high refractive index difference with SiOxNyHz core glass film has also been developed.

Acknowledgment

The author would like to thank Dr. H. Hoshino and Dr. N. Ando of Advanced Research Center, Hitachi Cable, Ltd.. The author also wish to thank S. Kashimura and S. Kawamata of Hitachi Cable, Ltd., and M. Shinohara and S. Koshihara of Hitachi Instruments Engineering co., Ltd., for their assistance in some experiments.

References

1. M. Kawasaki, M. Yasu, and T. Edahiro, "Fabrication of SiO_2-TiO_2 Glass Planner Optical Waveguides by Flame Hydrolysis Deposition, "Electron. Lett., Vol. 19, pp. 583-584, 1983.
2. Y. Yamada, M. Kawchi, M. Yasu, and M. Kobayashi, "Optical Fiber Coupling to High-Silica Channel Waveguides with Fiber-Guiding Grooves," Electron. Lett., Vol. 20, pp. 313-314, 1984.
3. K. Imoto, T. Asai, and M. Maeda, "Silica Glass Waveguide Structure and its Application to a Multi/Demultiplexers," ECOC'88, Sept. 1988, London.
4. K. Imoto, H. Uetsuka, and S. Kashimura, "High-Silica Guided-Wave Optical Devices," 41st ECTC'91, May, Atlanta, Georgia, pp. 483-488, 1991.
5. K. Imoto, H. Sano, and M. Miyazaki, "Guide-Wave Multi/Demultiplexers with High Stopband Rejection," Applied Optics, Vol. 26, no. 19, pp. 4214-4219, Oct. 1987.
6. K. Imoto, H. Sano, M. Miyazaki, and M. Maeda, "Conpensation for Fabrication-Induced Center Wavelength-Shift in Guide-Wave Multi/Demultiplexers," Applied Optics, Vol. 28, no. 10, pp. 1904-1909, May 1987.
7. K. Imoto, H. Sano, M. Miyazaki, Y. Takasaki, and M. Maeda, "Guide-Wave Multi/Demultiplexers with High Stopband Rejection," in Proc. Tech. Group Meet. IECE of Japan, 1987, pap. OQE87-7, pp47-53 (in Japanese).
8. H. Uetsuka, M. Kurosawa, and K. Imoto, "A Novel Method for Center Wavelength Tuning of Silica Waveguide Type Mach-Zehnder Multi/Demultiplexers," Electron. Lett., Vol. 26, no. 4, pp. 251-253, Febr. 1990.
9. K. Imoto, M. Nakamura, and H. Kunugiyama, "Optical Passive Components," HITACHI REVIEW, Vol. 72, pp. 51-58, 1990. 4.
10. K. Imoto and A. Hori, " New Technology for Reduction in Cost and Size of Silica Guided-Wave component," Electron. Lett., Vol. 28, no. 17, pp. 1665-1667, 1992. 8.
11. H. Sano and K. Imoto, " Analysis of High-Silica Optical Directional Coupler by Boundary Element Method," in Technical Digest, Annual Meeting of IEICE of Japan, 1987, paper345.

OPTICAL DEVICES FOR INTERCONNECTIONS

Hajime Imai
Fujitsu Laboratories Ltd.,
10-1 Morinosato-Wakamiya, Atsugi 243-01 JAPAN

The optical interconnection is the new important technology, and in the near future, intra/inter board connections will be done by the optical signals. This paper introduces the current device characteristics of lasers and photodiodes and discuss the future optical interconnection.

1. Introduction

The present optical devices such as lasers and photodiodes have been investigated aiming at the application to the trunk line transmission in the telecommunication field In this purpose the high performance of the devices is the key issue because of the large volume and long distance data transmission. For examples, the spectrum of the laser is single at 1.3 or 1.55 μm wavelength region and the modulation bandwidth is more than several giga Hz.

Figure 1 maps the optical signal transmission field on a chart of transmission distance versus bitrate[1]. Optical interconnection is large data throughput(high bitrate) but a rather short distance connection, while the trunk line transmission is toward the long distance and high bitrate. Therefore, in the optical interconnection, requirements to devices are different due to the short distance transmission. When we can not realize the high performance laser, the substitution is to use array devices.

In optical interconnection, one of the most important points is its cost. Optical interconnection is only to substitute electrical interconnection, so the increase in the cost due to the adoption of lasers, photodiodes, and related circuits should be minimized.

In this paper, we discuss the state of the art of lasers and photodiodes and the possibility of the application to interconnection.

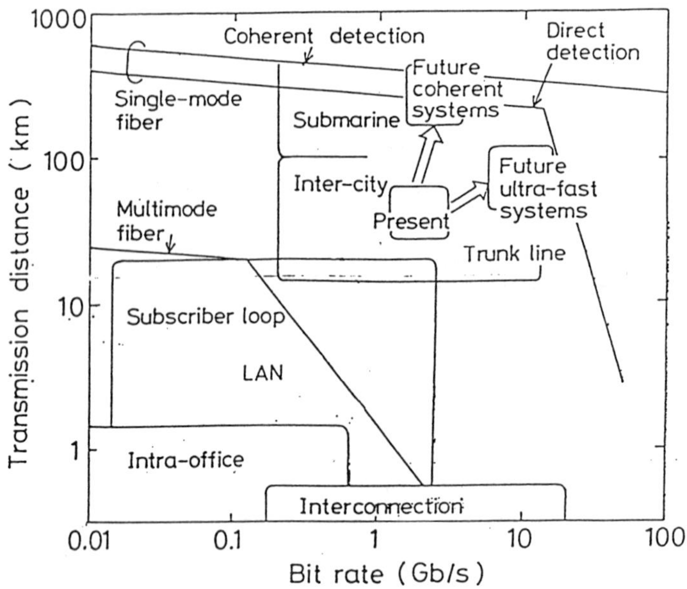

Fig.1 Optical signal transmission field[1]

2. Systems

Optical interconnection is divided into 3 categories, according to the scale of interconnect:

1. Inter equipment
2. Intra/inter board
3. Intra/inter wafer

Inter equipment connection is between machines and systems, and generally uses optical ribbon fibers. The remaining two forms of interconnection are similar in which they sometimes use optical waveguides for signal transmission.

In inter equipment connection, the system hardware architecture is almost the same as the conventional trunk line transmission, especially when the inter equipment interconnection is over short distance. Intra/inter board and wafer connection will be different because the transmitted signal does not always contain the full data length. This means that sometimes only the clock signal is distributed, because the clock signal should be sent to each port of LSI without variation of the delay.

The need of the optical interconnection is derived from another problem --- the wiring connection limit. Figure 2 shows the relation between the modulation bitrate and the transmission length. Coaxial cable for CATV has a large capacity but it is bulky with a diameter of about 10 cm --- far too large to be suitable for interconnection. Coaxial cable small enough for interconnection with a diameter of about 1 mm has a small capacity. When the modulation bitrate increases to 1 Gb/s, the possible transmission length of it is as short as 1 m. Optical fiber cable is flexible and has the large capacity and, although transmission is limited by optical loss and chromatic dispersion, its limit is large enough for optical interconnection(Fig.2).

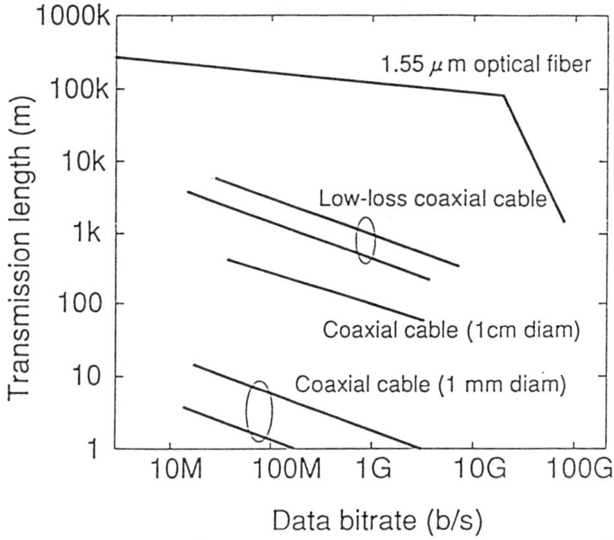

Fig.2 Transmission length limits in coaxial cables and optical fibers

The electrical strip-line is also in the same situation. However, as the clock frequency used in the present LSI is below 100 MHz, this problem does not become fatal for the interconnection on board or wafer. In the future, when the clock cycle increases up to 1 GHz, this becomes crucial.

When the optical interconnection is introduced into the real systems, the present devices commercially available will not be used. Because the threshold current of lasers is so high that it causes the large power consumption, the optical coupling system between laser and optical fiber is so large that it needs the large space, and the capacitance of the photodiode is so large that it limits the frequency response or photosensitivity. Therefore, much improvement in these devices is required.

The specification of optical devices can be summarized as follows:

Small power dissipation
High volume data transmission: parallel or serial
Small module size

3. Key devices

3.1 Lasers

According to above discussions, the important point of lasers is to reduce the threshold. Many studies have been performed for reducing threshold. The short cavity, the quantum well(QW) structure, and the strained QW structure are dominant methods. Recently the strained QW structure[2] was extensively investigated and low threshold current characteristics were reported. Figure 3 shows threshold current density and lattice strain of InGaAs/InP lasers.[3-6] The active region of this laser consists of an InGaAs well layer and InGaAsP barrier layer. The quantity of strain is set by changing the mole fraction of InAs of the well layer.

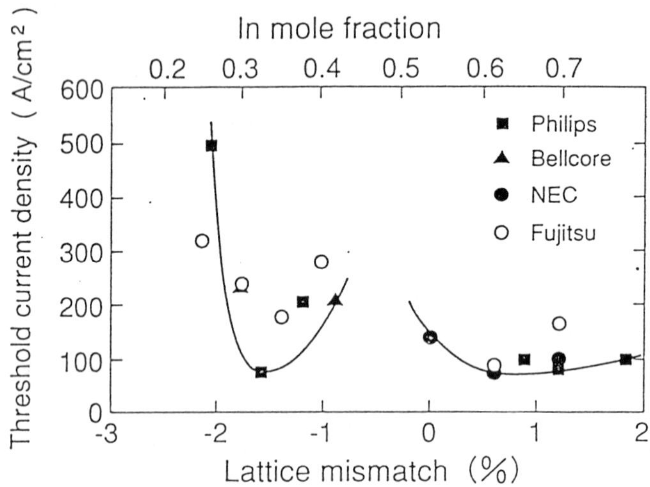

Fig.3 Threshold current density in strained-layer quantm well (QW) lasers[3-6]

Since it is difficult to estimate the strain, this figure plots the lattice mismatch. Threshold current density is expressed per well. When the active layer is stressed by lattice mismatch, the band structure is changed. For compressive stress, the degenerated valence band is released, the heavy hole band goes higher than the light hole band, and the heavy hole effective mass is decreased. For tensile stress, the degenerated valence band is also released. The light hole band goes higher than the heavy hold and the light hole effective mass is increased. Recombination occurs in the heavy-hole band for compression and in the light hole band for tension. It is still a question which stress is better for low threshold. From Fig.3, this is not easy to conclude, since each produces almost the same threshold. The difference exists in the lasing polarization, TE for compression and TM for tension.

Flat surface buried heterostructure(FBH) laser[7] shown in Fig.4 was fabricated in both compressive and tensile strained QW structures. Figures 5(a) and (b) show the L-I characteristics of both compressive and tensile strained QW lasers.[6] Both facets of these lasers are coated with 90% reflectivity. High temperature operation is also achieved and the maximum temperature of CW operation is 150 ℃ for compression and 120℃ for tension. For threshold current at 20℃, tensile stress is better. For the high temperature operation, compressive stress is better. More investigation is required to reduce the threshold at high temperatures.

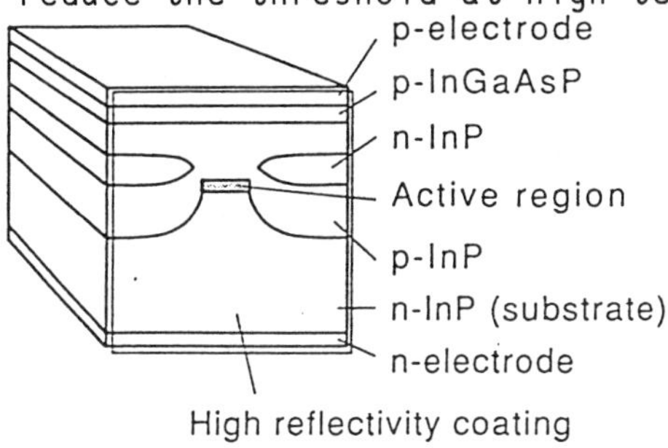

Fig.4 Flat surface buried heterostructure laser[7]

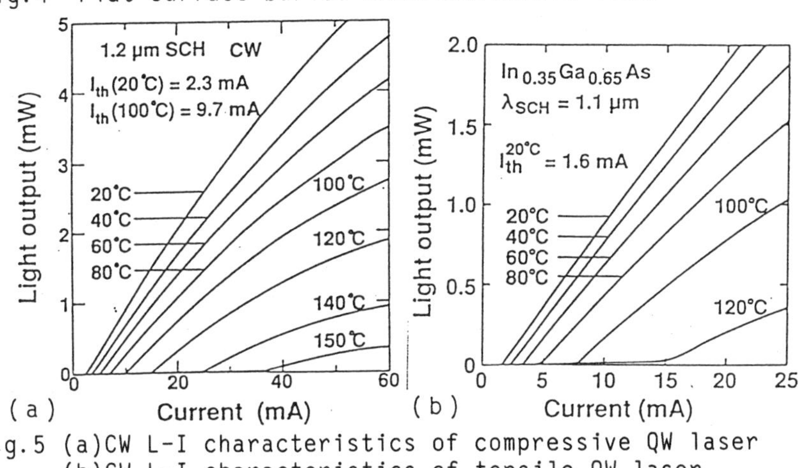

Fig.5 (a)CW L-I characteristics of compressive QW laser
(b)CW L-I characteristics of tensile QW laser

A high frequency response is also required. Recently 5 Gb/s modulation was reported by using tensile strained QW lasers.[8] The inferiority in the frequency response of QW lasers to that of bulk lasers is widely discussed because of the gain saturation effect in the QW structure. However in this modulation bitrate region, this effect does not seem to matter.

3.2 Photodiodes

For trunk lines, the received signal level is so weak that amplification keeping low noise level is inevitable in the receiver. For this reason, the avalanche photodiode (APD) is widely used. To reduce the excess noise, new materials are under investigation. When considering the optical interconnection, however, APD's are not always appropriate for photodiodes because an automatic gain control circuit and a DC/DC converter are necessary. The adoption of these circuits increases power consumption. Thus, other devices such as pin photodiode (PD) must be examined. The bias voltage for pin PD is 5-6 V, much lower than that for the APD.

The high speed response is dependent on its capacitance which is determined by the photosensitive area and the parasitic element. The capacitance of the photosensitive area corresponds to how small the area can be. Figure 6 shows a PD with back illuminated structure.[9] The diameter of the photosensitive area is 40 μm. This PD was a planar structure to reduce the surface leakage current. Two diodes, connected in opposite directions, are used. The photosensitive diode area is small, but the other diode remains large so its series resistance is small. This large diode does not affect the performance of the photodiode. Figures 7(a) and (b) show C-V and I_D-V characteristics. With 4-5 V bias, the capacitance is about 0.15 pF. The dark current under this condition is in the order of pA. The 3dB bandwidth is more than 8 GHz for 50 Ω load resistance.

Overall performance is determined by both the PD and the receiver circuit. To increase the receiver sensitivity, PD sensitivity or load impedance can be increased. In the latter case, it is necessary to reduce the capacitance. Since there is a limit to reduce the PD capacitance, the parasitic capacitance -- especially stray capacitance from PD mounting -- should be reduced. For optical interconnection, the structure should be simple, with capacitance as low as possible.

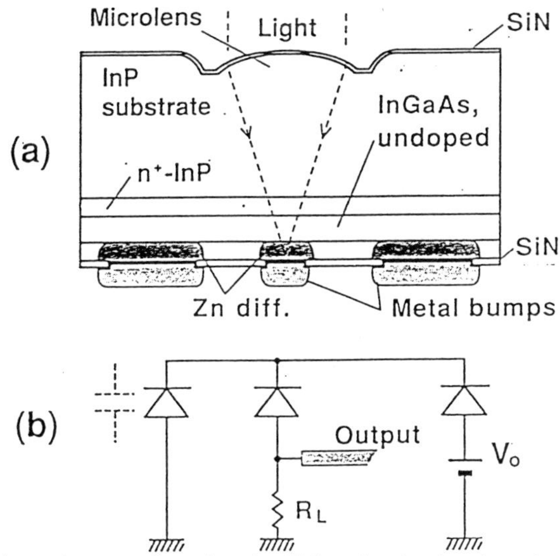

Fig.6 (a)Structure of a back-illuminated pin photodiode[9]
(b)Its equivalent circuit

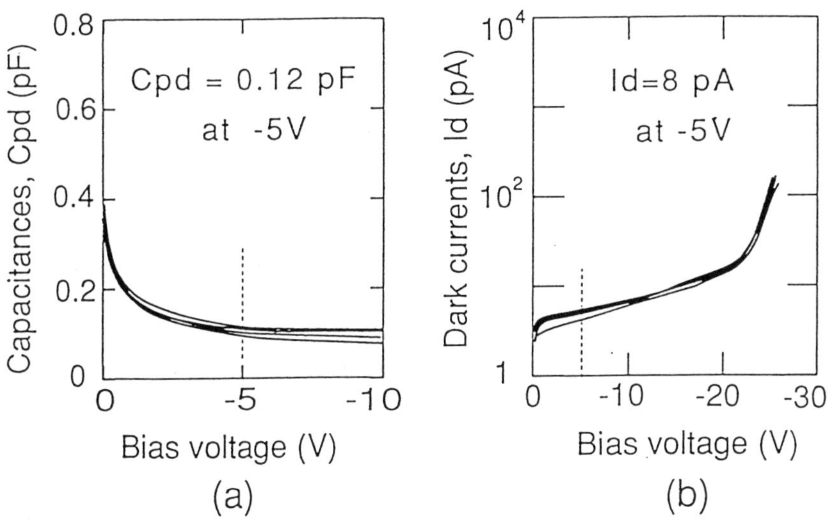

Fig.7 (a)C-V characteristics
(b)I_D (dark current)-V characteristics

4. Packaging and mounting

When packaging optical devices, optical coupling to optical fiber or waveguide is inevitable. For the trunk line transmission systems, the optical coupling system is

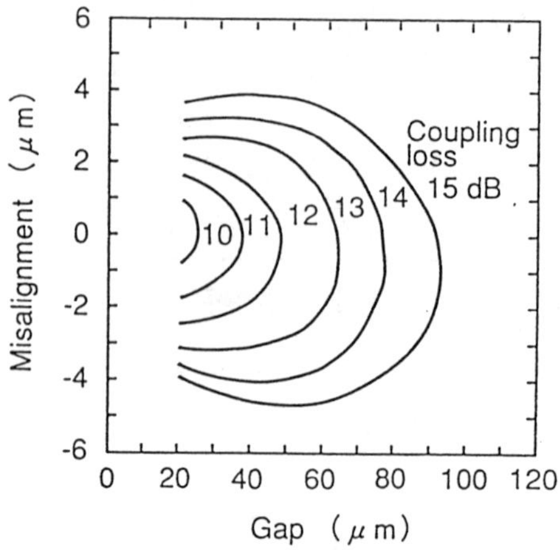

Fig.8 Variation of coupling efficiency for misalignment of optical fiber

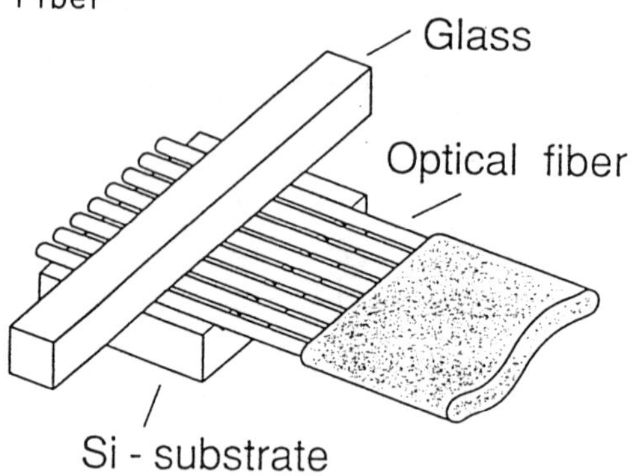

Fig.10 Coupling loss for a ribbon fiber

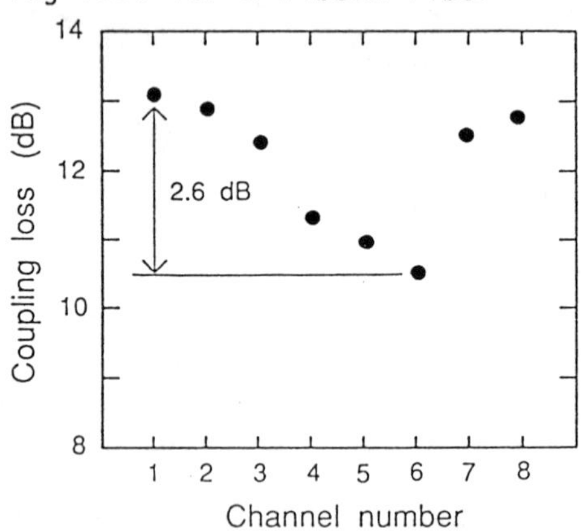

Fig.9 Tool for alignment of optical fibers

complicated such as confocal lens system. In this method, the coupling efficiency is as high as 2-3 dB, but fixing lenses to the package becomes difficult. For optical interconnection, simple coupling system is recommended. Figure 8 shows the measured variation of the coupling efficiency against the deviation of the single mode fused silica optical fiber. In this figure, the simple butt joint system is used. A laser used in this experiment is a 1.55 μm FBH laser as shown in Fig.4. The 3 dB tolerance is a few μm and the coupling efficiency is about 10 dB. In case of optical interconnection, the coupled optical power in optical fiber is not necessarily at high level, so this coupling efficiency is not a problem. Rather the simple coupling system is important. Butt joint is most simple. When we use array devices, the alignment of an optical ribbon fiber is a key issue. Figure 9 shows the mounting tool of the ribbon fiber with V-grooved Si substrate. By using this method each optical fiber is placed within 1 μm accuracy. Figure 10 shows the variation of the coupling efficiency using 8-channel laser array and 8-element ribbon fiber. The variation is 2.6 dB.

Figure 11 shows the photograph of the module. In this module, 4-channel driver LSI, 4-channel laser array, and 4-element ribbon fiber are mounted. The size of one laser is 200 μm long and 250 μm wide. 2.5 Gb/s modulation is demonstrated by using this module.[10]

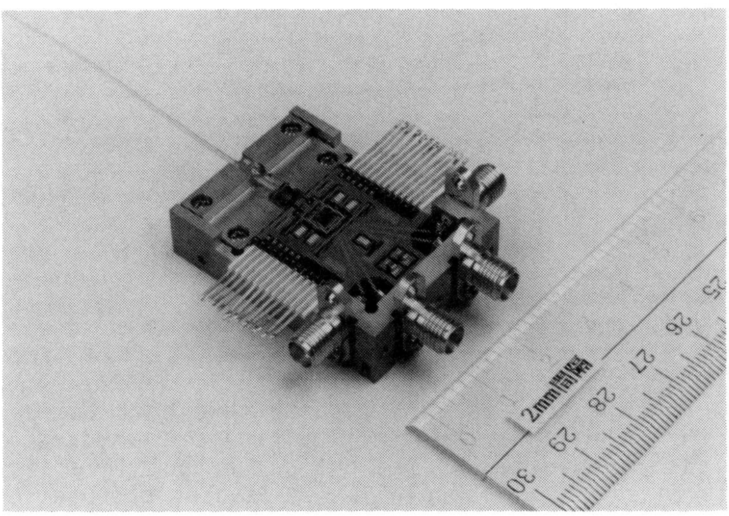

Fig.11 Photograph of a laser module[10]

5. Summary

Devices for optical interconnection were discussed. Present devices do not reach the level for inter/intra board or wafer interconnection. If intra board or wafer interconnection is possible, LSI layout is completely changed as shown in Fig.12. The 3-dimensional interconnection is shown. To realize this figure, the monolithic integration of optical devices of lasers, photodiodes, and waveguides should be investigated. After optical interconnection, the signal output is obtained freely from any position. This not only releases the present wiring limitation but also changes the LSI device design.

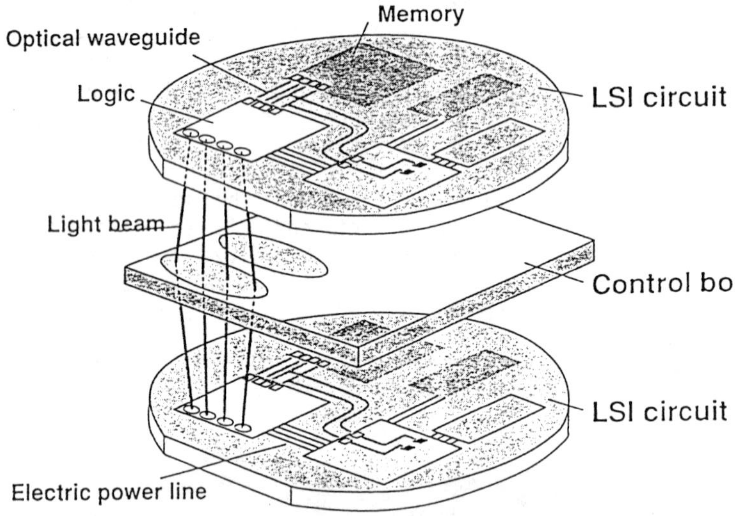

Fig.12 3-dimensional optical interconnection image

For this purpose, every laser should be able to operate under the electric output level which can be fed directly from logic or memory devices. Such output is in the range of a few mA and a few V, so that the low threshold lasers are foreseeable. The surface emitting laser with the microcavity is promising candidate.[11] The optical output level is not needed to be so high. The pin PD may be used because the DC bias level is 5 V, although it would be first necessary to improve the PD output in order to drive next stage devices.

The heteroepitaxy may provide other breakthroughs. Present heteroepitaxial growth on Si wafers results in many misfit dislocations. This is a hazard for long term reliability. Recently long term reliability of InGaAsP/InP lasers on Si substrate was reported.[12] InGaAsP/InP lasers may be suitable for the heteroepitaxial devices. Optical waveguide is another key issue. There are reports by using SiO_2 or organic materials. Lower loss is obtained for SiO_2 waveguide, but the fabrication on Si LSI is a coming problem. Semiconductor waveguides are promising, but electric isolation may be difficult.

The intra/inter board or wafer connection is the final goal for optical interconnection.

References

1. Wada O.: from correspondence.
2. Thijs P.J.A., Tiemeijer L.F., Kuindersuma P.I., Binsma J.J.M., and van Donogen T.: "High-performance 1.5 μm wavelength InGaAs-InGaAsP strained quantum well lasers and amplifiers", IEEE J.Quantum Electron., QE-27, pp.1426-1439 (1991).
3. Thijs P.J.A., Binsma J.J.M., Tiemeijer L.F., Kuindersuma P.I., and van Donogen T.: "Improved performance 1.5 μm wavelength tensile and compressively strained InGaAs-InGaAsP quantum well lasers (Invited)", ECOC/IOOC'91, Tu.A4.1, Paris (1991).
4. Zah C.E., and Bhat R., Pathak B., Caneau C., Favire F.J., Andreadakis N.C., Hwang D.M., Koza M.A., Chen C.Y., and Lee T.P.: "Low threshold 1.5 μm tensile-strained single quantum well lasers", Electron.Lett., 27, pp.1414-1415 (1991).
5. Yamada H., Kuroda N., Yamazaki H., Kitamura M., and Mito I. : "1.55 μm strained InGaAs/InGaAsP MQW LDs", IEICE Technical Report, OQE91-29, pp.67-72 (1991).
8. Yamamoto T., Nobuhara H., Sugawara M., Fujii T., and Wakao K.: "Low threshold operation of tensile strained InGaAs/InGaAsP quantum well lasers with simple separated-confinement heterostructures", SSDM'92, B-6(2), Tsukuba (1992).
6. Tanaka K., Nobuhara H., Yamamoto T., Machida T., Fujii T., and Wakao K.: "High temperature operation of low threshold strrained-layer InGaAs/InGaAsP multiple quantum well lasers", ECOC'92, Mo B2.4, Berlin (1992).
7. Ishikawa H., Yamakoshi S., and Isozumi S.: "Distributed feedback laser emitting at 1.3 μm for high-bit-rate systems", FUJITSU Sci.Tech.J., 22, pp.451-460 (1986).
8. Odagawa T., Nakajima K., Tanaka K., Inoue T., Okazaki N., and Wakao K.: "High-speed operation of strained InGaAs/InGaAsP

MQW lasers", 13th IEEE Semiconductor Laser Conf., H-3, Takamatsu (1992).
9. Makiuchi M., Hamaguchi H., Mikawa T., and Wada O.: "Easily manufactured high-speed back-illuminated GaInAs/InP p-i-n photodiode" IEEE Photonics Tech.Lett., 3, pp.530-531 (1991).
10. Rokugawa H., Ikeuchi T., Fujimoto N., Horimatsu T., Nobuhara H., Miura K.: "Multigigabit-per second optical parallel transmitter module using a 4-ch laser diode array with a wide temperature range operation", ECOC'92, Tu P1.11, Berlin(1992).
11. Soda H., Iga K., Kitahara C., and Suematsu Y.: " GaInAsP/InP surface emitting injection lasers", Japan.J.Appl.Phys., 18, pp.2329-2330 (1979).
12. Sugo M., Mori H., Itoh Y., and Sakai Y.: "InP-based optical devices on Si substrates", SSDM'92, S-II-12, Tsukuba (1992).

OPTOELECTRONIC PACKAGING OF MULTI-CHANNEL LED/PD ARRAY MODULES FOR HUNDRED Mb/s/ch PARALLEL OPTICAL INTERFACES

M. ITOH, *T. NAGAHORI, **H. KOHASHI, H. KANEKO,
H. HONMOU, ***T. UJI, and *M. FUJIWARA

Functional Devices Research Laboratories
*Opto-Electronics Research Laboratories
**R&D Planning and Technical Service Division
***Kansai Electronics Research Laboratory
NEC Corporation
4-1-1, Miyazaki, Miyamae-ku, Kawasaki, Kanagawa, 216 Japan

ABSTRACT

A novel optoelectronic devices packaging technology has been developed for high performance module operation and module package miniaturizing for high speed parallel optical interface multi-channel LED/PD array modules. Optical and electronic devices were mounted on a lateral point and the common submount surface, respectively. Flip-chip technology, using solder bumps, was employed for optical array chip bonding, in order to simplify the chip mounting. The AlN submount was used for both insulation and heat removal, in order to obtain stable module operation under simultaneous multi-channel utilization. The overall module package height was reduced to 7mm, which made it possible to attach the PWBs to a standard card rack with 0.5" spacing. A 12-channel 150Mb/s/ch 100 meter parallel optical transmission has been demonstrated.

1. INTRODUCTION

Parallel optical interfaces are very attractive for supercomputer and broadband switching systems, due to the advantages regarding large transmission capacity, long distance and small cable size, compared to parallel interfaces using metallic conductor electronic cables. Recently, monolithically integrated linear arrays of light emitting diodes (LEDs) and photodiodes (PDs) have been reported for hundred Mb/s/ch parallel optical interfaces[1-5]. When these LED/PD arrays are put into practical optical interfaces, they must be packaged in modules along with the necessary electronic circuits, in order to reduce inductance and capacitance, which might cause electronic crosstalk at high speed operation, as shown in Fig. 1 These optical and electronic devices cause much heat under simultaneous multi-channel operation, which might damage stable device characteristics. High thermal conductivity material is also required for device submount.

The optical modules are usually attached to printed wiring boards (PWBs), which contain many other electronic circuits. Assembly simplification for components, such as optical array chip and optical fiber array, and a compact module package size are also required, in order to reduce the fabrication cost and to increase components assembly density on the PWBs.

This paper reports optoelectronic packaging techniques for use on multi-channel LED/PD array modules, which meet the above requirements. The overall module package height was reduced to 7mm, which made it possible to attach the PWBs to a standard card rack with 0.5" spacing. The characteristics for 12-channel parallel optical transmission at 150 Mb/s/ch are also described.

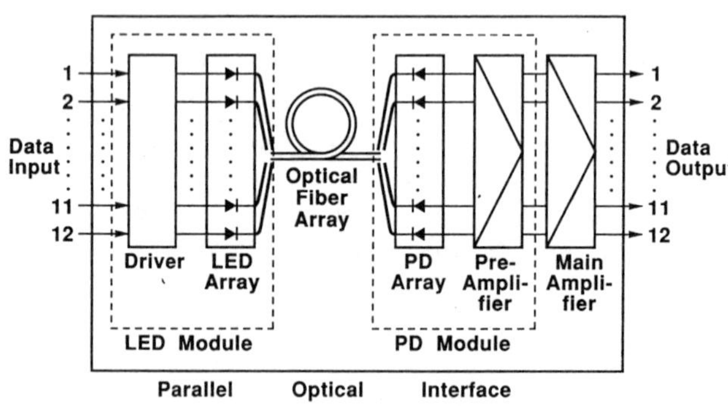

Fig.1 Module block diagram.

2. MODULE DESIGN

Figure 2 shows a cutaway view of the newly developed LED/PD array module. In order to meet the module requirements mentioned in the previous section, the LED/PD array and the driver/preamplifier circuits were mounted together on the common heatsink submount. AlN was employed as high thermal conductivity submount material. These optical device modules are usually attached to the PWBs along with many other electronic circuits. The fiber array must be assembled horizontally on the PWB, like electronic conductor cables, in order to allow attaching many PWBs to the card rack. When the LED/PD array is mounted at the submount top surface, the optical axis must be bent so that it may coincide with the optical fiber array optical axis. A 45° mirror block or beveled optical fiber end is used for optical axis bending. In this case, since precise cutting and polishing processes are needed, fabrication costs may increase. Furthermore, since spacing between LED/PD array and optical fiber array is increasing, coupling losses may be larger, compared with butt coupling.

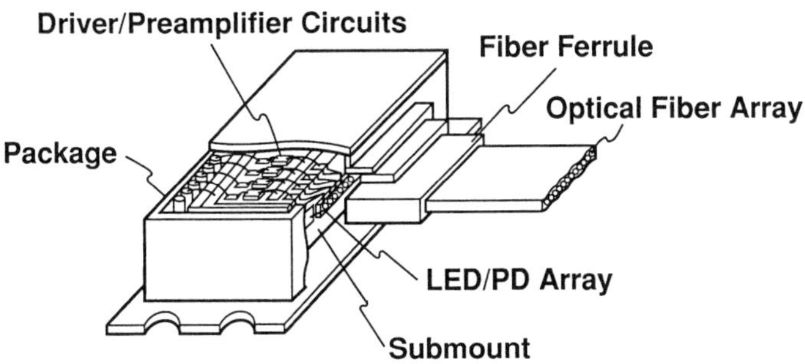

Fig.2 LED/PD array module package cutaway view.

For resolving these problems, the LED/PD array was mounted at a lateral point on the common submount. Figure 3 shows a submount diagram for the LED array. The LED array was mounted at a lateral point. LED lights were coupled directly to the closely attached optical fiber array, as shown in Fig. 4. This optical coupling also reduced the overall module package thickness to 7mm, which made it possible to attach the PWBs to a standard card rack with 0.5" spacing.

The driver/preamplifier circuits were hybridly mounted by

die bonding to the common submount top surface. When they are mounted at a lateral point, like an LED/PD array, large lateral space is needed for the submount. Therefore, package thickness increases. This device arrangement allows reducing the separation between electronic circuits and optical devices, as shown in Fig. 3. Consequently, inductance and capacitance, engendered by the electronic circuitry, are decreased, and electronic crosstalk between adjacent channels for high speed operation can be reduced.

For these high speed multi-channel optical modules, an optical and electronic device submount needs large electrical isolation and high thermal conductivity in order to achieve the low electrical crosstalk and stable signal transmission under simultaneous multi-channel operation. A small difference in thermal expansion coefficient between a submount and an optical device is also required, to reduce the thermal stress under high temperature bump reflow process. The AlN submount was employed for insulation, heat removal and low thermal stress. AlN is a suitable submount material for the InP LED/PD array chip, because the thermal expansion difference between AlN and InP is smaller than the other submount materials, such as Al_2O_3 or SiC.

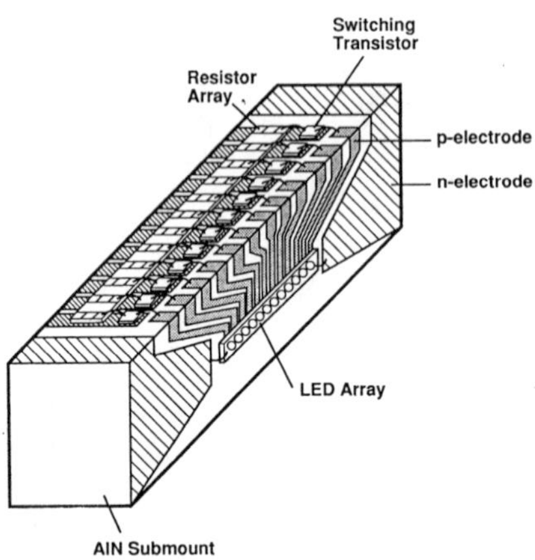

Fig.3 Submount diagram for LED array.

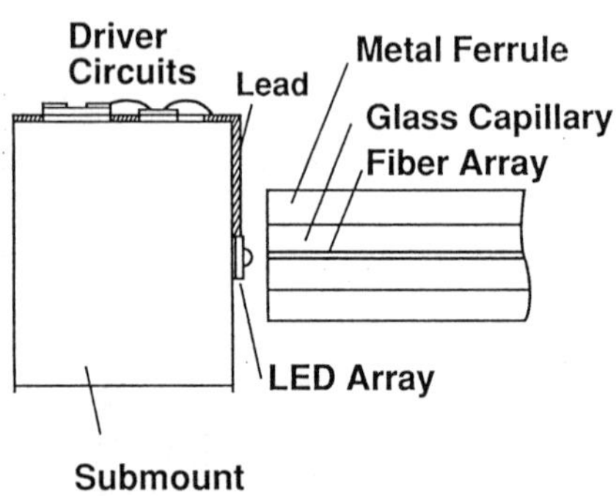

Fig.4 LED-fiber optical coupling diagram.

3. MODULE ASSEMBLY

3.1 Optical Chips

The LED array is a monolithically integrated 1×12 linear array, employing mesa-structure InGaAsP/InP 1.3 μm surface emitting LEDs [5-8] with 20 μm mesa diameter and 250 μm LED spacing. The PD array employs InGaAs/InP PIN PDs [5][6] with 80 μm diameter and 250 μm spacing, grown on a semi-insulating InP substrate. The optical array chip was fixed to the submount by flip-chip bonding. Flip-chip bonding using solder bumps is one of the effective techniques for both high precision and high density mounting for optical components. The self-alignment effect, inherent in bump bonding, can eliminate the requirement for precise chip bonding, as shown in Fig. 5.

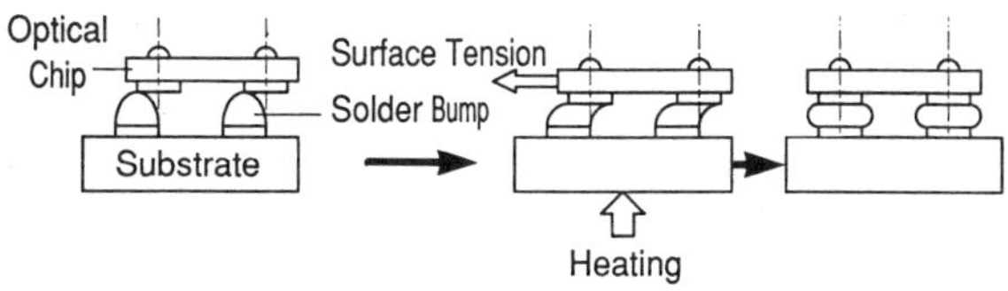

Fig.5 Flip-chip bonding Process.

Figure 6 shows the solder bump fanout array, after reflow on the submount. Figure 7 shows the LED array chip bonded solder bumps. The electrically isolated twelve p-electrodes on the optical array chip surface were directly bonded to individual submount fanouts by solder bumps. Perfect 12-channel connection between array chip and submount electrodes could be accomplished at one time. The self-alignment effect for the solder bump could allow using a simple bonding procedure, without precise chip alignment. Here, positioning accuracy was within about 3 μm. The accuracy would be higher as solder bump diameter becomes smaller.

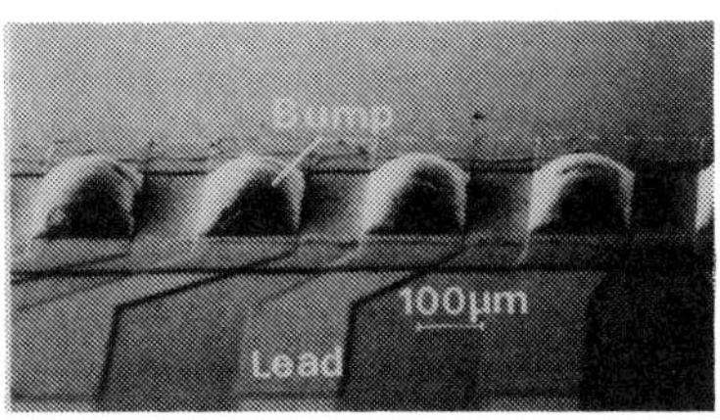

Fig.6 Solder bump array after reflow.

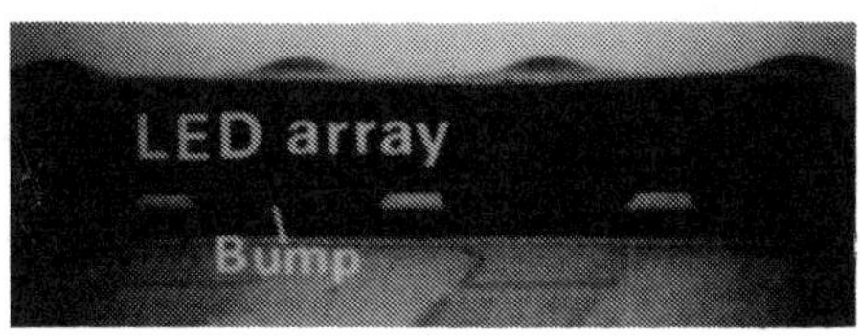

Fig.7 LED array bonded solder bumps.

Figure 8 shows the solder bumps formation and flip-chip bonding process. Process steps are as follows: (1)The Cr/Ni thin metal layer is deposited by evaporation. (2)Metal layer leads are patterned by photolithography technique. Polyimide dams are then fabricated on the leads in the same way. (3)The PbSn solder layer is deposited by electro-plating. (4)PbSn solder bump fanouts are formed by the reflow process. (5)The optical array chip is set on the solder bumps. (6)The chip is self-aligned and bonded.

Table 1 shows solder bump specifications. Reflow temperature was about 180 degrees centigrade.

Table 1 Solder bump specification.

Base Metal Thickness (μm)	Cr / Ni 0.1 / 2.0
Pad Size (μm)	150 × 400
Material	PbSn
Height (μm)	~ 50
Reflow Temperature (℃)	180

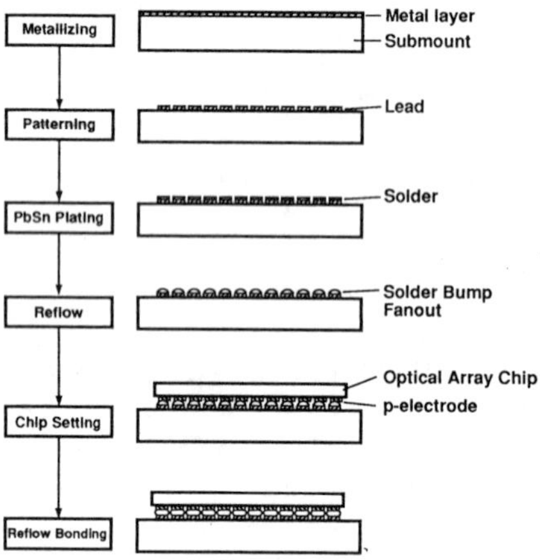

Fig.8 Solder bumps formation and flip-chip bonding.

3.2 Electronic Devices

The driver circuits were composed of switching transistors and resistors, as shown in Fig. 4. Transimpedance amplifier ICs were used for the preamplifier circuits. They were mounted by die bonding to the submount surface. Electronic connection between electronic circuits and optical array chips was made by wire bonding through Cr-Ni thin metal layer leads, as shown in Fig. 4. The submount was bonded to the metal package base, where there were 0.05"-pitch pins.

3.3 Optical Element

The optical fiber array, which had a 62.5 μm core diameter, 250μm spacing graded index 12-channel multimode optical fiber, was held by an inner glass capillary and outer metal ferrule, as shown in Fig. 9. The glass capillary was employed for the fiber substrate, in order to reduce the overall ferrule thickness.

The LED's lights were coupled directly to the optical fiber array, as shown in Fig. 4. The reasons are that optical axis alignment can be accomplished by monitoring optical power from the fiber output, and that large optical lights can be coupled without a coupling lens. Figure 10 shows the PD-fiber optical coupling. The PD was coupled to the output light from the optical fiber through a micro lens array, because, optical axis alignment can be easily accomplished by monitoring the output peak optical power from the fiber. Figure 11 shows a ball lens array. A sapphire ball lens array was employed for micro lens array fabrication simplicity. It had 250μm diameter and 250μm spacing, was covered with a glass block, and was adhered to the facet in the optical fiber array ferrule.

The fiber ferrule was fixed to the module package by YAG laser welding techniques, in order to obtain high reliability, after optical axis alignment. The LED/PD array module package dimensions, shown in the photographs in Fig. 12, were 15L×30W×7H mm, and 15L×45W×7H mm, respectively.

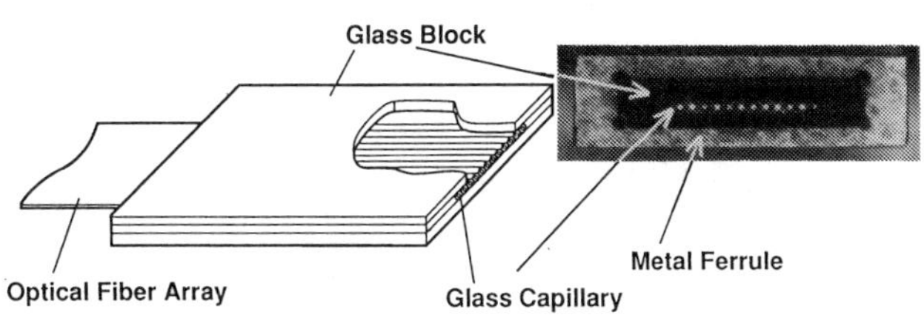

Fig.9 Optical fiber terminal.

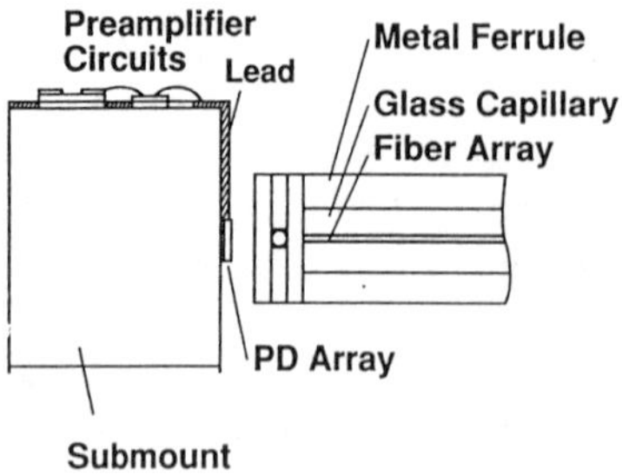

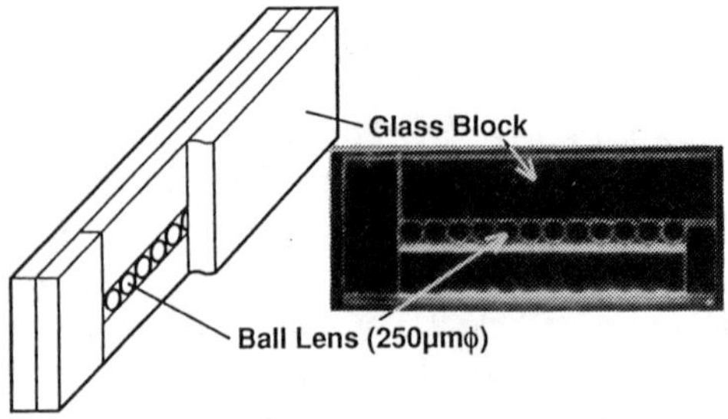

Fig.10 PD-fiber optical coupling. Fig.11 Micro lens array.

(a) LED

(b) PD

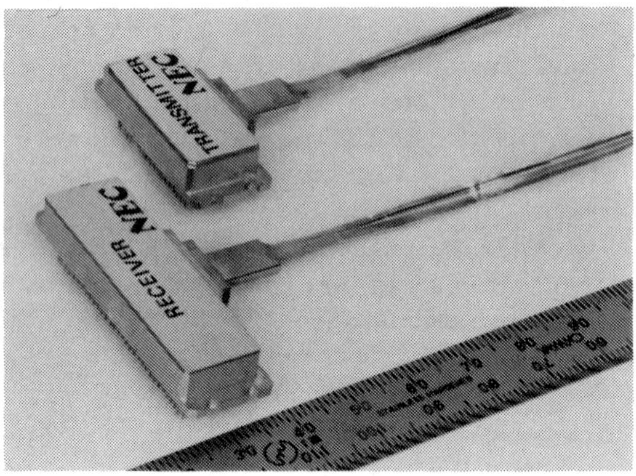

Fig.12 Array module photographs.

4. TRANSMISSION EXPERIMENTS

Figures 13 and 14 show coupling efficiency degradation (the difference between coupled power produced by optimum butt coupling and that produced by this coupling) for each channel in the LED and PD array, respectively. The coupling efficiency fluctuation for the LED array was ±0.2dB, and the average coupled power was -21.4±0.4 dBm at $15mA_{p-p}$ NRZ modulation. The coupling efficiency fluctuation for the PD array was ±0.2 dB. The DC optical crosstalk values between adjacent channels were small, at an average level of -39dB and -29dB for LED and PD modules, respectively. Table 2 shows the module features. Excess loss was about 0.8dB for the LED, and 1.2dB for the PD.

Figure 15 shows the parallel optical transmission experimental set-up. The upper eye pattern shows the LED module optical output, for 150Mb/s NRZ (2^{15}-1) pseudorandom modulation. The lower eye pattern shows the PD module output, after 100 meter transmission. Figure 16 shows the power budget in the transmission experiments. The transmission loss was about 2dB. The receiver sensitivity value for 10^{-9} BER with interference channels after 100 meter transmission was -28.3 dBm, including crosstalk power penalty. The loss margin for 100 meter transmission was calculated to be about 7 dB, even under a 12-channel simultaneous operation. A 12-channel 150 Mb/s/ch 100 meter parallel optical transmission, with 5dB power margin, was successfully achieved.

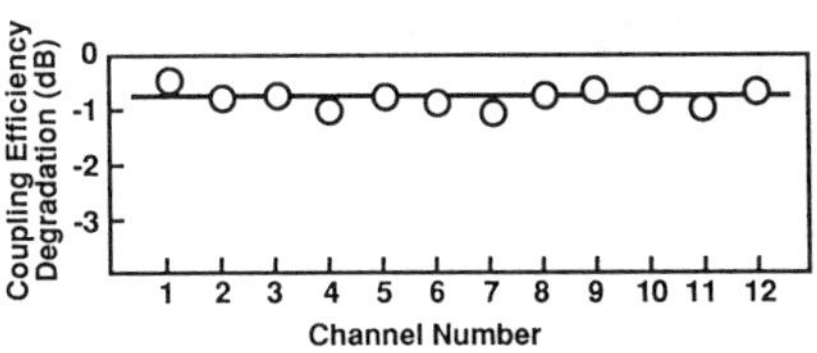

Fig.13 Coupling efficiency degradation for fiber array vs. individual LED channels.

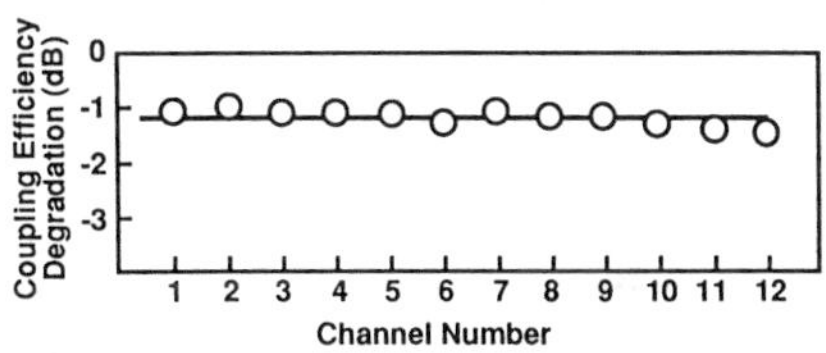

Fig.14 Coupling efficiency degradation for fiber array vs. individual PD channels.

Table 2 Module features.

	LED	PD
Optical Device	1.3 μm / 12 ch InGaAsP / Inp	InGaAs / InP
Optical Coupling	Butt Joint	Micro Lens Array
Electronic Circuits	Driver	Preamplifier
Excess Loss (dB)	~ 0.8	~ 1.2
Package Size (mm)	15L×30W×7H	15L×45W×7H

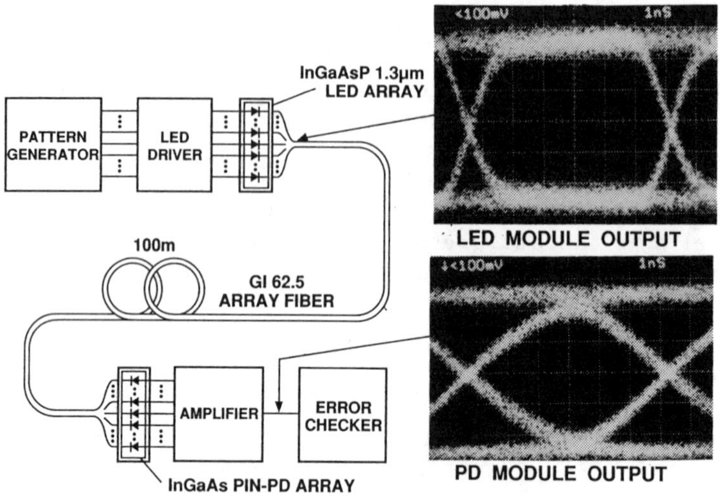

Fig.15 Parallel optical transmission experimental setup.

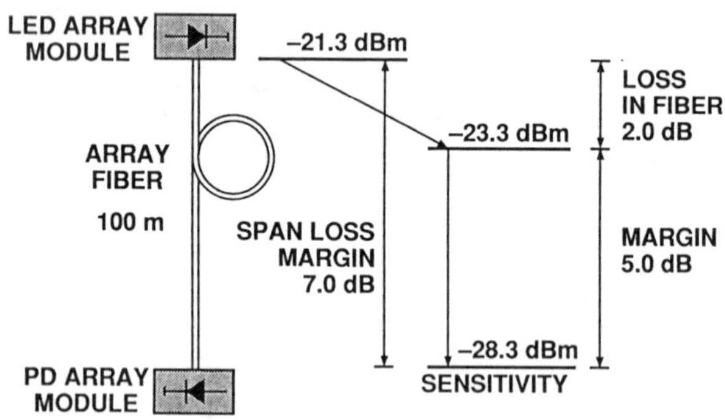

Fig.16 Loss budget.

5. CONCLUSION

A newly developed optoelectronic devices packaging technology for multi-channel LED/PD array modules has been demonstrated for high speed parallel optical transmission. Optical and electronic devices are mounted at a lateral point and at the surface of the common AlN submount, respectively. A multi-channel optical chip was bonded by solder bumps. A sapphire ball lens array and a glass capillary array were employed for optical fiber array terminal fabrication simplicity. The overall module package height was reduced to 7mm. A 12-channel 150 Mb/s/ch 100 meter parallel optical transmission, with small electronic crosstalk, has been demonstrated.

ACKNOWLEDGEMENT

The authors would like to thank Messrs S.Esho, K.Kobayashi, H.Goto, H.Abe, F.Yamamoto, J.Namiki, T.Ohkubo, T.Suzuki and K.Kaede for encouragement throughout this work.

REFERENCES

[1] S.R.Forrest, "Monolithic Optoelectronic Integration: A New Component Technology for Lightwave Communications", Journal of Lightwave Technology, vol.LT-3, 6, p.1248,(1985).

[2] Y.Ota, R.C.Miller, S.R.Stephen, D.R.Kaplan, C.W.Seabury, R.B.Huntington, J.G.Johnson and J.R.Potopowicz, "Twelve-Channel Individually Addressable InGaAs/InP p-i-n Photodiode and InGaAsP/InP LED Arrays in a Compact Package", Journal of Lightwave Technology, vol.LT-5, 8, p.1118,(1987)

[3]P.P.Deimel, J.Cheng, S.R.Forrest, P.H.-S.Hu, R.B.Huntington, R.C.Miller, J.R.Potopowicz, D.D.Roccasecca and C.W.Seabury, "Individually Addressable Monolithic 1×12 Light-Emitting Diode Array", Journal of Lightwave Technology, vol.LT-3, 5, p.988,(1985)

[4] M.G.Brown, P.H.-S.Hu, D.R.Kaplan, Y.Ota, C.W.Seabury, M.A.Washington, E.E.Becker, G.Johnson, M.Koza and J.R.Potopowicz, "Monolithically Integrated 1×12 Array of Planar InGaAs/InP Photodiodes", Journal of Lightwave Technology, vol.LT-4, 3, p.283,(1986)

[5] T.Nagahori, T.Uji, T.Torikai, I.Watanabe, T.Matsumoto, J.Hayashi, M.Itoh, H.Honmou, H.Kaneko and K.Kaede, "150Mb/s/ch 12-channel Optical Parallel Transmission using an LED and a PD Array", in Technical Digest of OFC'90, paper THC-3,(1990)

[6] K.Kaede, T.Uji, T.Nagahori, T.Suzaki, T.Torikai, J.Hayashi, I.Watanabe, M.Itoh, H.Honmou and M.Shikada, "12-channel Parallel Optical-Fiber Transmission using a Low-Drive Current 1.3-μm LED Array and a p-i-n PD Array", Journal of Lightwave Technology,

vol.LT-6, p.886,(1990)
[7]T.Uji and J.Hayashi, "High-power Operation of Mesa-structure 1.3-μm Surface-emitting LED's for Optical Fiber Communications", in Technical Digest of CLEO, 1985, paper THI-5
[8]T.Uji and J.Hayashi, "High-power Single-mode Optical-fiber Coupling to InGaAsP 1.3-μm Mesa-structure Surface-emitting LED's", Electronics Letters, vol.21, 10, p.418,(1985)

Section II. Microelectronic Packaging

AN OVERVIEW OF ADVANCED CERAMIC PACKAGING TECHNOLOGIES

Ananda H. Kumar and Rao R. Tummala

IBM Technology Products, Hopewell Junction, N.Y.

ABSTRACT

Although multilayer ceramic substrates (MCM-C) have been in use for over a decade, their widespread use has not yet materialized. There is, however, a general feeling that the time is now ripe for an explosive growth in the demand for these subsstrates. In this article we have briefly reviewed the ceramic materials that have been developed for MCMs and considered some of the important process challenges faced in their development.

INTRODUCTION

Ceramic packages for semiconductor devices have been in use from the time of the invention of the transistor. To-day ceramic packages are far outnumbered by their plastic counterparts, thanks to the latter's lower cost, greater dimensional tolerances which enables easier handling in automated assembly tools, and to the phenomenal improvements in their reliability. Single chip ceramic packages such as the dual -in -line packages (DIP), pin grid arrays (PGAs) and chip carriers (CCs) are still favored over their plastic counterparts in high reliability applications. Over a decade ago, ceramic packages won over a new niche application area viz. multilayer, multichip, substrates when IBM first introduced them in its mid-range and high - performance mainframe computers. Since then many others have adopted the multilayer, multichip, ceramic substrates in both mainframe and non-mainframe applications. However, their use has not developed to the extent once expected. But today these multilayer ceramic substrates are said are to be at another opportunity threshold, which if true, could mean an explosive growth in demand for them.

While there is now a clear trend in semiconductor development towards increasing levels of integration, made possible by decreasing feature sizes and the adoption of CMOS technology, there is no consensus on the direction of evolution in semiconductor packaging. For some, the ability to fabricate an entire system on a single chip (in effect, wafer-scale integration, WSI, at the chip level), spells doom for multichip substrates. Others contend that the growth in integration levels will be taken in stride and that, as in the past, several such chips will be utilized to provide enhanced systems, thus seeing no reason to abandon the multichip substrates. Indeed, if one were to judge from the number of recent articles written and conferences held on multichip technology, it would appear as if the world had just discovered them !

Multilayer ceramic substrates offer the following advantages over the printed wiring board or MCM-L substrates :

Higher circuit density.
Good thermal conductivity
Good hermiticity
Higher structural strength
Lower thermal expansivity
Excellent metal adhesion

Also MCM-C provides signal and power connections at the optimum locations in the package.

CERAMIC MATERIALS FOR FOR MCM-C

The selection of the ceramic for the multilayer substrate is governed principally by the following factors:
(i) Electrical design considerations (Dielectric constant, breakdown strength, wiring density, line resistivity etc.)
(ii) Conductor metal - its sintering characteristics and oxidation behavior at elevated temperatures.
(iii) Thermal design - the need, if any, for thermal dissipation through ceramic.
(iv) Thermal coefficient of expansion
(v) Mechanical properties
(vi) Sinterability at temperatures compatible with the conductor metal.
(vii) Cost of materials and fabrication.

Several ceramic compositions that have been used in multilayer substrate fabrication are briefly described below.

1. 99.6 % Alumina :

Many of the single chip substrates in use to-day are fabricated using as nearly pure alumina as possible, because of the higher thermal conductivity (1.5 x that of 96 % alumina compositions), higher strength, and superior surface finish that results. Foster et.al.[1] have reported on the fabrication of co-fired, multilayer, multichip substrates of pure alumina with 0.2 % magnesia (99.6 % alumina) and tungsten thick film metallization. The microstructure of these ceramics contains no glassy phase. The tungsten metallization is also devoid of the glassy phase traditionally added to promote adhesion to ceramic and to modify the thermal expansivity of the conductors, relying instead on the addition of titania to accomplish the same. Normal alumina powders having average particle size in the range of 5 microns and above cannot sinter to high density until at least 1650 °C. To obtain good sinterabiltiy at temperatures of about 1550 °C., they chose a reactive alumina powder having an average particle size of 0.6 microns. To obtain good dimensional control of +/-0.2 %, the particle size distribution in the powders had to be held to a very tight range. Magnesium oxide, added as magnesium aluminate (0.5 wt.%), acts as a sintering aid as well as grain growth inhibitor. Sintering at 1550 °C. yields a dense ceramic, which when polished produces a very smooth surface finish devoid of grain pullouts typical of alumina compositions containing glassy phase. The polished substrates were used as the base for fabrication of additional levels of wiring using thin film methods. The ceramic had a dielectric constant of 9.9, flexural strength of 640 MPa and thermal expansivity of 67×10^{-7} / ° C.

2. 90 - 94 % Alumina :

The common practice is to incorporate significant amounts of alumino-silicate glasses in alumina substrate compositions to promote sintering below 1600 °C, as well as to match the thermal expansivity of the co-fired metal features. In such compositions, the more readily available powders of particle sizes in the range of 5 microns are used. Figure 1 shows the shrinkage curves for pure alumina and and for alumina with glass added illustrating the beneficial effect of the glassy phase on sintering [2]. The microstructure of sintered substrates will show the presence of glassy phase at the interstitial sites and not as a continous grain-boundary phase. The glass does act as a reactive liquid phase at the sintering temperature

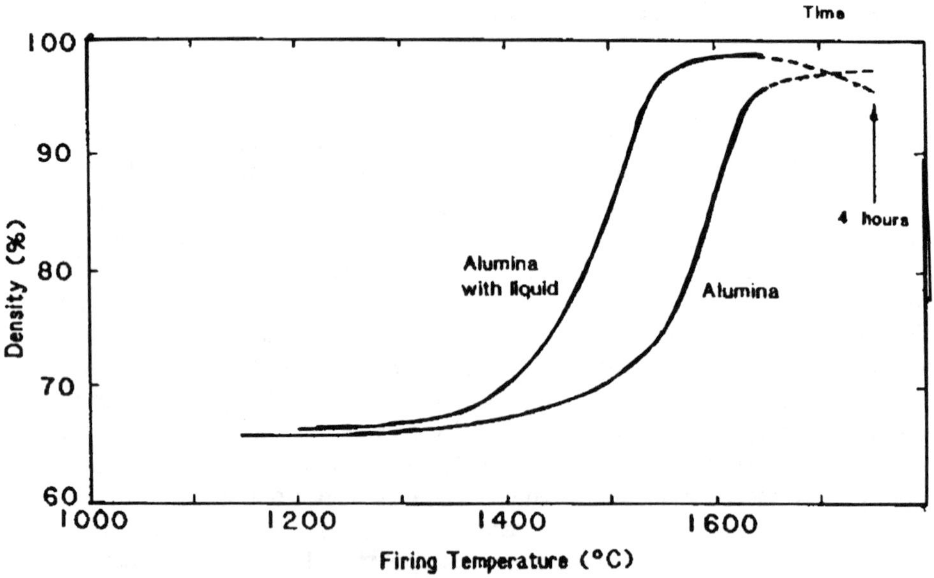

Fig. 1. Sintering of alumina (1 micron particle size) and alumina in the presence of glassy phase. [2]

promoting the sintering of alumina particles. For this reason the strength of the sintered ceramic is only slightly lower than that of the 99.6% material but the degradation in the thermal conductivity of the material is proportionately larger. Typical properties of the resulting ceramic are as follows ; dielectric constant = 9 - 9.5, TCE = 60 - 65 x 10^{-7} / °C and modulus of rupture in the range of 400 - 600 MPa.

3. Mullite Compositions:

Mullite, ($3Al_2O_3$ $2SiO_2$), is attractive as a ceramic for MCM-C because of the lower values of its dielectric constant (k= 5-6) and thermal expansion coefficient (TCE = 35x10^{-7}/ °C) compared to alumina compositions described above. It is also a very stable compound and can be sintered at temperatures in the range of 1500° - 1700 °C. Mullite was one of the first ceramics to be considered for MCM - C in the early days, but its low dielectric constant and low expansivity were not important factors in substrate material selection when chips were small and propagation delay was not a concern. On the other hand its inferior strength and thermal conductivity compared to alumina won the day for latter. The situation has turned around somewhat today, with chip sizes becoming large, and propagation delay being one of the important criteria in material selection. Also, the strength of mullite prepared from synthetic powders used to day is significantly higher than found in earlier mullite substrates. In some applications, the thermal conduction of the ceramic is not an important factor in thermal management. All these developments have made mullite an attractive candidate for MCM -C fabrication in place of alumina.

The earliest reference to the usefulness of mullite for multilayer substrate application was made by Arrhenius [3]. First reduction to practice is believed to have been by Gardner and Wilcox [4] who produced substrates from very fine mullite powder (0.03-0.1 micron) obtained by flame pyrolysis of a solution containing aluminum chloride and silicon tetra chloride in stoichiometric proportion. Mullite powders are now prepared by a sol-gel process which yield mullite ceramics having strength comparable to that of alumina.[5]

Kurihara et.al.[6] have described the development of mullite substrates for use in MCM-D applications. Utilizing mullite powders of nearly stoichiometric composition produced by the sol - gel process, they were able to obtain dense sintering at 1570 °C whereas ordinary mullite of this composition needs to be sintered in the range of 1600° - 1650 °C to achieve the same density. This was possible because of the very fine grain size (average particle size = 2.3 microns) and the presence of about 0.2 wt.% magnesia impurity which formed a glassy phase with trace amounts of alumina and silica. The substrates obtained were well suited for subsequent thin film wiring fabrication because they yielded a nearly pore free, fine grained and smooth surface upon polishing. Off - stoichiometric mullite powders tend to form a glassy phase at grain junctions which gives rise to porosity when polished. Kobayashi et. al.[7] have described the development of multilayer board of mullite with tungsten metallization. These large (160 mm - sq.), multilayer (36 layers) boards are used as carriers for 41 VLSI chips. Among the substrate's characteristics are its low dielectric constant (k = 5.9), thermal expansivity (TCE =3.5 $\times 10^{-7}$ /°C) closely matched to that of silicon, and reasonable flexural strength (300 MPa).

4. Glass-ceramics:

Glasses and glass-ceramics have been used as cross-over dielectrics from a long time. Although references to the suitability of making multilayer substrates with glass-ceramics as the dielectric material have been made as early as 1969 [8], clearly defined development efforts started only in the late 1970's. These efforts were directed towards developing multilayer substrates with cosintered , high conductivity metals such as copper, silver or gold, with ceramics of dielectric constants significantly lower than that of alumina and thermal expansion coefficient as closely matched to that of silicon as possible. The choice of the conductor metals (copper, silver or gold) dictates that the ceramic chosen for the substrate should be sinterable at temperatures no higher than about 1000 °C.

Two distinctly different approaches were pursued to develop ceramic compositions cosinterable with copper pastes. One approach was based on the development of suitable glass compositions whose powders had the ability to sinter to pore-free, glass-ceramic bodies. The choice of the glass compositions was also based on the properties desired in the resulting glass-ceramic viz. low dielectric constant, low thermal coefficient of expansion and high strength. Glasses in the MgO-Al_2O_3-SiO_2 and Li_2O_3-Al_2O_3-SiO_2 systems were successfully developed to fulfill all these requirements [9]. Others [10,11] developed ceramic and glass mixtures, wherein the nature and amount of the glass used were such as to obtain sinterabiltiy in the desired temperature range. These compositions contain nearly equal amounts of glass and ceramic powders in their starting mixtures.

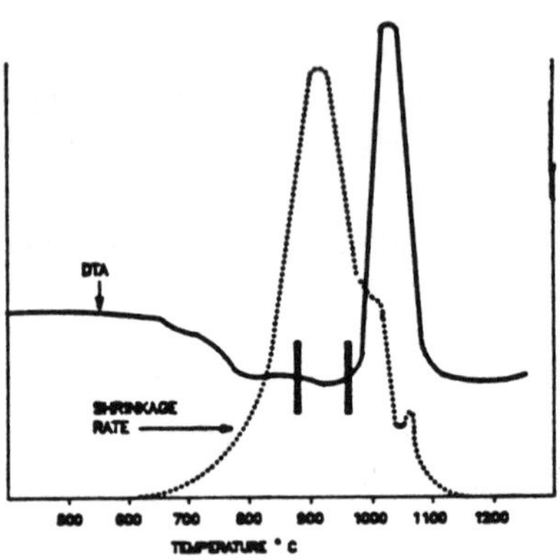

Fig. 3. DTA and shrinkage curve of a cordierite glass powder

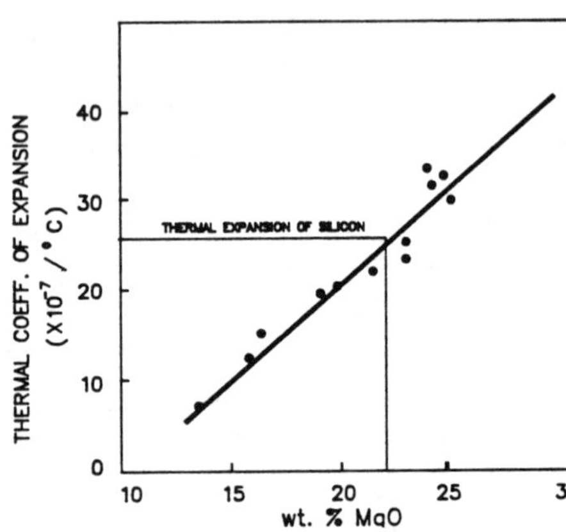

Fig. 4. TCE variation of of cordierite glassceramics.

The challenge faced in the first approach is the avoidance of the crystallization of the glass powder until after substantial densification is achieved through viscous coalescence. This is difficult because of the ease of surface nucleation on glass and the abundance of such surface nucleation sites in the fine powder. The $MgO-Al_2O_3 3-SiO_2$ glasses developed are situated in the cordierite composition region. The off-stoichiometry of their composition with respect to the crystalline phases of the composition field and the absence of known nucleating agents in the glasses, helped achieve good sinterabiltiy over a wide composition range. In Figure 2, the DTA and shrinkage curves of one particular glass are shown. It is seFigen that the full densification of the glass occurs prior to crystallization . The dielectric constant of the sintered glass-ceramics ranged from 5.0 - 5.5. The thermal expansion coefficients of the glassceramics were found to vary systematically with composition, as shown in Figure 3, making it easy to choose a composition that yielded a substrate having exactly the same TCE as silicon. The sintered glassceramics were nearly 99% crystalline, and more than 99% dense. The polished substrates were as smooth as silicon with very few small pores - making them ideally suited for fabrication of thin film wiring. Glass-ceramic -copper multilayer (66 layers, 127 mm. square) substrates with 2 levels of copper-polyimide thin film wiring are used in some of IBM's high performance, mainframe computers.

When mixtures of glass and ceramics are used to obtain the low temperature sintering ceramic, the nature and amount of the glass chosen is key to obtaining the desired sinterabiltiy. During the sintering process, the fluid glass could react with the ceramic constituent forming crystalline phases. In the composition (55wt.% alumina and 45 wt.% lead borosilicate glass) used to fabricate the 225mm X 225mm glassceramic substrates featured in NECs supercomputers, the reaction between the glass and the alumina produced anorthite microcrystalline phase that pervaded the entire network of the initially glassy grain boundary phase giving rise to a highly crystalline, non-porous, and strong body [10]. When cordierite powder and borosilicate glass reacted during sintering, the formation of cristobalite resulted in a body of very high thermal expansivity [11]. This reaction was suppressed by the addition of up to 5 Wt.% of alumina to the initial mixture. Use of lead borosilicate glass in the composition gave a relatively higher dielectric constant (k = 7.8) in the NEC glassceramic, whereas the Fujitsu glassceramic substrate had dielectric constant in the range of 5 - 5.5 from mixtures of alumina (40 - 50 wt.%) and simple borosilicate glass. Still lower dielectric constants can be obtained through the substitution of mullite or cordierite for alumina and through the controlled introduction of closed porosity into the structure. Many others have developed similar glassceramic substrates but NEC and Fujitsu are to be singled out for having actually deployed large, multilayer, glassceramic substrates in their high performance computers.

5. Aluminum nitride :

In applications where high thermal conductivity of the substrate is important, alumina and beryllia have been the ceramics of choice for decades. Lately, aluminum nitride is making important inroads into these applications. Like beryllia , aluminum nitride stands above most other ceramics in its high thermal conductivity. But unlike beryllia, its powder is not toxic. This allows one to consider aluminum nitride for multilayer substrate fabrication while beryllia could not be. In fact multilayer substrates incorporating cosintered tungsten conductors have been developed by several companies in the US and in Japan [12,13].

Aluminum nitride powder is synthesized either by direct nitridation of aluminum or by the carbothermal reduction of alumina in flowing nitrogen. Powders obtained from the latter have very fine (less than 2 microns average) particle size . In the absence of sintering aids this powder sinters to 93 % theoretical at 2000 °C. For substrate applications, certain alkaline earth, yittrium and rare earth metal compounds are added in small quantities as sintering aids. These form liquid phases at the sintering temperatures of 1300° - 1600° C promoting the densification of the aluminum nitride grains while also protecting them from oxidation. When calcium aluminate is used as a sintering aid, holding at 1800° C after full densification expels most of the calcium and restores a high thermal conductivity to the substrate [13]. The room temperature thermal conductivity of aluminum nitride substrate is in the range of 160-180 w / cm^2. Its dielectric constant (k = 9.2) is similar to that of alumina and thermal coefficient of expansion (TCE = 4.4 x 10^{-7} / °C) is close to that of silicon. The bending strength of sintered body is in the range of (400 -600 MPa). In addition to use in substrates, aluminum nitride is also used as caps and heat spreaders [7]. The worldwide market for aluminum nitride was estimated to be about $20M in 1990 and $250M in 1995.

MULTILAYER SUBSTRATE FABRICATION

Young [14] has given an excellent description of the MLC process. The generic processing scheme for all the materials discussed above are similar and is shown in Figure 4. In fact, 99.6 % alumina, 94-96% alumina, mullite and aluminum nitride MLC substrates with molybdenum or tungsten metallization can be sintered in the same furnaces and atmospheres by careful control of grain size and / or additives. In this paper we have briefly described some important process challenges involved in MLC substrate fabrication.

(i) Shrinkage and distortion control:

(a) Ceramic shrinkage control:

Precise control of the dimensions of the substrate and of the location of the circuit elements on its surface are of critical importance because of the need to assemble silicon chips with precisely defined and closely placed terminals on to the circuit elements on the substrate. The substrate, in turn, has to be connected to other module elements that have been precisely dimensioned to facilitate automated assembly processes. The larger the substrate, the more precise the shrinkage and distortion control needs to be because it is maximum deviation from the nominal position of a feature that dictates tolerances needed. Foster et.al [1] found this to be the case in their multilayer substrates. Reported values for shrinkage control in large, multilayer substrates are in the range of 0.1 to 0.2 %. This represents a remarkable achievement in materials and process control.

Two primary factors that determine shrinkage and distortion in ceramic, multilayer substrates are : (i) the ceramic sintering and shrinkage and (ii) the matching of the shrinkage regimes of the metallization pastes and of the ceramic. The overall shrinkage level is determined by the ceramic particle loading and distribution in the green laminate, while the paste shrinkage can be considered as a perturbation on the inherent ceramic shrinkage. Very close control over raw materials, slurry composition, mixing, green sheet casting and lamination conditions are needed to assure good shrinkage control and reproducibility. The reproducibility of the ceramic powder grain size and size distribution is either assured by the method of forming the powder or by prolonged ball milling until a stable distribution is obtained. The work of Foster et al (1) with ultrafine alumina powders is an example of the former. In the case of glassceramics, the glass cullet is reduced to powder of particle sizes in the range of 2-5 microns by prolonged milling. A major cause of porosity in the sintered body is particle agglomeration. This can be minimized by assuring good dispersion of the ceramic powder

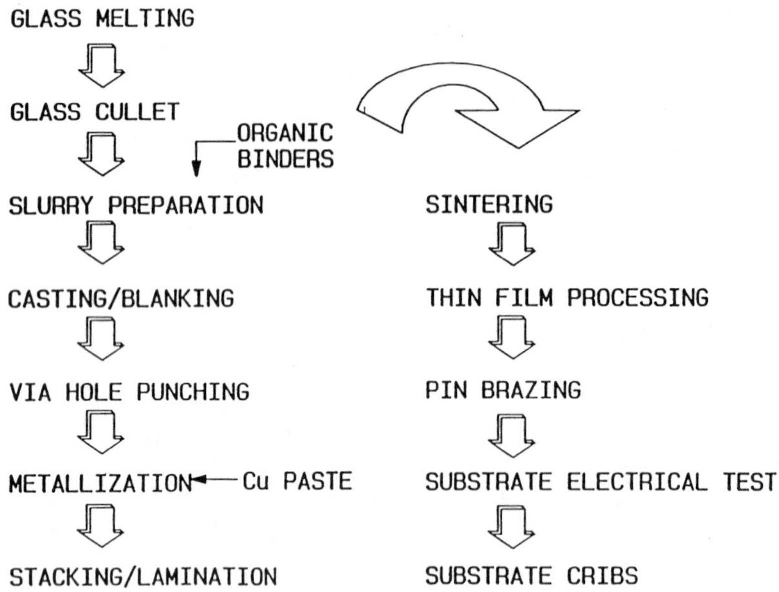

Fig. 4. Process flow for glassceramic substrates

in the slurry through the use of suitable surfactants. Alumina grains of particle size of 5 microns or larger tend to have an elongated shape which causes them to line up in the direction of slurry casting. This could give rise to anisotropy in shrinkage in the sheet which cannot be ignored. Glass and ultrafine ceramic powders tend to have isotropic shapes.

(b). Conductor paste shrinkage :

Bordia and Raj [15] have considered the consequences of the conductor sintering after the ceramic has set. More common is the problem where the conductor tends to sinter prior to the onset of sintering in the ceramic. Figure 5 shows the shrinkage curves for 96 %lumina by itself and for two forms of molybdenum pastes. The normally used molybdenum paste sinters at temperatures at or above the sintering temperature of the alumina body and thereby can easily accommodate the shrinkage of the ceramic. When an attempt was made to modify the molybdenum pastes to cause them to sinter dense using small amounts of palladium powder, known to be a very powerful sintering catalyst for molybdenum, the shrinkage behavior of this paste changed drastically with sintering occurring at temperatures well below the sintering temperature regime for the alumina body. As a consequence, the internal and external molybdenum circuit features would set to dimensions dictated by the unsintered green laminate. When, eventually, the latter started to sinter, the molybdenum circuit features tended to severely constrain the ceramic sintering or pop out of the ceramic, as in the case of the seal band shown in Figure 6 [16].

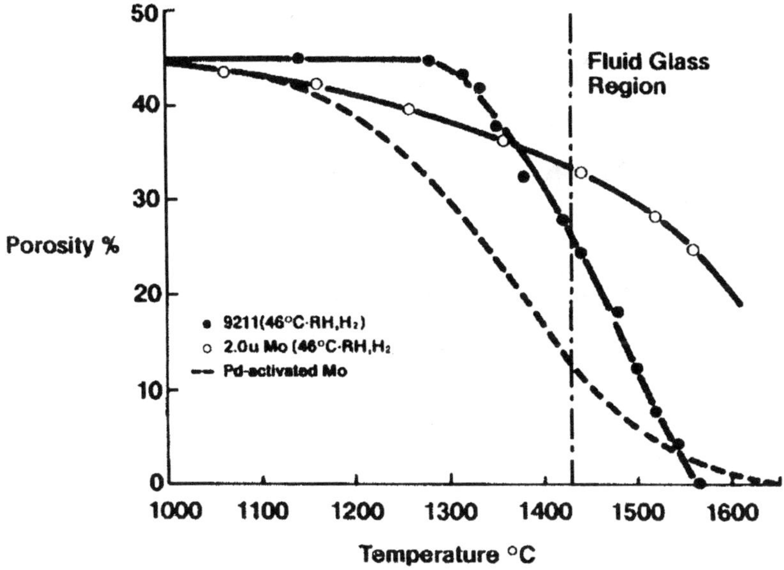

Fig.5. Shrinkage curves for alumina and molybdenum powders.

Similar problems were also encountered in the case of the cordierite glass-ceramic and copper pastes. The sintering regimes for the two were quite far apart. In addition, the degree of inherent shrinkage for the two were also different,but this was corrected by optimizing the solids content of the paste. The large degree of shrinkage modification of the copper paste could only be achieved by adding sintering retardants in the form of proprietary surface coatings on the copper particles. Sintering modification of copper pastes with solution - deposited titania surface coatings has also been reported [17].

ii) Sintering ambient:

Except in the cases where noble metal pastes are used to form the circuitry, the sintering ambient selection is very crucial to the successful fabrication of multilayer substrates. The commonly used metallizations viz.

Fig. 6. Delamination of molybdenum seal band from ceramic due to a mismatch in the sintering temperatures.

(molybdenum, tungsten and copper), are easily oxidized at elevated temperatures. Therefore a controlled atmosphere of low oxidation potential is necessary during sintering. This, in turn, causes the polymers used as binders to char leaving behind a carbon residue. The complete removal of carbon residue is essential for the optimum sintering of the substrate. Since carbon can only be removed by oxidation to gaseous products, certain degree of oxidizing condition should exist in the furnace ambient. The oxygen potential in the furnace ambient has to be controlled to prevent the oxidation of the base metal powders in the paste, to oxidize the carbon residues formed by the binder polymers, and to ensure that the ceramic components are not reduced. The required oxygen potentials are obtained by using mixtures of gases such as cracked ammonia, and hydrogen - water vapor. Wahl and Sohn [18] have studied the detailed mechanisms and kinetics of carbon removal from alumina green sheets in a wet hydrogen atmosphere by the reaction:

$$C(s) + H_2O\ (g) = CO(g) + H_2(g)$$

Figure 7 shows the various events that take place during the sintering of alumina - molybdenum multilayer substrates in a hydrogen - water vapor atmosphere. The earliest reactions are the expelling of volatiles in the metallizing paste and the laminate which is followed by the complete breakdown of the organic into carbonaceous residues at about 500 °C. This is followed by a period of carbon oxidation by the water vapor in the sintering ambient. This reaction occurs faster the more elevated the temperature. Since there is hardly any sintering of the alumina below about 1200 °C., the gaseous reactants and products can easily traverse through the laminate facilitating complete removal of carbon at around 1100 °C. This is followed by the melting of the glass in the ceramic composition. The molten glass causes the sintering of alumina particles, and penetrates the still porous molybdenum circuit features cementing a bond between the two. The presence of water vapor in the sintering ambient at these elevated temperatures is essential to prevent the reduction of silica in the glass.

When copper is the conductor in the multilayer substrates, the sintering temperatures for the substrates are typically in the range of 800° - 1000 °C., with measurable densification of the ceramic occurring at temperatures of

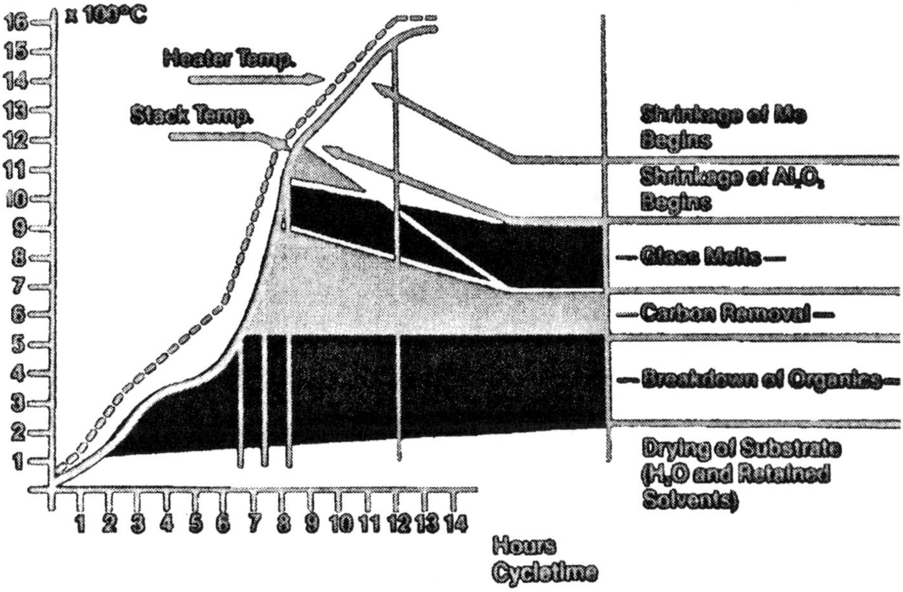

Fig. 7. Events occuring during the sintering of alumina substrates.

about 700° C. The challenge then is to remove all of the carbon before it gets trapped in the sintered ceramic. At these temperatures it is a race between complete carbon removal and ceramic densification. Fortunately, there exists a range of oxygen potentials which is oxidizing to carbon while being reducing to copper as shown in Figure 8 [19]. The rate of carbon oxidation by water vapor is relatively slow requiring prolonged hold at the reaction temperatures to ensure the complete oxidation of carbon. To shorten the hold period, certain metal compounds can be added to catalyze the carbon reaction with water vapor [20].

Presence of even very small amounts of carbon is detrimental to the properties of the substrate. In one case it was found that carbon levels of about 200 PPM in the glassceramic reduced its modulus of rupture by nearly half and the insulation resistance to zero [21]. This is consistent with the view that the carbon residue settles at the surfaces of the particles of glass, preventing their dense sintering and creating a conducting path between the grains. Short of a complete breakdown in the insulation resistance, the dielectric constant of the glassceramic increases with the increased carbon content. Entrapped carbon can also cause localized areas of unsintered copper leading to opens [22].

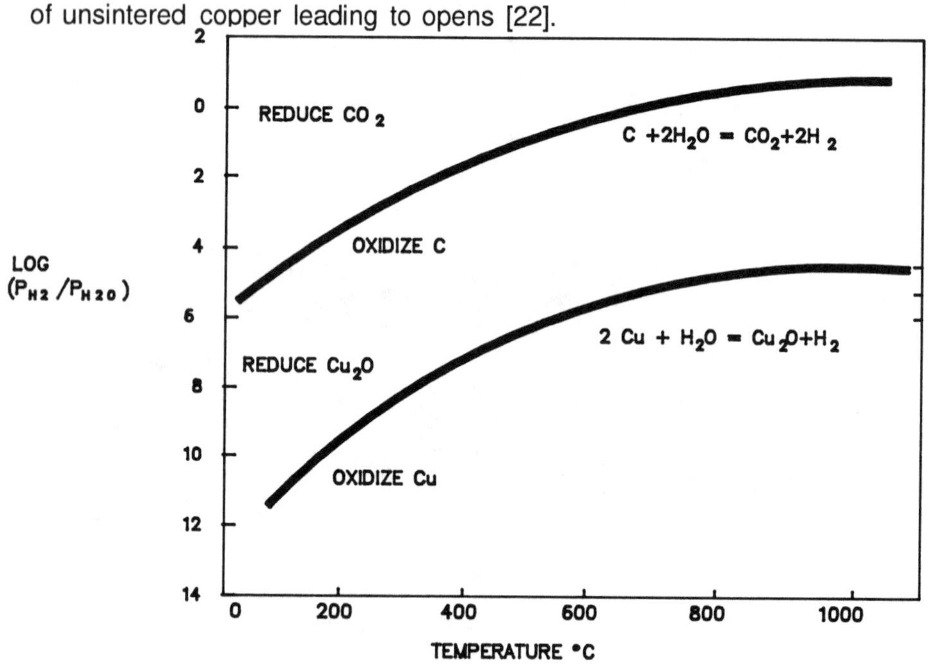

Fig. 8. Oxidation potentials for copper and carbon.

iii) Thermal expansivity mismatch :

The multilayer ceramic substrate can be regarded as a metal - ceramic composite structure. Its structural integrity can be undermined if there exists large disparities in the thermal coefficients of expansion of the metallic and ceramic parts. For example, in the case of 96% alumina / molybdenum substrates, the molybdenum vias (TCE of Mo= 55×10^{-7}/ °C) are surrounded by the ceramic of higher thermal expansivity (TCE ceramic = 70×10^{-7}/ °C). This TCE disparity sets up hoop stresses in the ceramic around the vias, Figure 8. If the vias are sufficiently close to each other (less than 100 microns apart), there would be cracking of the ceramic in between. In one case, ceramics of lower thermal expansivity were formulated by substituting mullite for some of the alumina in the ceramic composition to successfully overcome this problem.(Figure 9) [23]. Other approaches involved adding glass or ceramic powders to the conductor paste to modify its thermal expansivity. In the case of the glassceramic / copper substrates, the copper metallization has a much higher thermal coefficient of expansion than the glassceramic. Here , the via would pull

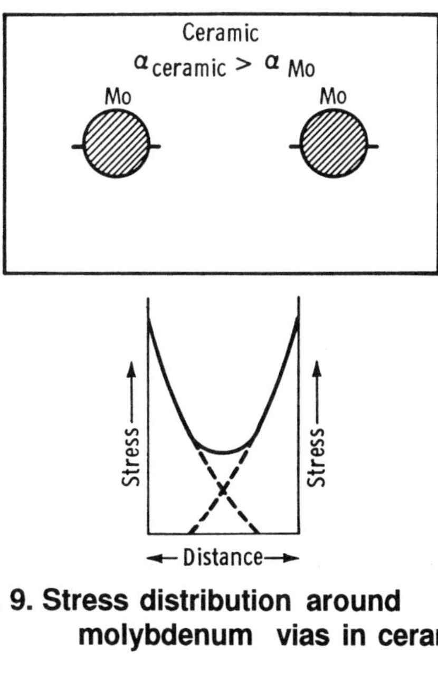

Fig. 9. Stress distribution around molybdenum vias in ceramic.

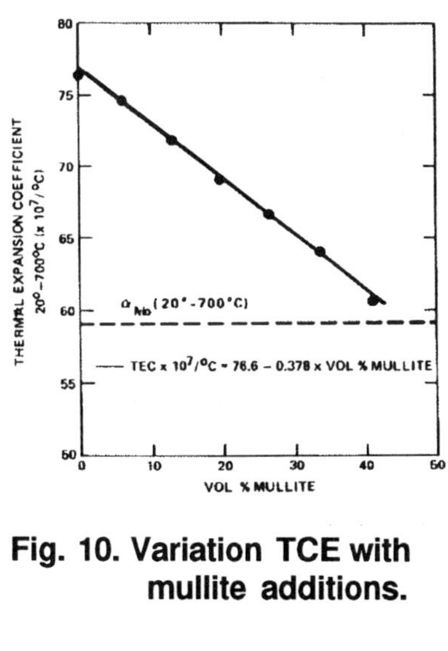

Fig. 10. Variation TCE with mullite additions.

away from the ceramic side walls if the adhesion between the two is marginal. Once again the problems associated with this can be solved by suitably modifying the metallization paste.

REFERENCES

[1] B.C. Foster, F.J. Bachner, E.S. Tormey, M.Occhionero and P.A. White ," Advanced Ceramic Substrates for Multichip Modules with Multilevel Thin Film Interconnections", IEEE Trans CHMT, Vol.14(4),pp. 1991,

[2] J. S. Reed in Introduction to The Principles of Ceramic Processing, pp. 463-464, Ed. J.S. Reed.

[3]. G. Arrehinius, ' Monolithic Ceramic Interconnecting Structures', U.S.Patent 3,423,517, 1969.

[4]. R. A. Gardner, and D.L. Wilcox, ' Mullite Package for Integrated Circuits ' U.S.Patent 3,857,923, 1974.

[5]. M. G. M. U. Ismail, J. Am.Ceram.Soc., Vol.70,1987.

[6]. T. Kurihara, M. Horiuchi, S. Takeuchi, S. Wakabayashi, ' Mullite Substrate for Thin Film Applications ', Proc. 40th. Electronic Technology and Component Conference, pp.68-75, 1990.

[7]. F. Kobayashi, Y.Wanatnabe, Y. Yamamoto, A. Anzai, T. Takaheshi, T. Daikoku,and T.Fujita, ' Hardware Technology for Hitachi M-880 Processor Group, Proc..41st. Electronic Component and Technology Conference, pp.693-703,1991.

[8]. D.L. Wilcox, ' Ceramic for Packaging', Solid State Technology, pp. 40-60,1991.

[9]. A.H. Kumar and R.R. Tummala, ' State- of-the - Art,Glass-ceramic / Copper Multilayer Substrate for High Performance Computers', Intl. J. Hybrid Microelectronics, 14(4),pp.137-150, 1991

[10]. Y.Shimada, Y. Kobayashi, K. Kata, M. Kurano, H. Takamizawa, ' Large-scale,Multilayer Glass-ceramic Substrate for Supercomputers', Proc.40th Electronic Component and Technology Conference, CHMT, p.76,1990.

[11]. K. Niwa,and Y. Takeda,' Multilayer Glass-ceramic Composite Circuit Board for FUJITSU VP-200 Series, Fujitsu Science and Technology Journal, v.27(2), pp.179 -186, 1991

[12]. J. K. Hagge, 'State-of - the - Art, Multichip Modules for Avionics',IEEE CHMT, v.15(1), pp.29-42, 1992.

[13]. K. Kuramoto, H. Taniguchi, and I. Akso,'Sintering properties of High Purity Aluminum Nitride Ceramics', in Advances in Ceramics, v.26,

American Ceramic Society Publication, pp.107-119, 1989.
[14]. W. S. Young, 'Multilayer Ceramic Technology' in Ceramic Materials for Electronics, Ed. R.C. Buchanan, Marcel Dekker Inc.., N.Y., pp.403-424, (1986).
[15]. R. K. Bordia and R. Raj , ' Sintering Behavior of Ceramic Films Constrained by a Rigid Substrate ' J. Am.Ceram. Soc., v.68(6), 287, 1985.
[16]. A. H. Kumar Unpublished work.
[17]. V.P. Suita, 'Meatl-oxide Coated Copper Powder' U.S.Patent # 4,671,928 , 1987.
[18].D.R. Wall and H.Y. Sohn, ' Removal of Carbonaceous Residue with Wet Hydrogen in Green Sheet Processing for Multilayer Module', J. Am.Ceram.Soc., 73(10)2944, 1991.
[19]. R. N. Master, L. Wynn Herron and R.Tummala, ' Co-firing Process for Glass -Ceramic / Copper Multilayer Substrates', Proc. Electronic Technology and Component Conference,CHMT, pp. 5-9, 1991.
[20]. L.W. Herron, A.H. Kumar, R. master and R.Nufer ,' Method for Removal of Carbonaceous Residues from Ceramic Structures.S. Patent 4,627,160, 1986.
[21]. K. Niwa, N. Kamahera, H. Yokoyama, K. Kurihara, 'Multilayer Ceramic Board with Copper Conductors', in Multilayer Ceramic Devices, Advances in Ceramics, vol.19, J.R.Blum and R. Cannon, eds.American Ceramic Society Publications, pp.41-48, 1986.
[22] W. Borland and V.P. Suita, ' Matyerails Interactions in the Firing of Copper Thick Film Multilayer Ceramics', Matls. Res. Soc. Symp., Proc., vol. 108, pp. 287-300, 1988.
[23]. A. H. Kumar and J.B. Niklewski,' Alumina-Mullite Ceramics for Multilayer Substrates', Bull. Am. Ceram. Soc., 58(7), 1179-1181,1979.

PERSPECTIVES OF MULTICHIP MODULE PACKAGING TECHNOLOGY

W. Radlik, Siemens AG, Munich, Germany

Abstract

Multichip module (MCM) packaging technology provides most significant performance gains to complex electronic systems. Thus the employment of MCMs fabricated by lamination technology has been the only way to make the H90 series mainframe computers feasible. Currently these modules are under volume production.

As the level of packaging density and signal transmission speed is advancing steadily, thin film MCMs are commonly expected to take over for the laminated and ceramic types in system packaging. This has triggered now the development of thin film MCMs based on metal/polymer multilayer structures. The implementation of a test vehicle and a demonstrator module is reported.

1. Introduction

The rapid evolution of IC technology yields a considerable increase of performance especially regarding elevated clock frequencies, advanced minaturisation and extended functionality. But most of these performance gains do not translate onto the system level, because conventional IC packaging and interconnection technology turns out as a major limiting factor. Mainly it is the poor packaging density and related phenomena like significant chip-to-chip delay, signal distortion and noise, that make all the on-chip benefits ineffective again.

The impact of this packaging hurdle on future trends in microelectronics is exactly resumed by the following statement: "In the next decade, more performance strides will be made through improvements in semiconductor packaging than through semiconductor technology"[1].

Today multichip module (MCM) technology is a most promising packaging approach to meet system level requirements. MCMs - that are multiple bare dice directly attached to a common high density multilayer interconnection substrate. For enviromental protection, the assembly may be encapsulated as a whole.

There are three basic types of MCMs, which can be classified by feature size and substrate material - among some more criteria. MCMs derived from an upgraded pc-board technology at least are referred to as MCM-L. The L indicates a laminated type of substrate using materials like FR4. The D extension is assigned to those MCMs, in which high density interconnection is achieved by deposition of thin film metal/polymer multilayer structures on Si or Al_2O_3 ceramic substrates (MCM-D).

Concerning the L and D versions of MCM technology there are major activities at Siemens. To its tightest definition the MCM-C cofired ceramic approach is not pursued inhouse. Nevertheless high volumes of ceramic thick film multilayer hybrids are currently manufactured. The concentration on the D- and L-types is somehow corresponding to latest BPA data on trends in worldwide MCM consumption[2]. The analysis predicts the cofired ceramic MCMs becomming far outstripped by their deposited and laminated counterparts up to the year 2000.

2.1 MCM-L Requirements

The development of the H90 generation of Siemens mainframes forced the introduction of the MCM-L technology. This step to MCM-L has been driven clearly by performance with respect to the system configuration, which is characterized by features like 512 Mb main storage and computing power rated at 40 Mips and 24 ns cycle time. The hardware design of the H90 processor board provides nine subunits for the implementation as MCM. Each of them is holding 16 device locations equal in size, which can either be assembled with a single LSI chip or with nine MSI chips alternatively (Fig.1). At least a total of more than 60000 gates is typically integrated on the entire module.

Interconnecting all these chips requires about 3500 signal lines. To manage this complexity and to achieve the 175 µm pitch required for the LSI chip accomodation, the MCM basic layout has been specified to 15 layers, 80 µm

Fig.1: H90 MCM-L populated with 108 MSI and 4 LSI chips
(original size100mm x 100mm)

conductor linewidth and 250 μm pitch.

All the conductor layers are applied successively to the frontplane of a rigid carrier, which gives mechanical stability to the whole module. The material chosen for this supporting function is fiberglass reinforced polyimide. For the external connections of the MCM there are provided 1850 copper pads at the backplane of this substrate. Contact to the front is achieved by through hole vias.

2.2 Lamination Technology

Each conductor layer is a dual compound of the actual conducting plane and the dielectric underneath. This assembly is fabricated by laminating 18μm of Cu foil onto the preceding surface by means of an adhesive film, which directly serves as the dielectric. It is consisting of a 25μm polyimide core sandwiched between two adhesive acryl coatings. The total thickness of 75μm is matched to the overall dielectric constant k=3.7 to keep the line impedence at 52Ω.

Signal feedthrough between adjacent conductor layers is achieved by vias of 80μm diameter. Typically 6000 vias per layer are necessary for complete interlayer connection. The process of via formation is subdivided into three major steps: (1) photolithographical removal of Cu at the via locations, (2) laser drilling of the dielectric layer, (3) metallisation of the vias.

After the first step, the Cu layer is directly taken as a mask for the laser drilling process. Hole generation within the dielectric is achieved by the UV radiation emitted from an KrF excimer laser, which induces the photodecomposition of the polymer material. The ablation process stops automatically as soon as the bottom Cu layer is reached due to the high reflectivity of Cu. The main advantage of this laser drilling procedure is to get always contaminant-free bottom Cu surfaces. Any kind of conta-mination may result in poor contact resistance at this point.
As a preliminary step to via metallisation electroless Cu deposition is done covering entirely the substrate surface as well as the interior parts of the vias. The final metal plate up of the vias is intregrated then into the semi-additive formation of the conductor pattern. The total conductor thickness attained at least is 30μm according to the line resistance specified at 13Ω/m.
Subsequently a flash of Sn is applied on top of the Cu features as an interim protection against etchant attack during the removal of the plating base. Afterwards this resist is stripped off again. Exemplary for the results a SEM via cross section is shown in Fig.2. A slightly different treatment is necessary for the top layer, which is provided for the die attachment. Here the Sn resist is replaced by a solderable Sn-Pb coat remaining permanently on the surface.

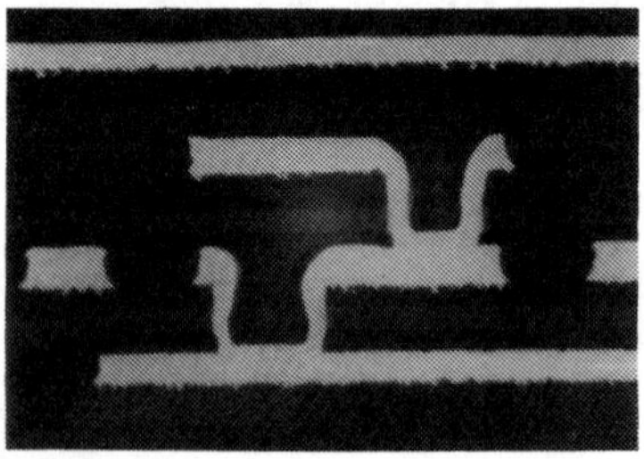

Fig.2: SEM micrograph of via cross-section

2.3 Die Attachment

Chip-to-substrate interconnection has to fit into the frame of managing 143μm on-die pitch and cooling power dissipation at a level of 20W per chip. Thus the technology of the choice is flip TAB - that is tape automated bonding of the dice in face-down position. The major benefit of this approach is that the full chip backplane area can be utilized for heat removal. Consequently the heat sink is directly attached to the reverse side of the chips. The heat sink employed is a solid Cu plate, which is also part of the module housing. To achieve a maximum heat tranfer an excellent thermal contact is required between chips and the cold plate. This is guaranteed by elastic elements in the interstice of chip and substrate pressing the chips to the heat sink (Fig.3). Any chip displacements arising from this adaptation are compensated by the TAB spider leads.

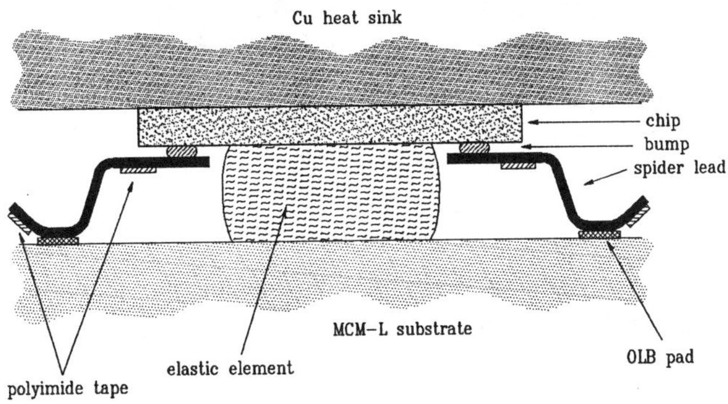

Fig.3: Principle of Flip TAB die attachment

The chips are received from the vendor as standard types. Thus modifications become necessary concerning the assembly by TAB. That is particulary bumping the chips and bonding them to inhouse fabricated TAB tapes.
The bumps are obtained by electrodeposition of 18µm Cu cores onto the bonding sites. The core surface is overcoated with Ni then and finally an Au layer is applied to ensure bondability. As bonding of the inner leads to the bumps is done by thermocompression, Sn-coating of the Cu spider is applied before. Due to this interface metallurgy the thermode bonding pressure required is less than 10 cN and the average peel forces resulting come to 28 cN for the LSI circuits.
Joining the outer leads to the module pads is a normal soldering procedure following the TAB package excision from the tape and the lead bending operation. A part of the completly assembled MCM-L is shown in Fig.4.

2.4 Benefits of the H90 MCM-L

Comparing the H90 MCM-L to its conventional PCB equivalent reveals impressively the drastic size reductions achieved (Fig.5). Although the same

Fig.4: Close-up of the MCM-L top layer

functionality is implemented in both of these assemblies, the PCB approach is pushed to its limits. So the line delay of 500 ps emerges as a hurdle, the circumvention of which makes switching over to MCM technology inevitable. Accordingly the delay is reduced to 150 ps within the MCM-L version.

Fig.5: Comparison of conventional PCB and equivalent MCM-L rendering

In the mid-1990s the performance level required for the next generation mainframes will increase tremendously. Thus major challenges will arise even for the MCM-L technology when the conductor line width is shrinking down to 40 μm and substrate size is blown up to 200mm x 200mm. Furthermore the number of layers is expected to grow from 15 to 40.

3. Thin film MCMs - the upcomming generation

MCM-D technology however provides maximum performance gains to meet future demands for high density interconnection. This is due to MCM-D

specific advantages: Wiring densities from 200 to 400 cm/cm² according to 10μm, 20μm or 25μm conductor trace design rules translate into minimum interconnection length, which is corresponding to superior electrical performance. In particular signal fidelity and propagation time are close to the on-chip standards. Concerning mechanical aspects the use of silicon substrates makes a perfect thermal expansion match to the dice attached.
On the other hand various infrastructural drawbacks boost production costs, pushing back the technological merits of MCM-D and impeding widespread acceptance thereby. Actually this may be true for current costs ranging from \$7/cm² to \$15/cm². As MCM-D technology is moving up the learning curve, the MCC projects the costs comming down approximately to the level from \$3/cm² to \$6/cm². Furthermore the MCC expects MCM-D becomming even more cost effective at increasing wiring densities than the L and C types(3).
All this is the typical scenario of any emerging technology. At least the pursuit of performance drives the efforts to MCM-D break through, which is predicted to be in about two or three years(4).
The basic construction of MCM-Ds is a multiple stack of alternating thin film metal and dielectric layers deposited onto alumina or silicon substrates. Typically five conductor layers are required: Two signal layers in x-y orientation sandwiched between power and ground, and the top layer for die attachment. A schematic cross section is given in Fig.6, which also shows some features of the conception Siemens is following.
Firstly copper and aluminum are the conductor metals under consideration, both of them with specific pros and cons concerning resistivity, adhesion properties and processing ease. Prefering the conformal via type is related to cost reduction. On the other hand this requires excellent planarizing interlevel dielectrics. At least well proven wire bonding technique is provided for chip to substrate interconnection.

3.1 MCM-D Substrate Metallurgy

Copper is the conductor material of choice, if emphasis is laid on low line resistivity. The formation of the copper conductor lines is done by a standard semi-additive process, that is electroplating of copper into the photo-image defined trench structure of a photoresist mask. As a preliminary step the generation of a plating base by sputter deposition of approximately 150nm

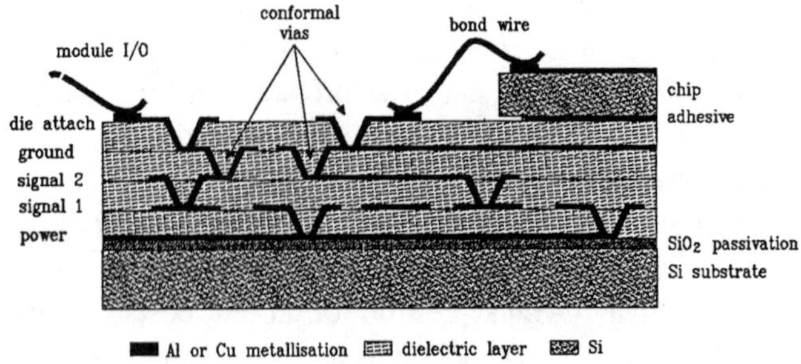

Fig.6: Basic construction of thin film MCMs

copper onto the preceding dielectric layer is required. Due to the cured state of the polymer organics found here, adhesion of the copper is a critical matter arising from the poor chemical interaction at the copper/polymer interface(5). In this case sufficient increase in adhesion is achievable by sputtering few nanometers of Ti prior to copper deposition.

At the polymer/copper interface in turn attention must be paid to the polymer solvents, which might attack the copper chemically. To prevent such corrosive processes, a nickel overcoat is applied onto the copper conductors. Additionally the nickel overcoat suppresses interdiffusion effects and ensures selectivity needed for the wet etch removal of the plating base. Compared with pure copper surfaces, nickel-plating improves the adhesion to polymers.

All in all semi-additive processing of copper conductors appears to be rather complex and by that more costly. If superior electrical conductivity is not demanded, the use of aluminium is favourable, because conductor delineation is simplified considerably. So aluminium is sputter deposited in one step to full layer thickness and patterned subtractively. Due to the self passivation of aluminium surfaces no protective coating is required.

To our experience there are also some slight drawbacks of aluminum processing to be taken into account: Due to its isotropic character, the wet etch back procedure generates losses in linewidth by undercutting. This effect is only tolerable, if the maximum conductor thickness does not exceed 2 μm. Furthermore undercutting is less pronounced if spray etching is done instead of dip etching.

Improvement of adhesion is a concern for the aluminum metallisation too. An effective solution is the introduction of titanium-clad aluminium layers.

3.2 Interlevel Dielectrics

The properties of the interlevel dielectrics are of crucial importance for the overall MCM-D performance. So it is with the dielectric constant k, which is the determining factor of signal speed within the MCM. The value of k should be as low as possible, because signal propagation delay is a function of the square root of k. Polyimides and benzocyclobutenes (BCBs) are the materials selected by this criterion according to their low k-values at 3.5 and 2.8 respectively. Among all the various types of polyimides especially the photoimigable formulations offer remarkable gains of processing ease. Thus only polyimides of this group will be subject of further interest here.
Processing of BCB requires some more steps again due to hard-masking and dry etch patterning of this non-photosensitive resin. Therefore BCB is found to be superior to polyimides concerning k-value, planarization, adhesion to metal and water uptake.
Good planarization is essential for MCM processing and performance. If the topografic structure of each level is not smoothed out, surface roughness is shaking up as the number of layers increases. From that difficulties arise mainly in lithography and impedance control lowering the fabrication yield at least. The effect of lower planarization is demonstrated in Fig.7. The profiles of the conductor traces orthogonally cut by the cross-section clearly

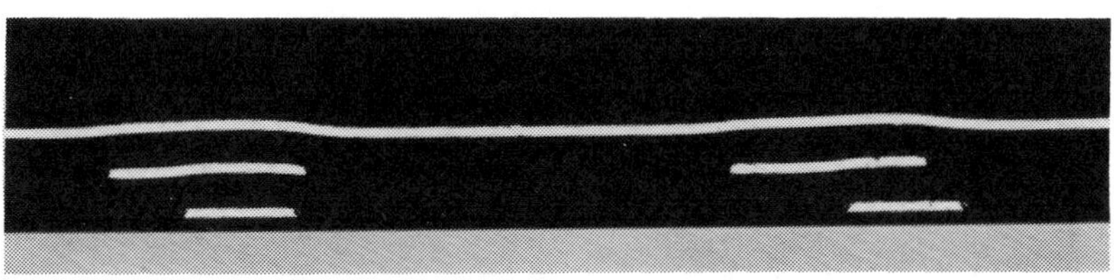

Fig.7: Surface deformations due to poor polyimide planarisation

translate into a dent-like surface deformation. The polyimide employed is Probimide 300, the degree of planarization achieved is about 60%. A better result is achieved with the XU13005 BCB-type dielectric. Nearly no deformation of the top layer are recognizable in the vicinity of the buried traces due to the excellent planarization of the BCB polymer (Fig.8).
The ability to smooth out surface topograhy is substantially affected by the volume shrinkage, the polymer is undergoing during the polymerization

process. As it is with the BCB, a very low shrinkage (2%) is the key to good planarization.

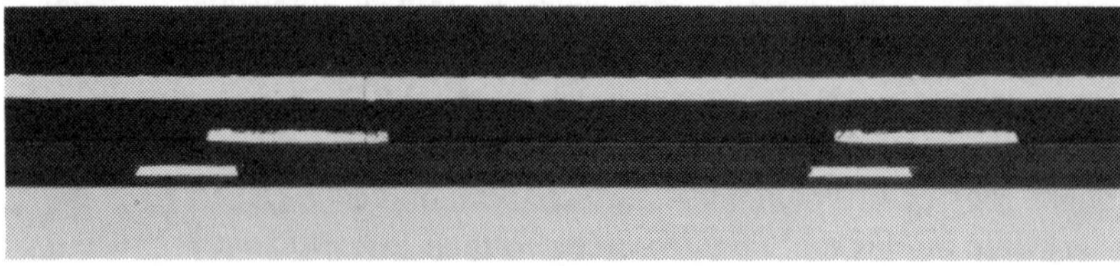

Fig.8: Similar detail as shown in Fig.7, now excellent planarisation achieved by BCB-type dielectric

Another problem related to polymer densification is the induction of mechanical stress within the multilayer system. To get some quantification of this effect, stress measurements are done for 7 μm Probimide 300 and XU13005 films. Both materials are applied to equivalent Si wafers and cured before following the manufacturer's recommendations. For the polyimide a tensile stress of about 41.9 MPa is detected then whereas the value obtained for BCB is only at 23.2 MPa.
Depending on the UV exposure conditions unisotropic polymer shrinkage may occur, which causes geometrical distortions of the features patterned before curing. Mainly deformations become apparent in the vicinity of steps, which somehow take a cusp-like contour. This effect can be extremly pronounced as it is at the upper polymer edge of the via cross-section shown in Fig.9. The polymer involved is Pyralin 2732.In consequence trouble may arise on such items due to the poor step coverage achieved with deposition of the next layer.

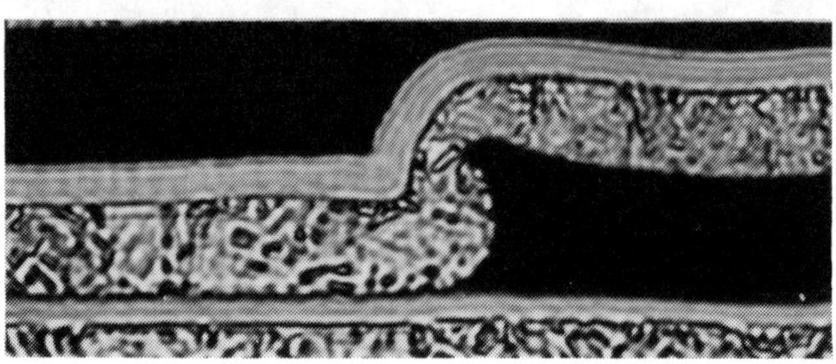

Fig.9: Step distortion due to unisotropic polyimide shrinkage

As unisotropic shrinkage is related to the mechanism of light absorption within the precursor film, tuning the UV wavelength to optimal film transmission or increasing the exposure dose minimizes the disturbing phenomenon.

All the difficulties above can be circumvented almost completly, if non-photosensitive BCB-type dielectrics are used.

In addition to the discussion before, no dramatic differences in adhesion properties are found between the polymers under consideration here. The application of organic adhesion promotors has a positive effect anyway. Tape tests qualitatively exhibit the tendency of better adhesion with BCB. This is valid especially for the adhesion of BCB to itself due to a partial cure procedure. Unlike polyimide BCB films withstand full MCM processing even at 80% polymerization. Now chemical bonding is made possible between subsequent BCB layers thus forming monolithic structures. To achieve full polymerization at least, hard curing is done with the application of the final BCB layer.

Sputter etching is a proven method to increase self-adhesion of polyimides. This kind of surface treatment reveals to be sensitive to the parameters power and time. Too much power and/or too long duration may reduce adhesion again due to surface degradation.

3.3 Test Vehicle

Fabrication of test vehicles is an appropriate form of probing the technology. To this end a four-layer configuration holding the following test structures has been designed:

- Signal lines: simple parallel and double layer X-Y arrays, pitch ranging from 50 to 200 μm;
- Via chains: staircase and meander types, via diameter from 20to 75μm;
- Ground: solid and mesh pattern planes;
- Capacitors for monitoring dielectric parameters and detection of polymer pin-hole defects.

The materials selected as substrates are alumnia (4"x4") and silicon (4" oxidized wafer). Conductor traces are formed employing the semi-additive copper process. The overall thickness attained for signal lines is at 7μm including the bottom titanium adhesion layer and the top nickel barrier. Roughly adjusting the signal line impedance to 50Ω then requires a dielectric film thickness of about 10μm.

Fig. 10 gives a cross-sectional view of the Pyralin 2732 implementation of the module. The planarization achieved is still sufficient although the surface topography shows some waviness.

In contrast to Fig. 9 step deformations due to unisotropic polymer shrinkage are diminished by cutting out non g-line contributions of the UV radiation spectrum. Thus good metal coverage is ensured along the entire via profile.

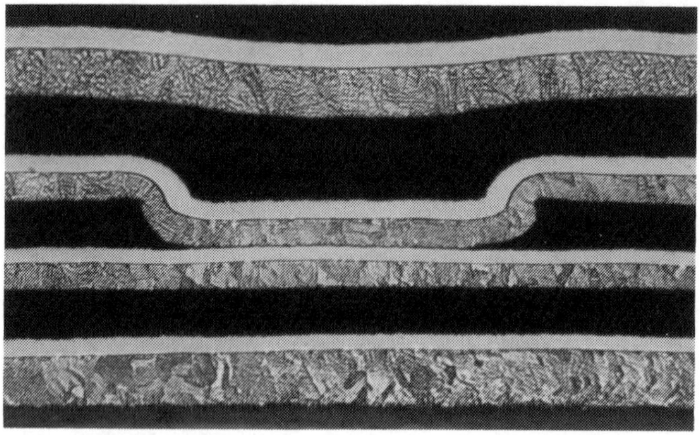

Fig.10: Cross-section of the 4-layer test vehicle

Concerning ground planes the mesh pattern type turns out to be superior to the solid plane design. On one hand this is due to better adhesion, because adhesion of the polyimide to large area metal features is still a weak point. On the other hand the capacitive load of grid structures is considerably lower than that of solid planes. Thus signal propagation is positively affected.

Similar results are obtained for the Asahi Pimel and the Probimide 300 polyimides. Processing the BCB version of the module confirms all the merits the material is promising. Furthermore the specific BCB cure procedure contributes considerably to processing efficiency. So less time elapses on BCB curing, because only a single soft bake of 30 minutes is required per layer and a final hard cure of 60 minutes after completion. This partiallly compensates the loss of processing ease due to the dry etching procedures. According to moderate temperature levels of 250°C maximum treatment of the thin film structures is more gently at least.

3.4 Memory Module

To verify the maturity of the present thin film MCM approach, memory modules holding nine 1 Mb DRAMs and their allocated 330nF decoupling capacitors are fabricated on the base of Si substrates. Three layers are provided for circuit wiring including the top layer facilities for component attachment. A linewidth of 50μm and 100μm pitch is specified for the signal traces. In contrast to the test vehicle an Al metallisation is chosen embedded in XU13005 or in Probimide 300 as well. Epoxy die bonding and thermosonic Al wire bonding are used for IC assembly. Capacitors are solder joined to copper-coated pads. Fig.11 gives a close up overview of the demonstrator module.

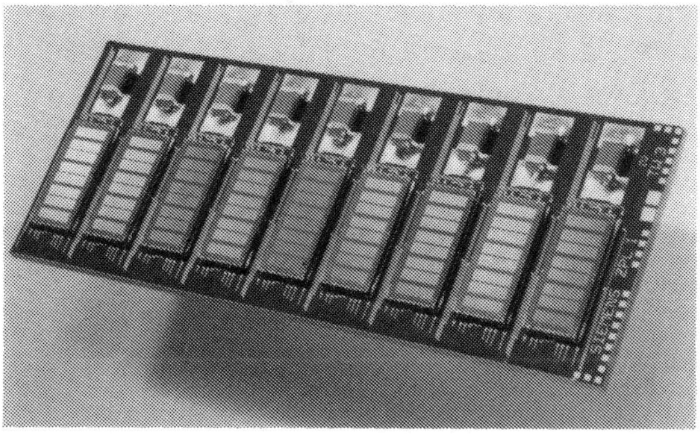

Fig.11: MCM-D demonstrator holding nine 1Mb chips (original size 20mm x 50mm)

Module testing is done in two steps. Before chip assembly the substrates themselves are probed for opens and shorts. To detect any failure event after completion of the MCM, all memory cells are successively addressed, loaded and read out again. Due to its simplicity this procedure is not capable of complete fault diagnosis of course. This would require more sophisticated debugging tools like boundary scan or e-beam probing, which however are still at their infancy concerning MCM testing[6].

4. Outlook

The potential thin film technology offers to MCM construction has not been fully exploited so far. That's not meant to refer to packaging density, which more likely will be limited by heat dissipation problems than by restrictions of thin film feature size. But in particular it is concerning the integration of passive components, especially of decoupling capacitors, within the MCM-D multilayer structure.

Mainly indispensable resistors and capacitors are still placed among the dice

on the module top. As an undesired effect the module size is blown up again. Furthermore the parallel employment of different assembly processes like wire bonding and soldering, is more costly at least.
Whereas the fabrication of buried thin film resistors should raise no problems, this is not with capacitors. Breaking point is the low k dielectric material required by the stripline characteristic of the conductors, which let capacitors grow beyond any reasonable dimensions.
Replacing the capacitor dielectric by materials providing k values in the magnitude of 1000 or above would be an appropriate solution to shrink down capacitor size again. Unfortunately materials, which meet the high k demand, e.g. ferroelectrics, do not fit into thin film processing up to now.
It is a challenge addressing material science to make such dielectrics more compatible to thin film MCM fabrication processes.
With respect to the future it should be emphasized that only thin film technology will be capable of providing planar optical wave guides, semiconducting and microsystem components integrated into the MCM substrates of tomorrow.
Apart from that there is still a couple of commonly known hurdles impeding widespread MCM acceptance. Whereas the breakthrough concerning technological problems like housing, test and repair is emerging today, full availabilty of known good dice cannot be expected even for the next future.
Nevertheless at Siemens MCM-D technology is on the way now from the basic delvelopment stage to product realisation.

References

(1) T.L. Ritter "Challenges Facing the MCM Industry in the next Decade", Proc. of the Techn. Program: National Electronic Packaging and Production Conf., pp. 161-168, June 1991

(2) BPA Technology and Management Inc. "Use of MCMs will Divide", Electronic World News, p. 35, March 1992

(3) M.C. Tucker " Meeting Global MCM Manufacturing Challenges", Solid State Technology, pp. 101-104, June 1992

(4) J.D. Mosley "Lack of Standards Impedes Design Issues", EDN, pp. 35-42, January 1992

(5) Y.N.Kim et al. "Adhesion and Interface Investigation of Polyimide on metals", Jounal of Adhesion Sience Technology Vol.2, pp. 95-105, 1988

(6) D. Kapenske, C.Talbot "Testing and Diagnosis of Multichip Modules", Solid State Technology, pp. 24-26, June 1991

RECENT PROGRESS IN CERAMIC SUBSTRATES FOR MICROELECTRONIC PACKAGING

Koichi Niwa
FUJITSU LABORATORIES LTD.
Morinosato-Wakamiya, Atsugi, Kanagawa, Japan

INTRODUCTION

Many kinds of ceramic materials are used as microelectronics packaging substrates. High-speed computer and supercomputer systems require a low dielectric constant ceramics to achieve a high speed signal propagation between integrated circuit chips, which are mounted on ceramic circuit boards. Chip carriers used for the computer systems require a high thermal conductivity ceramics, e.g. AlN, for heat dissipating chips. MCM (Multi-Chip Module) is now getting a very important position in computer packaging technology. A low thermal expansion and low dielectric constant ceramics, or high thermal conductivity ceramics and metals are applied as the base substrates of MCM. Composites of ceramics and organic resins are being applied to the substrates for flip-chip bonding. In this paper recent progress in ceramic substrates is reviewed.

PACKAGING IN HIGH SPEED COMPUTER SYSTEM

Three major computer companies in Japan, FUJITSU, HITACHI, and NEC, have applied ceramic materials as circuit board insulators. NEC has already applied ceramics to circuit board prior to FUJITSU and HITACHI. FUJITSU started to apply ceramics to a circuit board material[1)] in 1991.

Figures 1 and 2 show progress in propagation delay and number of gate of logic LSI applied to FUJITSU's high-speed computer system. The M-190 (1977) used 100-gate logic array LSIs, 14 years later, the M-1800 (1991) used about 15000-gate logic array LSIs, a more than 100 fold increase in integration. The propagation delay of logic LSI has decreased dramatically in 14 years, about 1/10 delay time has been realized. According to the rapid progress of logic LSI's, the package style was affected and changed as shown in Table I. The packages for M-190 and M-380 are QFP (Quad Flat Package, Ceramic Multilayer Package) with fin for forced air cooling. The

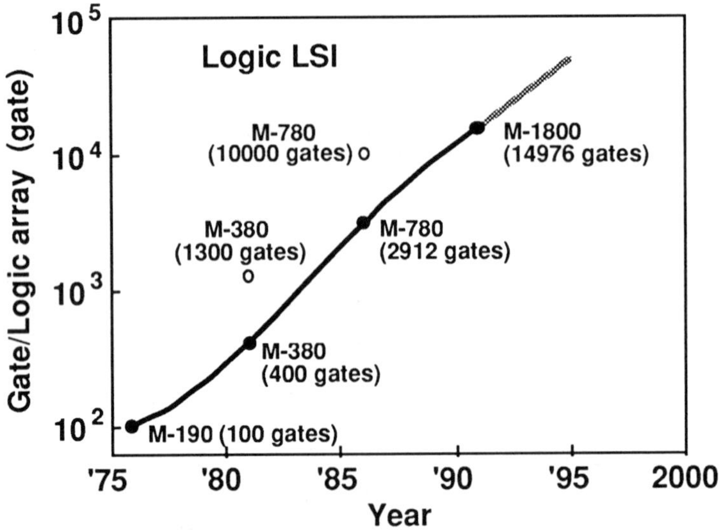

Fig.1 Progress in gate of logic LSI.

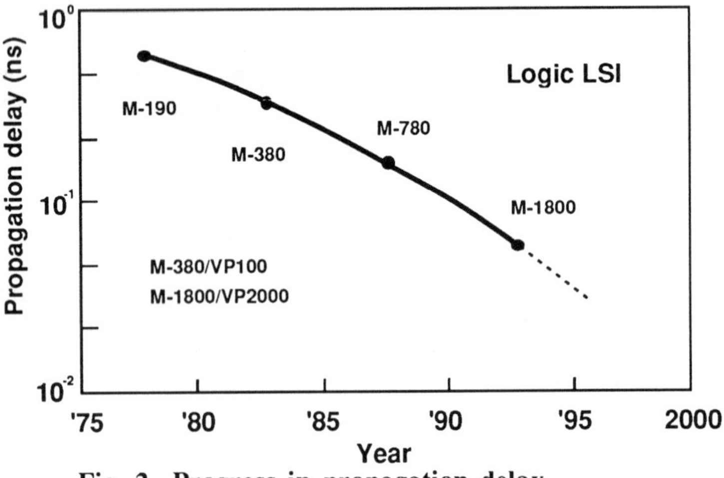

Fig. 2 Progress in propagation delay.

power consumption of an M-380 chip is 2.7 W, however, for the M-780 this increases to 8.5 W, which exceeds a maximum capability of forced air cooling[2]. In the M-780, chips are cooled by an impinging jet of water from the nozzle in FTC (Flexible Thermal Conductor) as shown in Fig. 3. The package (chip carrier) for M-780 is also QFP and is the same in the M-380 and M-190, but major difference is no fin structure, and pin count is about 2 times higher than M-380. Ceramic material for M-190, M-380, and M-780 packages is alumina. The chip is directly die attached to a molybdenum alloy plate for good heat dissipation. For M-1800 chip carriers, AlN is applied as PGA (Pin Grid Array) package[3]. AlN is used because of its high thermal conductivity and low thermal expansion. TAB (Tape Automated Bonding)

Table I. Package styles

M-190	M-380/VP400	M-780	M-1800/VP2000

84-pin QFP	84-pin QFP	180-pin QFP	462-pin PGA
12.7 x 12.7 mm^2	12.7 x 12.7 mm^2	22.5 x 22.5 mm^2	17 x 17 mm^2
4.0 x 4.0 mm^2	4.5 x 4.5 mm^2	8.9 x 8.9 mm^2	13 x 13 mm^2
Alumina + Mo plate	Alumina + Mo plate	Alumina + Mo plate	Al N
Wire bonding	Wire bonding	Wire bonding	TAB

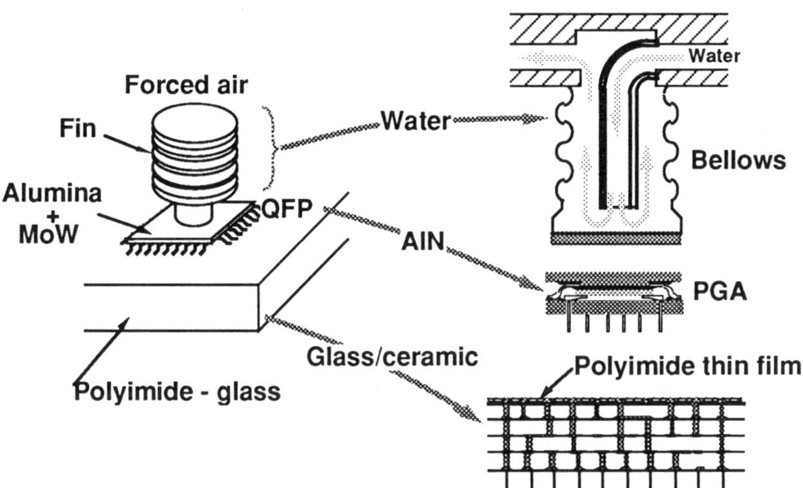

Fig. 3 Computer packaging.

was applied for inner lead bonding in the PGA. Wire bonding has been used in M-190 and M-380 for inner lead bonding.

Cross sectional views of the recent chip carriers for high speed computer systems are reviewed in Fig. 4. NEC's chip carrier is called FTC (Flipped TAB Carrier) and is used in the ACOS-3800. FTC uses Cu/W alloy cap for heat dissipation. Chip is die attached on Cu/W alloy cap. The substrate material is alumina. Resistors and capacitors are formed in the substrate[4]. HITACHI uses a chip carrier of AlN and mullite/glass composite[5]. AlN is applied for a water-cooled heat sink. The top of the chip carrier is called micro fin whose shape is looked like fin. The water jacket, also made of AlN, has fins which mate with micro fin. No thermal compound is needed because the close contact between micro fin and water jacket gives low thermal resistance. Inner and outer lead connections are flip-chip solder joints.

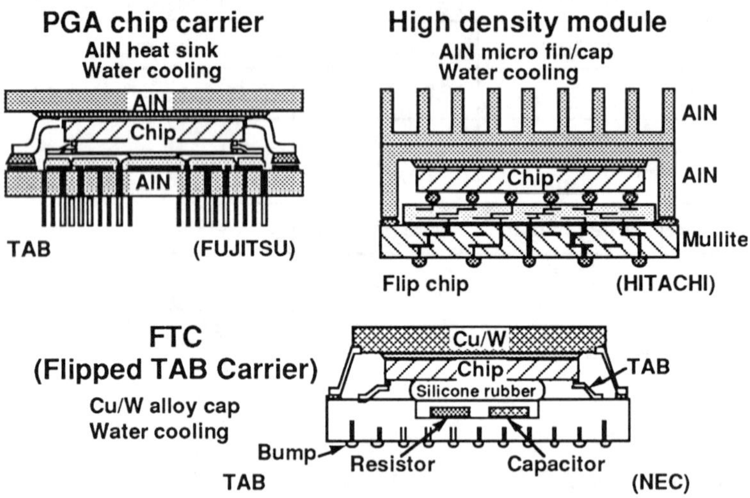

Fig. 4 Cross sectional view of recent chip carriers.

The circuit board material adopted to M-1800 (FUJITSU) is a glass/ceramic composite which can be fired at low temperatures as low as 1000°C and has a low dielectric constant[6]. The total number of layers is 51. On the top and bottom of the substrate[7] polyimide layers are formed. M-880 (HITACHI) uses a mullite/glass composite because of its low dielectric constant and low thermal expansion. Total number of layers is 44 including a distribution layer at the top, G/V and signal layers. The conductor used in this multilayer substrate is W[5].

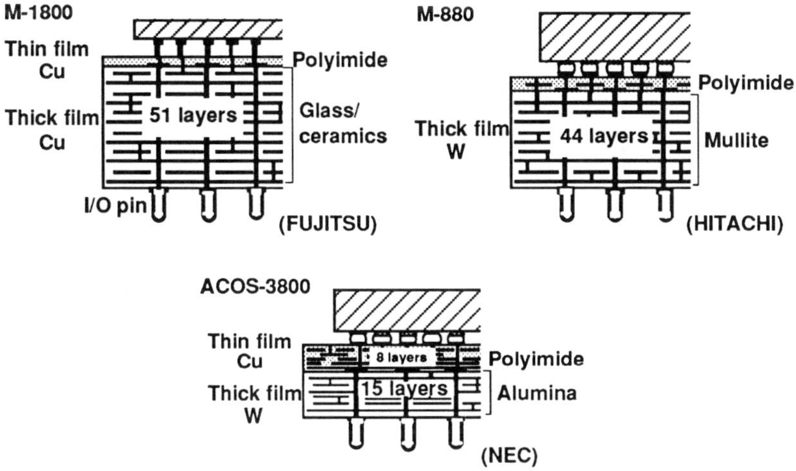

Fig. 5 Circuit boards used in main frame computer system.

SX-3/SX-X (NEC) uses an alumina multilayered with a copper-polyimide thin film layer on the top of the board[4]. Copper-polyimide layer is 8 and W-alumina layer is 15. Low dielectric constant layers are formed using polyimide and fine signal lines are patterned by photo-lithographic process. The three circuit boards are shown in Fig. 5.

PACKAGING TECHNOLOGY FOR GENERAL PURPOSE

AlN PACKAGE

High thermal conductivity AlN is useful for dissipating high heat from LSI chips. AlN was first applied as the chip carriers of high-speed computers, and is going to be adopted to the packages for high power LSIs such as LSI of one chip CPU by CMOS or ECL device.

Figure 6 shows latest QFP and PGA type AlN packages. The QFP shown in Fig. 7 is developed by FUJITSU as the package for MUSE decoder of HDTV (High Definition TV)[7]. This QFP uses two ceramic materials, AlN and mullite. The MUSE decoder chip is die attached to AlN, and aluminum fin is attached on the back of the AlN. Mullite is applied as a cap of QFP because thermal expansion coefficient of mullite is the same as that of AlN and mullite is cheaper than AlN. Another one shown in Fig. 6 is developed

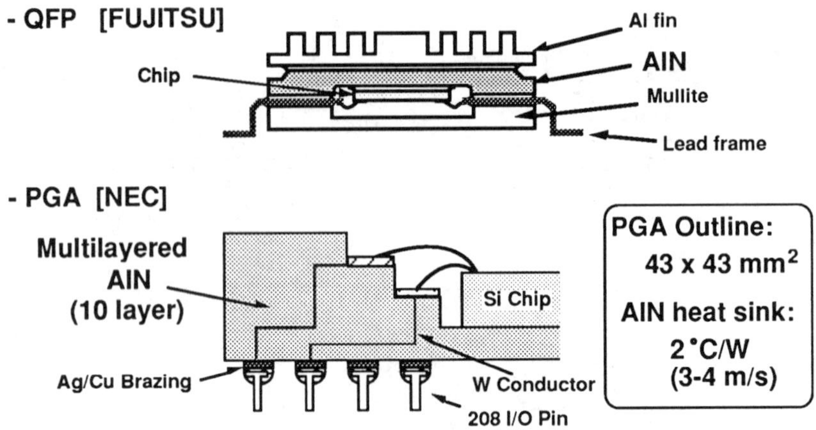

Fig. 6 QFP and PGA type AlN packages.

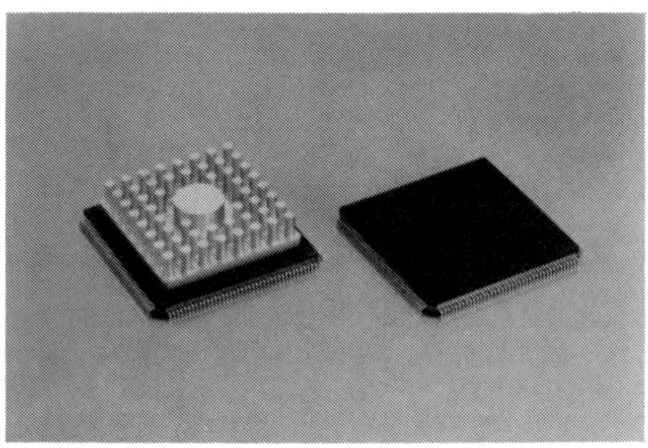

Fig. 7 PGA type package developed by FUJITSU.

by NEC. This PGA type package has multilayers of AlN and W conductors[8]. The thermal conductivity of AlN for this PGA is 220 W/mK. Thermal resistance is about 2°C/W at an air velocity of 2~3 m/s. I/O pins are brazed with a Ag/Cu alloy. The pin strength is 7 kgf/pin and the sheet resistivity of the W conductor is 7.2 mΩ /sq. This PGA is reported to be capable for dissipating high power of more than 40 W.

AlN is also applied as a base substrate for COW(Chip On Wafer). Figure 8 shows an example of COW made by Toshiba[9]. Toshiba developed a DSP(Digital Signal Processor) using a polyimide-copper thin film circuit on a 64×50×0.2 mm^3 Si wafer. The Si wafer is mounted on a 83×67×2 mm^3 AlN substrate with Ag-paste adhesive. After curing of adhesive a small amount of curvature (8 µm) was observed. If alumina is applied as the substrate, the substrate curvature becomes 28 µm. This result caused by the difference of thermal expansions of AlN and alumina with Si wafer. This COW package shows thermal resistance of 40°C/W(junction-air) at an air speed of 2 m/s.

- Chip on Wafer [Toshiba]

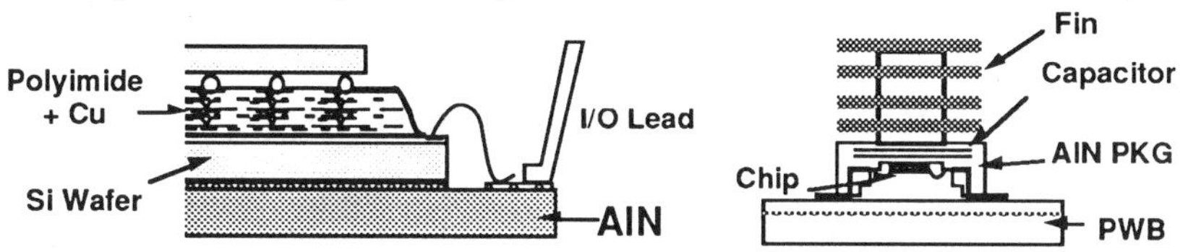

Fig. 8 AlN package developed by Toshiba.

METAL SUBSTRATE

A thin film wiring substrate has been developed by Sumitomo Electric Industries Ltd. Basic structure of the thin film wiring substrate is shown in Fig. 9 [10]. Metal plate for the base substrate is a Fe-42Ni alloy or Cu alloy.

- Thin film wiring substrate [Sumitomo Electric Ind.]

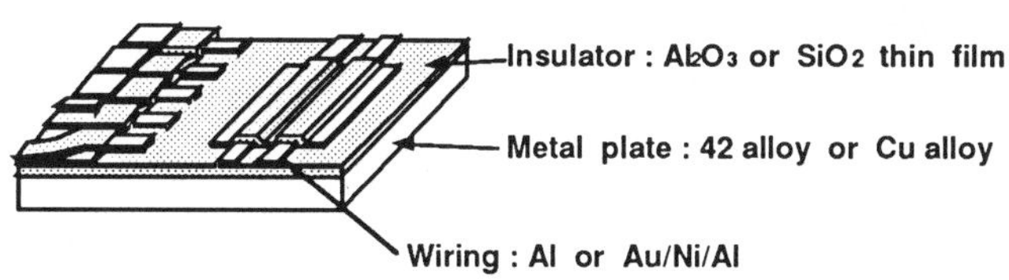

- MMIC [Mitsubishi Electric]

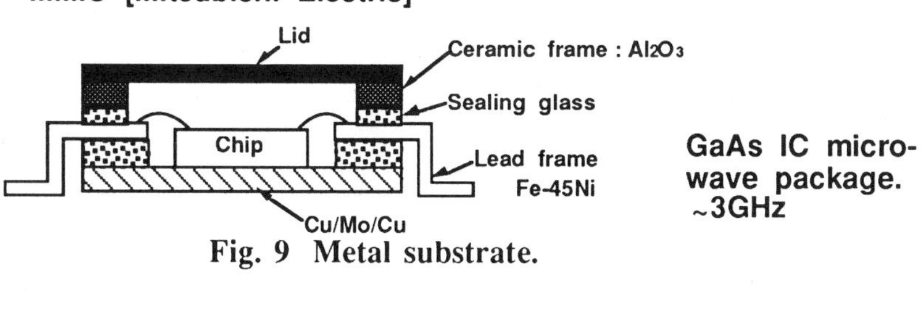

Fig. 9 Metal substrate.

Thin film (thickness: several ~ over 10 μm) of alumina or silica is deposited on the base metal as an insulator. Wiring materials used are Al and Au/Ni/Al. This substrate has a low thermal resistance and high heat dissipation due to its metal with thin ceramic layer structure. This substrate is used for an LED driver module which operates at 125 MHz.

A small size and simple structure package, called MMICs, has been developed by Mitsubishi Electric Co.[11]. The cross sectional view is shown in Fig. 9. Cu/Mo/Cu clad, Kovar, and Cu/W can be used as base metal. The lead frame metal in this package is Fe-45Ni alloy and lead frame is glass-sealed. This package features are brazing-free assembly and high thermal conductivity (194 W/mK, Cu/Mo/Cu clad). This MMIC was developed for GaAs monolithic microwave integrated circuits device which can be operate at up to 3 GHz.

CERAMIC/RESIN COMPOSITE STRUCTURE

A high density packaging of chips is now also required for organic printed circuit boards. The major problem in the organic board assembly is strongly caused by larger thermal expansion of the organic board materials, about 3 to 4 times higher than that of silicon. To decrease large thermal expansion, several approach has been tried by mixing the core materials with ceramics, or coating the organic board with ceramics. Examples of ceramic/resin substrates are shown in Fig. 10. These substrates are thought to be suitable for a 2-4 M DRAM substrates and FAX page printer application[12,13].

- Ceramic composite Copper-clad laminate [Hitachi Chemical]

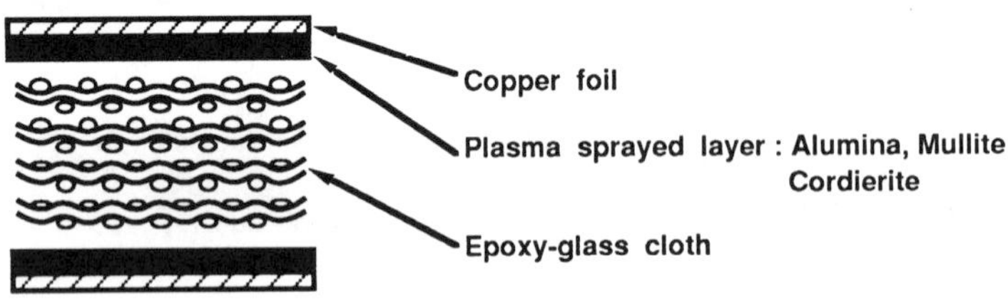

- CERACOM [Ibiden]

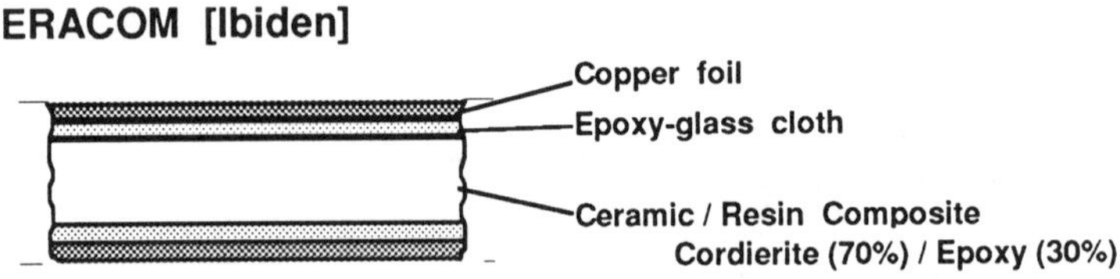

Fig. 10 Ceramic/Resin composite substrate.

THERMAL VIA

To increase heat dissipation, metal vias are formed in the glass-ceramic or polyimide layer. OKI developed a MCM for B-ISDN[14]. This MCM uses low temperature cofirable ceramics with a dielectric constant of 4.8 and a thermal conductivity of 3 W/mK. The chips are die attached(face up bonded) to ceramics. Heat flow from the chips goes through low thermal conductivity ceramics. Thermal via is employed between the chip and the fin. OKI reduced thermal resistance 40% by using thermal vias. Toshiba has also reduced the thermal resistance of copper-polyimide MCMs[15]. Thermal vias are formed in the polyimide layer on an alumina substrate. An aluminium fin is attached to the back of the alumina substrate.

SUMMARY

This paper summarizes recent trends in packaging technologies for computer and general applications. At the chip carrier level, TAB or solder bump is applied for chip bonding, and chip carriers are connected to circuit board with pins or solder bumps. A high thermal conductivity cap of AlN or Cu/W is used for water cooling technologies. At the circuit board level, only solder connection (flip-chip bonding) is applied. Although various circuit board materials have been used, requirement for the circuit board materials is almost the same, a low dielectric constant for high-speed signal propagation. Ceramic circuit boards made it possible to achieve a high gate density on circuit board(Fig. 6). FUJITSU M-1800 realized 10 times higher gate density than M-780 which used organic printed circuit boards. Substrates for general purpose packaging can be divide into three or four types: (1) AlN packages, QFP, PGA, and COW, (2) Metal substrates, thin film, lead frame, (3) Ceramic/resin composite substrates, (4) Substrates with thermal vias. The major purpose of these substrate is to deduce materials or process cost. The requirements for these general-purpose substrates are, however, almost the same as those of high-speed computers. The big difference in high speed computer circuit boards is total wiring length, which is reflected in the number of multiple circuit layers. Computer systems use over 50 layer-board for LSI mounting. General purpose packages have about 10~20 layers. Most of them are designed for the simple package called MCM(Multi-Chip Module). MCMs are likely to be popular because of their low cost and high performance.

REFERENCES

1) K. Niwa, and Y. Takeda "Multilayer Glass-Ceramic-Composite Circuit Board for FUJITSU VP2000 Series" FUJITSU Sci. Tech. J. **27** 179-186 (1991).
2) H. Yamamoto, Y. Udagawa, and T. Okada "Cooling and Packaging Technologies for FACOM M780" in Japanese FUJITSU **37** 124-134 (1986).

3) K. Ohno, K. Ooami, H. Itoh, and K. Suyama "Semiconductor Devices for FUJITSU VP 2000 Series" FUJITSU Sci. Tech. J. **27** 169-178 (1991).
4) A. Dohya, T. Watari, and H. Nishimori "Packaging Technology for the NEC SX-3/SX-X Supercomputer" IEEE. ECTC'90 Proceedings 525-533 (1990).
5) T. Fujita, S. Ishihara, S. Ogihara, G. Toda, and Y. Watanabe "Multilayer Mullite-Glass Ceramic Substrate for Main Frame Computers" ISHM Proceedings '91 187-190 (1991).
6) A. Kaneko, K. Seyama, and M. Suzuki "LSI Packaging and Cooling Technologies for FUJITSU VP2000 Series" in Japanese FUJITSU **41** 12-19 (1990).
7) NIKKEI NEW MATERIALS, Feb. **17** 51-59 (1992).
8) Shibuya, Y. Kurokawa, and Y. Shimada " Highly Thermal Conductive AlN PGA Package" IMC'92 Proceedings 285-290 (1992).
9) T. Sudo, S. Kimijima, O. Shimada, and N. Iwase "A Silicon-Based Mutichip Module with Co-Fired Aluminum Nitride Package" IEICE Trans. **E74** 2323-2330 *(1991).*
10) T. Takikawa, K. Harada, and S. Yamada "Reliability of Thin-Film Wiring Substrate with Ceramic Thin Film as Insulating Layer "Proceedings ISHM'91 483-487 (1991).
11) H. Yamada, T. Fusayasu, M. Takada, M. Kohara, and M. Nunosita "Small-Size Package for Monolithic Microwave Integrated Circuits" IMC'92 Proceedings 281-28 (1992).
12) H. Hasegawa, N. Okano, and M. Inoue "Ceramic Composite Copper-clad, Laminate IMC'92 Proceedings 47-50.
13) S. Saito "Ceramic Substrates for Flip Chip Bond" in Japanese Electronic Ceramics **22** 40-46 (1991).
14) T. Hanada, S. Iida, M. Yamada, and J. Utsunomiya "Development of a High Speed High Density Multi-Chip Module for B-ISDN" ISHM'91 Proceedings 196-200.
15) M. Nii, J. Omori, T. Nakazawa, T. Sasaki, J. Kobayashi, S. Kimijima, and Y. Fukuoka "Evaluation of Copper/Polyimide Thin Film Substrate for Multi-Chip Module" IMC'92 Proceedings 259-264 (1992).

RECENT ADVANCES IN HERMETIC EQUIVALENT IC PACKAGING - A REPLACEMENT FOR CERAMIC HERMETIC PACKAGING OF IC DEVICES

C. P. Wong, AT&T Bell Laboratories, P. O. Box 900, Princeton, NJ 08542-0900

INTRODUCTION

A modern electronic device is a complex 3-dimensional structure that consists of millions of components for each single IC chip. These components are densely-packaged in a multilayer structure with different ultrafine, submicron metallized conductor lines separated by dielectric insulating layers. The advances of the IC technology have had great technological and economic impact on the industry throughout the world. The rapid growth of the number of components per chip, the rapid decrease of device feature dimensions, the steady increase in IC chip size, and the ever-increasing number of input/output (I/O) interconnects have imposed stringent requirements, not only on the IC physical design and fabrication, but also on the IC encapsulants and packaging (see Figure 1). For the past couple of decades, we have seen the number of components per chip double for every 18 months. The increase of integration in ULSI technology has resulted in the miniaturization of the device size which has reduced the propagation delay due to higher density packaging and interconnection. As a result, a modern advanced device operates at a faster speed (~50 -100 MHz or greater), consumes more power, and consequently dissipates more heat during operation[1]. This complex and delicate device requires effective encapsulation and packaging to ensure its long-term reliability. The device encapsulant requires not only excellent electrical and physical properties, but also suitable mechanical properties for the hostile and extreme temperature cycling requirements. Low stress encapsulants are the preferred choice for microelectronic packaging. Silicone-based materials, with their low modulus and excellent electrical properties, are one of the best encapsulants[2]. AT&T has been using these silicone materials to encapsulate all their bipolar, MOS and hybrid IC devices since 1969[3]. There are billions of silicone-encapsulated devices in service with excellent reliability throughout these years. In fact, there is not a single circuit failure that could be attributed to the encapsulant. In this paper, we will describe the general chemistry of silicones (both room temperature cure elastomer and heat curable silicone gels), their reaction mechanism, electrical performance and their applications to electronic packaging as replacement for hermetic ceramic packaging.

WHY DO IC DEVICES WITH PASSIVATING LAYERS NEED ENCAPSULATION?

The purpose of encapsulation is to protect the electronic IC devices and prolong their performance. At the final stage of IC fabrication, a thin layer of passivating layer is normally deposited onto the finished fragile IC. However, moisture, mobile ions, (such as: sodium, potassium, chloride, fluorides), UV-VIS (ultraviolet-visible), alpha-particle radiation, and hostile environmental conditions, are some of the possible modes of degradation or interaction which could negatively affect device performance or lifetime. Silicon dioxide, silicon nitride and silicon-oxy-nitride, which are commonly used as passivation layers, are excellent encapsulants. These passivation layers are known to have excellent moisture and mobile ion barriers of the devices. As for the sodium ion barrier, silicon dioxide is inferior to silicon nitride. However, the use of phosphorous-doped (a few weight percent) silicon dioxide has greatly improved its mobile ion barrier property. A thin layer (approximately 1-2μm thickness) of one of these dielectric materials is deposited uniformly on the finished device, except on the bond pad areas where the passivating material is etched out for interconnection bonding. Due to the "edge effect" of passivation of IC device after wire bonding interconnection, we need the protection from the bond pad area (see Figure 2). In addition, the passivation layers (such as inorganic oxides and nitrides) are not 100% pinhole or crack-free. Thus, to ensure the device reliability, a conformal coating of organic encapsulant is usually used for device encapsulation. Epoxies, polyimides, polyxylylene (Parylene®), silicone-polyimides, and silicones (elastomers and gels) are usually used for this application. Typical properties of these potential encapsulants and their general advantages and disadvantages are listed in Tables 1 and 2[2-5].

Table 1. Physical Properties of Some Potential Encapsulants

Encapsulants	TCE* (ppm/°C)	Modulus (psi)
Epoxy	40 - 80	$1 - 5 \times 10^6$
Polyimide	3 - 80	1×10^6
Parylene	35 - 40	0.4×10^6
Silicone-Polyimide	5 - 100	0.4×10^6
Silicone Gel	200 - 1000	0 - 400

*TCE = Thermal Coefficient of Expansion

Table 2. Overview of Potential Encapsulants

Encapsulants	Application	Advantages	Disadvantages
(1) Epoxies	•Normal Dispensing or Molding	•Good Solvent Resistance •Excellent Mechanical Strength	•Non-Repairable •High Stress •Marginal Electrical Performance
(2)Polyimides	•Normal Dispensing (spin-coat)	•Good Solvent Resistance •Thermally Stable (~500°C)	Non-Repairable •High Stress

(3) Polyxylylene (Parylene®) (®Trademark from Union Carbide)	•Thermal Deposition (reactor)	•Good Solvent Resistance •Conformal Coating	•Thin Film Only •Non-Repairable
(4) Silicone-Polyimides	•Normal Dispensing	•Less Stress vs. Polyimide •Better Solvent Resistance vs. Silicone	•Higher TCE vs. Polyimides •Thin Film Only
(5) Silicones (RTV, Gel)	•Normal Dispensing	•Good Temp. Cycling •Good Electrical •Very Low Modulus	•Poor Solvent Resistance •Low Mechanical Strength

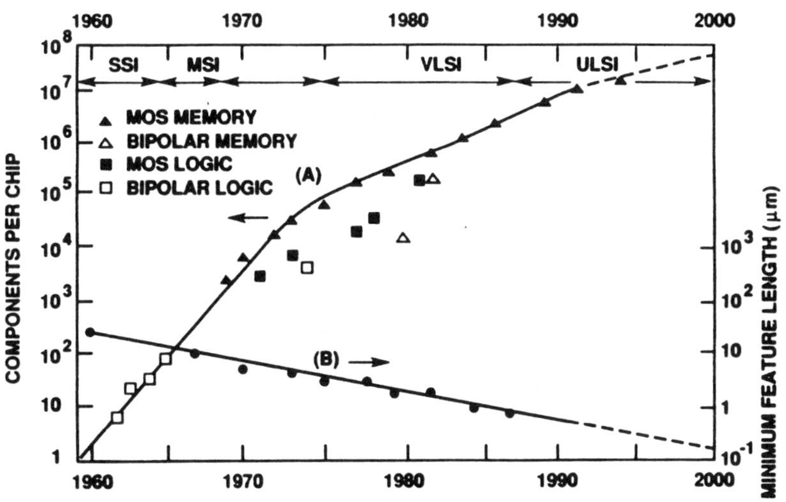

Figure 1: IC Technological Trends

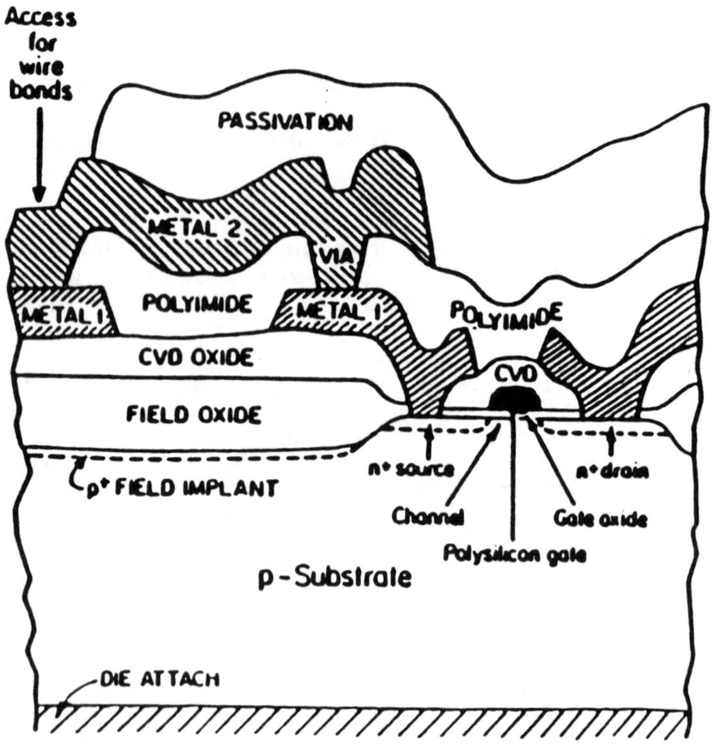

Figure 2: "Edge Effect" of a Passivated IC Device

GENERAL CHEMISTRY OF SILICONES (ELASTOMERS AND GELS)

The basis of commercial production of the silicones is that chlorosilanes are readily hydrolyzed to give disilanols which are unstable and condense to form siloxane oligomers and polymers. Depending on the reaction conditions, a mixture of linear polymers and cyclic oligomers is produced. The cyclic components can be ring opened to linear polymers and it is these linear polymers that are of commercial importance. The linear polymers are typically liquids of low viscosity and, as such, are not suited for use as encapsulants. These must be cross-linked (or vulcanized) in order to increase the molecular weight to a sufficient level where the properties are useful. Two methods of cross-linking are used: those which can be classified as condensation cures and those which are addition cure systems. For electronic applications, only the room temperature vulcanized silicone which uses alkoxide condensational cure (with alcohol as byproduct), and platinum catalyzed additional heat cure vinyl and hydride silicone systems (with virtually no byproduct) are suitable for device encapsulation. RTV silicone elastomer, and organosiloxane, cures by a moisture initiated-, catalyst-assisted process. It is one of the most effective encapsulants used for temperature cycling and moisture protection of IC devices (see Figure 3).

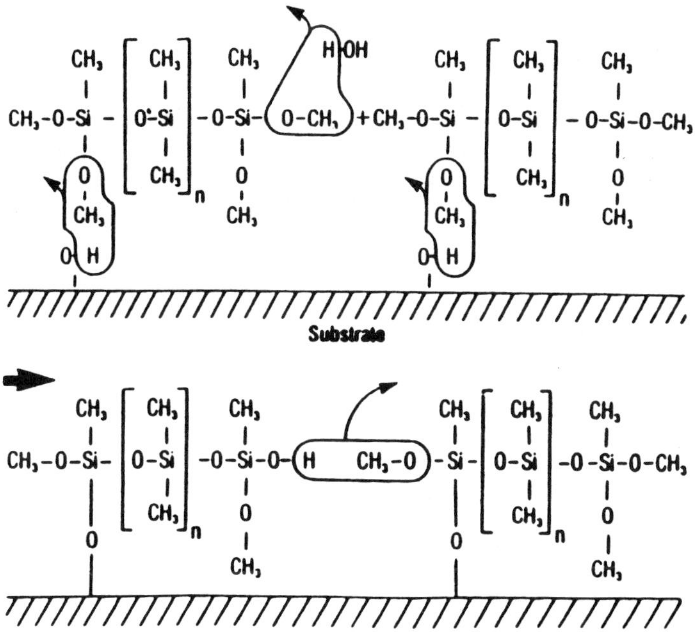

Figure 3: Room Temperature Vulcanized (RTV) Silicone: Condensation Cure Mechanism and Reaction with Coated Substrate Mechanism

Since World War II, silicones (organosiloxane polymers) have been used in a variety of applications where properties of high thermal stability, hydrophobicity, and low dielectric constant are necessary, eg., as encapsulants of conformal coatings for electronic components. In 1969, it was demonstrated that room temperature vulcanized (RTV) silicones exhibited excellent performance as moisture protection barriers for IC devices and a number of different RTV silicones have since been adapted for use in the electronics industry.[2-4]

Ionic materials, whether from the device surface or from the encapsulation materials of the environment, affect the electrical reliability of encapsulated IC devices. For this reason, the silicones are subjected to intense purification. The concentration of Na^+, K^+, and Cl^- mobile ions is less than a few ppm, and alpha particle emission is less than 0.001 alpha/cm^2 hr. Thus it offers excellent alpha particle shielding for eliminating soft error in dynamic random access memory devices, such as 64K, 256K, and megabit chips. The drawbacks of RTV silicone as an IC encapsulant are both its poor solvent resistance and weak mechanical properties. Highly fluorinated alkyl substituted siloxanes have shown an improvement in solvent resistance. For example, a recently developed silicone material with high cross-linking and high filler loading system seems to significantly improve the solvent resistance of the silicone encapsulant.[8]

Heat curable silicone (either elastomer or gel) has become an attractive device encapsulant. Its curing time is much faster than the RTV-type silicones. Besides, heat curable silicone gels tend to have slightly better thermal properties than conventional RTV silicone. With its excellent jelly-like (very low modulus) intrinsic softness, silicone gel is a very attractive encapsulant in wire bonded large chip size IC devices. The two part heat curable system which consists of the vinyl and hydride reactive functional groups and the platinum catalyst additional cure system provides a fast cure system without any byproduct (see Figure 4 for cure mechanism). This solventless type of heat curable silicone gel have increased usage in electronic applications[8-12].

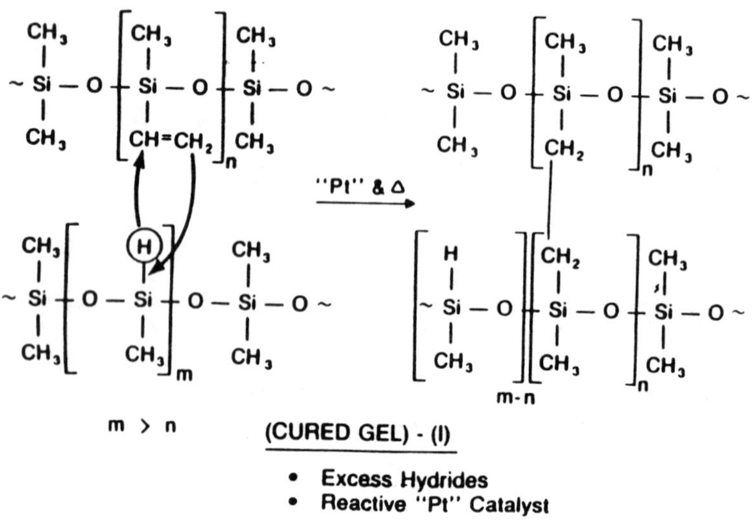

Figure 4: Heat Curable Silicone: Hydrosilation Additional Cure Mechanism

EXPERIMENTAL

Curing Of Encapsulants

In order to optimize each material property, the complete cure of the material is essential. Various analytical methods are used to determine the complete cure of each material. Differential Scanning Calorimetry (DSC), FT-IR and microdielectrometry provide quantitative measures of the curing process.

Differential Scanning Calorimetry (DSC)

DSC measures exothermic and/or endothermic heats of reaction associated with the material curing reaction. This provides quantitative information regarding the enthalpic changes in each material. A DSC thermogram is a plot of the energy supplied to the material as a function of temperature. In addition to providing measures of the heats of reaction based on the areas under thermogram peaks, additional information is readily obtained. Included are the glass temperature and kinetic parameters such as apparent activation energy, reaction rate and reaction order (see Figures 5 and 6).

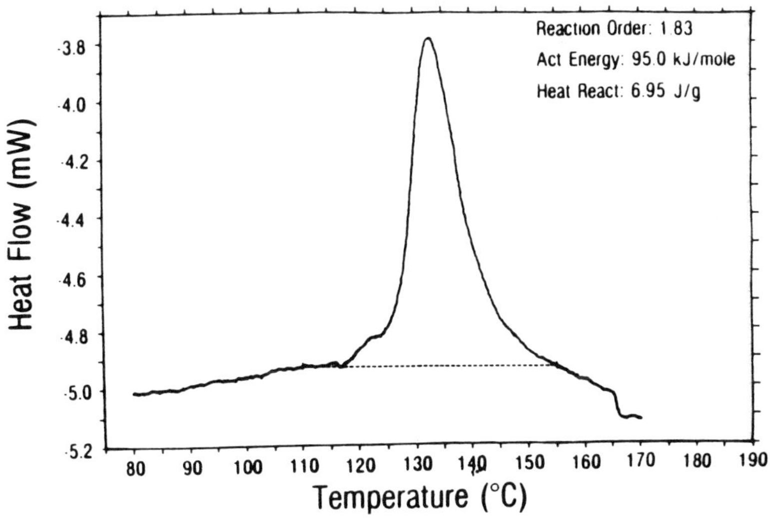

Figure 5: Differential Scanning Calorimetry (DSC) of a Silicone

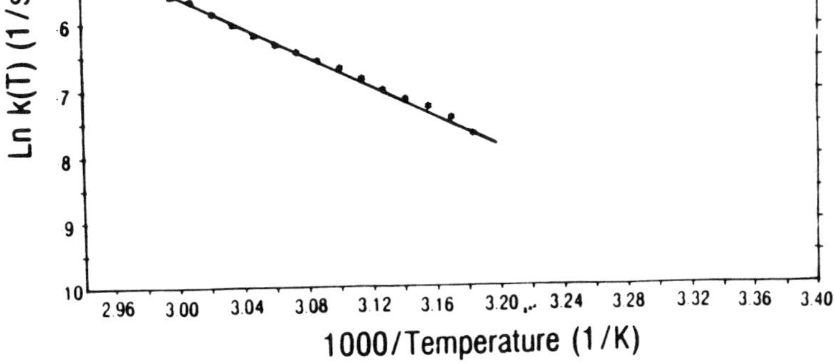

Figure 6: DSC Reaction Kinetics of a Heat Curable Silicone

Fourier-Transform Infrared (FT-IR)

FT-IR is a very sensitive tool for measuring the vibrational energy associated with the reactive functional groups of an encapsulant such as heat curable silicones (see Figure 7). The strong absorption of Si-H at ~2100cm^{-1} is characteristic of this uncured silicone. Upon the hydrosilation cure, the decrease of Si-H absorption at the initial cure stage could be easily measured by its peak height and peak area integration which provides a useful process tool to quantify the material cure. When the Si-H absorption stability, it is a good indication of the complete cure of the material. Figure 8 illustrates this. Similar information can be obtained for other types of encapsulation material systems.

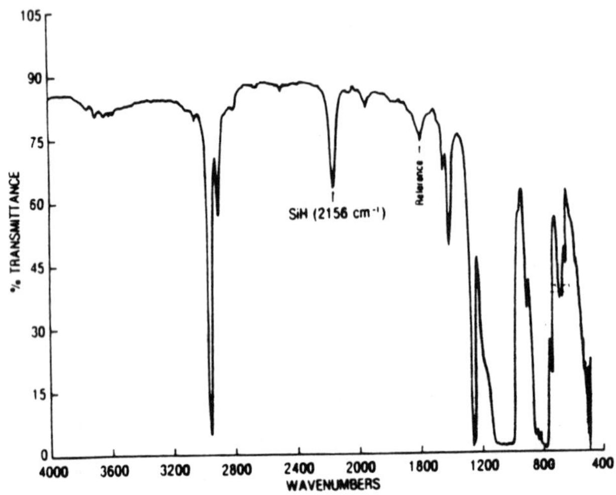

Figure 7: FT-IR of an Uncured Heat Curable Silicone

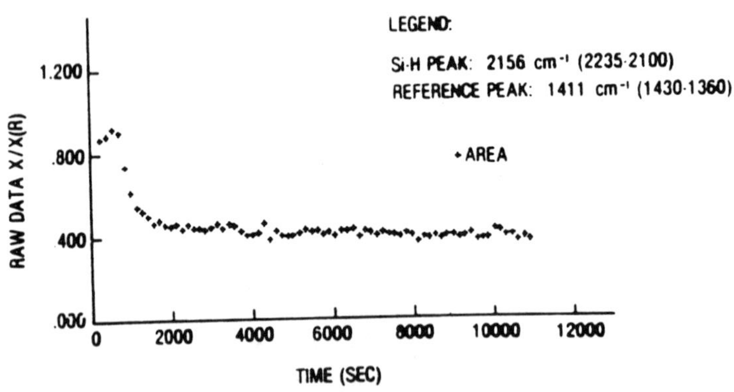

Figure 8: FT-IR of a Heat Curable Silicone Reaction Kinetics

Microdielectrometry

Microdielectrometry is another recently developed technique that can be used to quantitatively characterize encapsulant cure. Instrumentation available from Micromet Instruments utilizes a miniature IC sensor to measure the loss factor (E") of an encapsulant over a wide range of frequency (0.05 Hz to 10 KHz). In general, a thin layer (~20 mil thickness) of freshly prepared encapsulant material is coated on the miniature IC sensor and placed inside a programmable oven set at a constant temperature (e.g., 120°, 150° or 175°C). The loss factor (E") is then measures at various frequencies (e.g., 0.05, 1, 100, 1000, 10000 Hz) periodically during the curing of the material. Typical results of the microdielectric loss factor measurement are shown in Figure 9. When all the storage and loss factors (E' and E", respectively) become essentially constant, the curing process is considered to be complete.

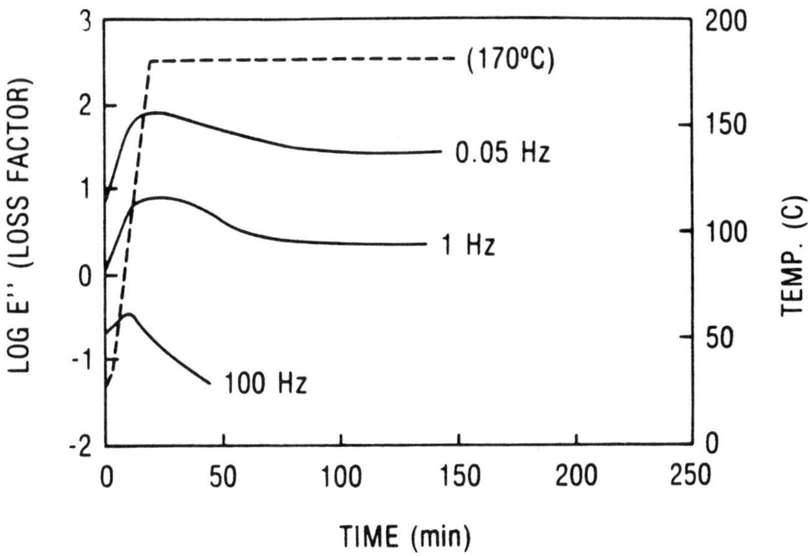

Figure 9: Microdielectrometry of a Heat Curable Silicone

Cure Study of Heat Curable Silicone

Heat curable silicone (either elastomer of gel) has become an attractive device encapsulant. Its curing time is much shorter than the RTV-type silicone. Furthermore, heat curable silicones also tend to have slightly better stability at elevated temperatures than conventional RTV silicone. However, the adhesion of the vinyl-hydride heat curable system is inferior to RTV system. With its excellent jelly-like (very low modulus) and intrinsic softness, silicone-based gel is a very attractive encapsulant in encapsulating wire-bonded large chip size IC devices. The two-part heat curable system which consists of the vinyl and hydride reactive functional groups, and the platinum catalyst additional cure system provides a fast cure system without any byproduct (see

Figure 4 for cure mechanism). To formulate a low modulus silicone gel, a vinyl-terminated polydimethylsiloxane with a moderate low viscosity range from 200 to a few thousand centipoise (cps), and a low viscosity (range from a few cps to 100 cps) mono-or multi-functional hydride-terminated polydimethylsiloxane are used in the formulation. The low viscosity hydride resin usually blends in with the higher viscosity vinyl resin to achieve a desire mixing ratio of part A (only vinyl portion) and Part B (hydride plus some vinyl portion). The key to formulate a low modulus silicone is the deliberate undercrosslinking, low reactive functional groups silicone system. This solventless type of heat curable silicone-based gel will have an increasing use in electronic applications.

TEMPERATURE HUMIDITY BIAS (THB) TESTING

THB testing was performed with a triple track resistance measurement. In this type of triple track testing, we grounded the two outer tracks and biased the center track. Then, we measured the leakage current change of the center conductor line. Good encapsulant shows very low leakage current with long testing time. Typical test conditions are 85°C, 85% RH and 10-180 V bias[13-19]. Of all the potential device encapsulant, silicone is one of the best THB electrical performance of all materials. However, other high purity encapsulants such as silicone-polyimide (such as Occidental Chemical's 2010M or Ablestik's Rely-Imide 1326), benzocyclobutene (BCB), polyimide such as Hitachi's PIQ8140 and AT&T's Modified Silicone Elastomer (MSE) and parylene also are have relatively good electrical performance (see Figure 10).

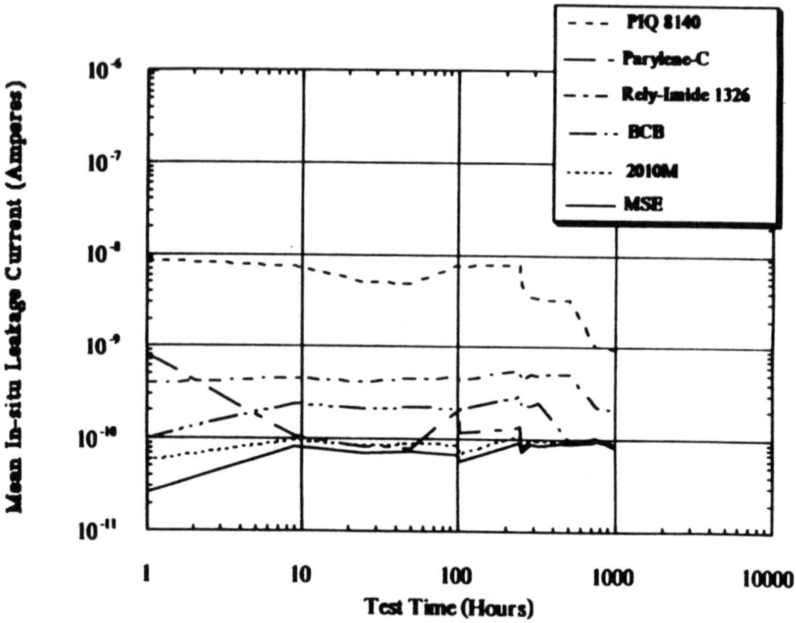

Figure 10: Temperature Humidity Bias (THB) Accerating Testing of Potential Encapsulants

TEMPERATURE CYCLING TESTING

To be an effective encapsulant, the material must possess excellent temperature cycling result. The siloxane backbone of the silicone provides good thermal stability (continual use temperature ~150°C) among all potential elastomers. Polyimides, silicone-polyimides, and BCB have better thermal stability than silicone. However, their modulus of elasticity is larger. With their intrinsic low modulus, silicone-based polymers have shown excellent performance in temperature cycling. Silicone samples were cycled from -40°C to +130°C with 20 minute dwell time at each temperature extreme show excellent results versus other potential encapsulants. Results of these temperature cycling data are clearly demonstrated on Figure 11.

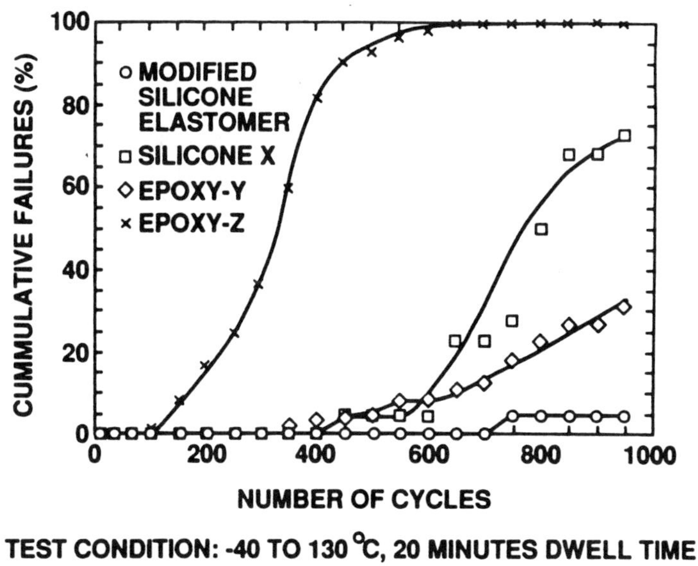

Figure 11: Temperature Cycling Test Results of Modified Silicone Elastomers

RESULTS AND DISCUSSIONS

Epoxy (plastic) molded devices have shown excellent reliability, and they are well-documented in the literature[6,7]. However, recent advances in hermetically equivalent packaging have been centered in the use of soft silicone gel. This section is meant to review some recent developments in this timely topic. A spectrum of numerous potential encapsulation materials exist. However, for high reliable IC device encapsulant, only a few select encapsulants are potential candidates. Table 2 selects six promising materials and presents advantages and disadvantages of these encapsulants. Two important mechanical properties play an extremely strong role in overall reliable functionality of the encapsulated device. The two properties are thermal coefficient of expansion (TCE) and modules of elasticity (E). These properties give rise to detrimental thermal cyclical stress as

$$\text{stress} = k\int_{T_1}^{T_2}(\Delta\text{TCE}) \cdot E \cdot dT,$$

where k is a constant, dT is the difference of the cycle temperatures (T_1 and T_2). Table 1 presents these physical data for some potential encapsulants. One will note that comparing the modulus of elasticity of silicone gel to the other results in orders of magnitude differences. Although the TCE for gels is higher, the resulting stresses will be substantially lower with the use of silicone gels (see Table 1) (as long as the volume is not restricted).

As we all know, high performance silicone gel possesses excellent electrical and physical properties for IC protection. With their intrinsic low modulus and soft gel-like nature, silicone gels have become very effective encapsulants for larger, high I/O, wire-bonded VLSI chips. Recent studies indicate that adequate IC chip surface protection with high- performance silicone gels in plastic packaging could possibly replace conventional ceramic hermetic packaging[8-12]. General Motors (Delco) have used silicone gels for over 10 years, and so has British Telecom. In each case, simple low-cost packages have been used to achieve a reliability performance better than available with simple epoxy molding compounds. However, recent advances in low stress ultrapure epoxy compounds has improved the material reliability[20]. The advantages of epoxy are cost, and the ability to surface-mount those packages on simple epoxy/glass circuit boards. Reliability performance with silicone gels and elastomers has been extensively studied by AT&T Bell Labs[3-5] and by Hitachi[11]. The corrosion protection results have been quite exciting. Hitachi has used gel-encapsulated devices in their super-computer systems. In addition, IBM and UNISYS have explored the performance of the gels and also reported good results.

Perhaps the real strength of the silicone gel encapsulation is for the future use in multichip module (MCM) packaging. MCM packaging is becoming an essential approach of high density packaging not only for meeting high operation speed requirements, but also for improving system integration[21,22]. This is true especially when package size and weight are critical driving factors. Ceramic packaging technology is severely stretched by the increases in the size and complexity of the packages. It is difficult to provide hermeticity at reasonable cost, and the match of the expansions of the lids to the substrates is demanding. Because of the ever-increasing large sizes, up to 4 inch square, pressure seals are very expensive and technically difficult, if not consistently impossible. The use of silicone gel or elastomer can be a simple solution. Plastic molding is not desirable when the chips must be changed to either get working systems, or to provide the necessary changes in performance. Here, the silicone-based gel encapsulation is useful even if the performance of molded plastic is adequate. The ability to change or salvage a package is worth far more than the possibly slight increased cost of premolded plastic packages needed for the use of encapsulating gels. The recent packages of Rockwell, for example, suggest the use of gel encapsulants to eliminate the need for ceramic packaging[22]. In addition, the use of silicone gel as a large VLSI chip passivation could provide a stress-relief buffer coating of the post-molded packages and achieve hermetically equivalent reliability. Clearly, the use of multichipping instead of wafer scale integration will rely on the application of silicone-based gel encapsulation[21-27].

CONCLUSION

We believe that the high performance silicone-based polymers possess an excellent IC device protection and have a potential to replace the ceramic hermetic package of IC. The reliability without hermeticity or hermetic equivalent packaging of IC device could be achievable with the high performance silicone encapsulant.[7] However, the combination of the multilayer inorganic passivating materials and the high performance polymeric organic coatings may be the key to achieve the reliability without hermeticity[28-29].

REFERENCES

1. Tummala, R. R., Rymaszewski, E. J., Ed., "Microelectronics Packaging Handbook", Van Nostrand Reinhold, New York, New York (1989).
2. Wong, C. P., "Polymers for Electronic and Photonic Applications", Ed., Academic Press, San Diego, CA (1992).
3. White, M. L., "Encapsulation of Integrated Circuits", Proc. IEEE, Vol. 57, p. 1610 (1969).
4. Wong, C. P., "IC Device Encapsulants and Polymeric Reliability", *Polymers for Electronic Applications*, Chapter 3, p. 63, CRC Press (1989).
5. Wong, C. P., "Overview of IC Device Encapsulants as Device Packaging", *Journal of Electronic Packaging, American Society for Mechanical Engineering*, Vol. 111,97 (1989).
6. Manzione, L. T., Lando, D. J., "Polymers in Electronics Packaging", AT&T Technical Journal, Vol. 69, No. 660 (1990).
7. Kinjo, N., Ogata, M., Nishi, K., Kaneda, A., "Epoxy Molding Compound as Encapsulation Materials for Microelectronic Devices", Advances in Polymer Science, Vol. 88, 1 (1989), Springer-Velag, Berlin.
8. Lin, A. W., Wong, C. P., "Encapsulant for Non-hermetic Multichip Modules Packaging", IEEE Trans. on Components, Hybrids, and Manufacturing Technology, Vol. 15, No. 4, 510 (1992).
9. Wong, C. P., Segelken, J. M., Balde, J. W., "Understanding the Use of Silicone Gel for Non-hermetic Packaging of ICs", IEEE Trans on Components, Hybrids and Manufacturing Technology, Vol. 4, 419 (1989).
10. Balde, J. W., "The Effectiveness of Silicone Gels for Corrosion Precaution of Silicon Circuits: The Final Report of the IEEE Computer Society Packaging Committee Special Task Force", IEEE Trans. on Components, Hybrids and Manufacturing Technology, Vol. 14, No. 2, 352 (1991).
11. Otsuka, K., Ishida, H., Utsumi, Y., Miwa, T., Shirai, Y., "Advantages of Silicone Gel for Packaging Devices with Very Large Scale Integration", Chapter 21, in "Polymeric Materials for Electronic Packaging and Interconnection", Lupinski, J. H., Moore, R. S., Eds., American Chemical Society Symposium Series, Vol. 407, p. 240, (1989).

12. Wong, C. P., "Electrical Performance and Reaction Kinetics of Silicone Gel", Journal of Materials Research, Vol. 5, No. 4, 795 (1990).

13. Mancke, R. G., "A Moisture Protection Screening Test for Hybrid Circuit Encapsulants", IEEE Trans. Components, Hybrids and Manufacturing Technology, Vol. 4, No. 4, 492 (1981).

14. Jaffee, D., Soos, N., "Encapsulation of Large Beam-Leaded Devices", IEEE Proc. Electronic Components Conf., 213 (1978).

15. Wong, C. P., "High Performance RTV Silicones as IC Encapsulants", The International Journal for Hybrids and Microelectronics, Vol. 4 (2), 315 (1981).

16. Wong, C. P., Maurer, D. E., "Improved RTV Silicone for IC Encapsulants", National Bureau of Standards, Special Publication 400-72, Semiconductor Moisture Measurement Technology, 275 (1982).

17. Wong, C. P., "Improved Room-Temperature Vulcanized Silicone Elastomers as Integrated Circuit Encapsulants", "Polymer Materials for Electronic Applications", American Chemical Society Symposium Series, No. 184, 171 (1982).

18. Wong, C. P., "Thermogravimetric Analysis of Silicone Elastomers as Integrated Circuit Device Encapsulants", American Chemical Society, Organic Coatings and Applied Polymer Science Proceedings, Vol. 48, 602 (1983).

19. Wong, C. P., Rose, D. M., "Alcohol Modified RTV Silicone for IC Device Packaging", IEEE Trans. on Components, Hybrids and Manufacturing Technology, Vol. 6, No. 4, p. 485 (1983) and references therein.

20. Yamada, T., et al., "Low Stress Design of Flip-Chip Technology for Si on Sil Multichip Modules", Proc. Int. Symp. on Electronic Packaging, Orlando, FL (1986).

21. Gates, L. E., Ward, T. G., "Seal and Chip-on-Board Circuit Protection", IEEE Proc. 41st Electronic Components and Technology Conference, 198 (1991).

22. Hagge, J. K., "Ultra-reliable Packaging for Silicon-on-Silicon WSI", IEEE Proc. of 38th Electronic Components Conference, 282 (1988).

23. Neugebauer, C. A., "Comparison of Wafer Scale Integration with VLSI Packaging Approaches", IEEE Trans. on Components, Hybrids and Manufacturing Technology, Vol. 10, No. 2, 184 (1987).

24. Tai, K. L., Wong, C. P., "Challenge of the High Density Packaging in the 1990's", Proceedings of the 6th Microelectronic Conference, 26 (1990), Tokyo, Japan.

25. Wong, C. P., "High Performance Silicone Gel as Integrated-Circuit Device Chip Protection; Cure Study and Electrical Reliability", Chapter 19, in "Polymeric Materials for Electronic Packaging and Interconnection", Lupinski, J. H., Moore, R. S., Ed., American Chemical Society Symposium Series, Vol. 407, p. 220, (1989).

26. Lin, A. W., Wong, C. P., "Encapsulant for Non-hermetic Multichip Modules Packaging", The IEEE Proceedings of the 41st Electronic Components and Technology Conference, 820 (1991).

27. Wong, C. P., "Recent Advances in the Application of High Performance Siloxane Polymer in Electronic Packaging", Proceedings of the 6th Electronic Materials Conference, Society for the Advancement of Material and Process Engineering, Vol. 6, 806 (1992).

28. Wong, C. P., "Passivating Organic Coatings with Silicone Gels: The Correlation Between the Material Cure and Its Electrical Reliability", Material Research Society, Proceedings of Electronic Packaging Materials, Vol. 154, 195 (1990).

29. Wong, C. P., "Recent Advances in IC Passivation and Encapsulation: Process Techniques and Materials", Chapter 4 in "Polymers for Electronic and Photonic Applications", Wong, C. P., Ed., Academic Press, San Diego, p. 167 (1992).

CERAMIC PACKAGING FOR TELECOMMUNICATIONS

David W. Johnson, Jr. and John Thomson
AT&T Bell Laboratories, Murray Hill, NJ 07974

ABSTRACT

Ceramic packages for integrated circuits are commonly used where requirements include high reliability, high IO count or large thermal loads. The advantages and shortcomings of ceramic packages for housing VLSI chips are reviewed. The various materials available to fabricate ceramic packages are listed along with how their properties meet ideal packaging demands. Designs of ceramic packages are reviewed with examples used in telecommunications. Finally, the use of a cofired multilayer ceramic package is described for a high speed laser transmitter housed with an optical isolator and a HIC for associated control. The fabrication of this high profile package is described in detail.

INTRODUCTION

Ceramic packages are part of the modern day fabric of electronic technology.[1,2] Just as in other sectors of electronic technology, ceramic packages are used in various facets of telecommunications products. Therefore, many portions of the technology are common to both general electronics and telecommunications applications.

This paper will review the needs for ceramic packages, the materials utilized, and then describe some examples of the types of packages used in the telecommunications industry. Finally, the processing steps used to fabricate a ceramic laser transmitter package will be described. This package is unique to telecommunications since it is used with optical fiber transmission media.

Applications of Ceramic Packages

Some form of packaging is needed for silicon integrated circuit technology. The reasons are: 1) protection of the circuitry from the external environment; 2) interconnection of chips; 3) removal of heat from the chips and 4) the provision c power and ground. Clearly, ceramic packages are not the only vehicle for provid ing this function. Table I shows that ceramic packages offer advantages such a hermeticity and high heat transfer while polymer packages offer the advantages o low dielectric constant and low cost.

Table I. Ceramic vs. Polymer Packages

Ceramic	Polymer
Hermetic	Low K
High Heat Transfer	High Toughness
Low Thermal Expansion	Low Cost

Table II lists some of the conditions which favor the utilization of ceramic pack ages.

Table II. Conditions Favoring Ceramic Packages

- High Thermal Loads
- Demanding Electrical Performance
    - Many Signal Layers
    - High Pin Count
    - Multiple Chips
- High Reliability
    - Military Specifications
    - New Designs

Clearly, because of the higher costs of the ceramic packages, these applications ar usually the most technologically advanced. The advantage of ceramic materials i thermal performance relates to their high thermal conductivity. The advantages i electrical performance relate to the ability to make cofired multilayer ceramic pack ages containing many layers with a complex pattern of 3-dimensional electrica interconnection. Ceramics are favored for high reliability applications largel because of their mechanical strength and hermeticity.

Ceramic Packaging Materials and Properties

Table III outlines some of the properties a package designer desires for ceramic packages.

Table III. Desirable Properties for Packaging Materials

- Hermeticity
- High Thermal Conductivity
- Low Dielectric Constant
- High Mechanical Strength
- Low Thermal Expansion
- Low Sintering Temperature
- Low Cost

Clearly, not every one of these properties can be satisfied by any one ceramic material. Therefore, it is necessary to understand the properties of common ceramic packaging materials. Figure I shows some common ceramic materials used for packaging and demonstrates qualitatively how well they perform in the various property categories.

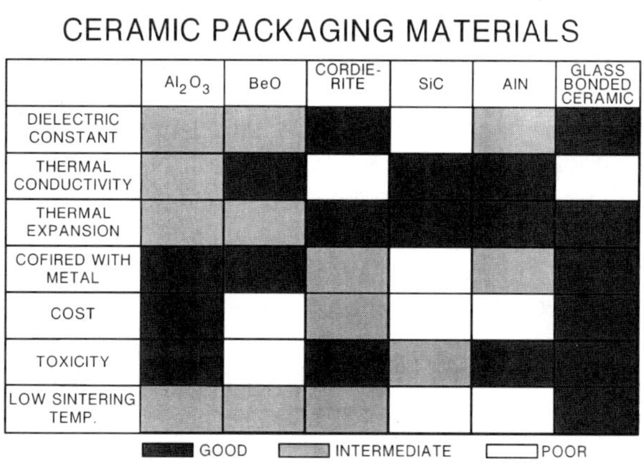

Figure 1 Comparison of material properties for some ceramic packaging materials

Aluminum oxide is the most broadly used material for ceramic packages. It is easily cofired with metals such as tungsten and molybdenum, is relatively low cost and offers no toxicity hazards. However, its performance is mediocre in some other categories. Its dielectric constant of about 9 is higher than some of the best ceramics. Even though its thermal conductivity is 10 times higher than some other ceramic materials, it is still 10 times lower than the best materials. Finally, its thermal expansion is somewhat higher than typical semiconductor chips and its sintering temperature is in the range of 1600°C.

In many respects, berylium oxide can be a direct replacement for aluminum oxide with a clear advantage of about 10 times better thermal conductivity. However, its utilization is severely restricted by the toxicity hazards associated with its production. These hazards have limited utilization and maintain it as a high cost alternative.

Cordierite, a magnesium alumina silicate, is an example of a ceramic material with a low dielectric constant (about 5). It also has a thermal expansion match which is close to that of semiconductors. However, there is little experience in cofiring this material with metal and its sintering temperature is rather high and of a narrow range.

Silicon carbide and aluminum nitride are two materials which have high thermal conductivities, approximately the same as that of berylium oxide. Silicon carbide suffers from a high dielectric constant (20-40) and could be perceived as posing a toxicity hazard since the electronic grade frequently uses BeO as a sintering aid. Therefore, of these two materials, aluminum nitride appears to be the one which will gain wide acceptance. It has a dielectric constant close to that of aluminum oxide, but has a thermal expansion coefficient significantly lower and closer to that of semiconductors. While in its relative infancy, this material has been successfully cofired with tungsten metal in multilayer package applications. Its current high cost is associated with its high sintering temperature (1800-2000°C), its high raw materials cost and its low volume production. It is anticipated that the cost will be reduced as the usage increases.

As can be seen from Fig. 1, glass bonded ceramics offer many advantages for packaging. These are typically made by mixing ceramic powders with glass frit. The sintering temperature can be made low and variable by the use of the appropriate glass composition and the ratio of glass to ceramic. As evidenced by the recent introduction of these materials for the new IBM thermal conduction module [1], low dielectric constant and low thermal expansion can be achieved simultaneously

Furthermore, these materials can be cosintered with copper metallization leading to low cost and low resistivity. The singular shortcoming of glass ceramics is the low thermal conductivity which is due to the intrinsic low thermal conductivity of the glass phase. Therefore, utilization of these materials where high thermal loads are anticipated demand other engineering solutions to heat removal.

Ceramic Package Designs

Ceramic packages are made either as simple pressed parts or as multilayer cofired devices.[3] The pressed packages generally accept a single chip and have the lead frame sealed in glass under the lid. Two common names for these packages are CERDIP and CERQUAD; the former having a dual in-line attachment and the latter having attachment pins on all 4 sides. These packages do not make use of multilayer technologies and will not be discussed further here.

The multilayer cofired packages are one of the first applications of multilayer ceramic technology [4] and makes use of buried metal lines and vias to make electrical connections between the chip and the external pins or pads.[5] For VLSI applications, ceramic multilayer packages are sub-categorized according to the mounting technology. The two major mounting categories are through-hole mounting and surface mounting. For through-hole mounted packages, common names are dual in-line packages (DIP) and pin grid arrays (PGA). Common names for surface mounted packages are chip carriers and ceramic quad flatpacs. Surface mounted packages can be either leaded or leadless.

Figures 2 & 3 demonstrate 2 examples of ceramic packages for VLSI applications. Fig. 2 shows a PGA package housing a multichip module. The module contains a digital signal processor (DSP) with 8 SRAMS on a silicon substrate. The chips are solder bumped to the silicon substrate and the substrate is wire bonded to the ceramic package. The pin grid array is on a 100 mil pitch. Fig. 3 shows two forms of quad flatpacs. In this case, the leads are in a gull wing format for surface mounting and are on a 20 mil pitch.

Figure 2 An example of a pin grid array package containing a DSP chip and SRAM chips

Figure 3 Examples of surface mounted ceramic packages using gullwin leads on 0.020" pitch

Multilayer ceramic packages are well developed for housing VLSI chips as the previous examples have shown. However, the rapidly expanding field of digital optical telecommunications has spawned a need for packaging transmitters and receivers. Because the number of IOs (inputs and outputs) are small, a simple metal can with a few glass bonded leads can often suffice.[6] For more demanding applications, ceramic packages can offer advantages. This is particularly true in those cases where it is desired to house electronic hybrid integrated circuits (HIC) within the same package as the transmitter and receiver. Multilayer ceramic packages can easily provide the IOs necessary for these applications. Such packages have been designed and built[7,8] and material property considerations made.[9]

Fig. 4 shows a package designed to have a laser transmitter operating up to 10 Gbits along with a thermoelectric cooler to provide a constant temperature for the laser, an optical isolator and a HIC designed for temperature control and laser driving functions.[10]

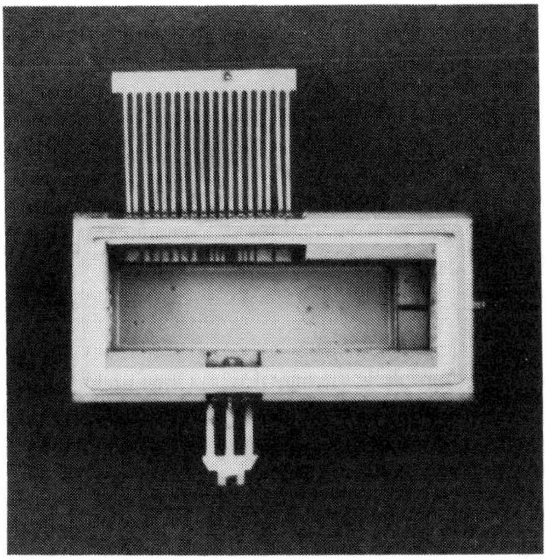

Figure 4 A completed ceramic optoelectronic package designed to house a laser transmitter, a thermoelectric cooler, an optical isolator and a hybrid integrated circuit

The dimensions of this package are 53 mm long x 28mm wide x 15 mm hig Fabricating a package of this height is a particular challenge. Its large size makes ideal to demonstrate some of the multilayer processing steps used.

Fabrication of a Laser Transmitter Package

The processing steps to fabricate this laser transmitter package and other cerami packages by the multilayer method are shown in Table IV.

Table IV. Ceramic Package Fabrication Steps

- Prepare Slurry
- Cast Tape
- Punch Blanks
- Punch Via Holes and Cavities
- Fill Vias
- Print Conductor Traces
- Stack and Laminate
- Cut or Groove
- Sinter
- Electroplate and Braze

For this application, the active aluminum oxide powder was mixed with sma amounts of transition metal oxides in order to provide opacity in the final produc This opacity is needed to prevent stray light from entering the transmitter packag and the fiber. These powders were mixed with a polyvinyl butyral using MEK an toluene as solvents. This was cast into a tape approximately 17 cm. wide in a sma laboratory tape caster shown in Fig. 5.

Figure 5 A laboratory tape caster with doped Al_2O_3 tape

The bulk of this 20 layer package could be made with tape about 0.5 mm (.020 in) thick. From this tape 12.7 cm (5 in) sq. sections of tape were cut in a blanking press which also punched a registration hole in each corner. These registration holes provide for alignment of the tape for each of the subsequent processing steps.

Via holes for electrical interconnection between layers and the major cavity cutout were punched into individual layers using a numerically controlled punch. The via holes were filled with a tungsten paste which was forced through a template on a screen printing apparatus. A tungsten paste was then used to screen print the internal traces as well as the external and internal interconnection pads. This package design contains 2 sets of external pads to which lead frames were ultimately brazed. These can be seen in Fig. 4. The longer array of pads provide the signal information to the hybrid integrated circuit and the short array of 3 pads are for the high speed transmission line bringing the data bit stream into the package. The screen printing step also provided the metallization for brazing the copper tungsten base and cover to this package. Some of these layers with the associated screen printings are shown in Fig. 6.

Figure 6 Sheets of Al_2O_3 tape with screen printed tungsten

The layers were then stacked onto a laminating frame. Fig. 7 shows a partially laminated stack on the left, revealing some of the internal interconnection traces and a fully completed stack on the right.

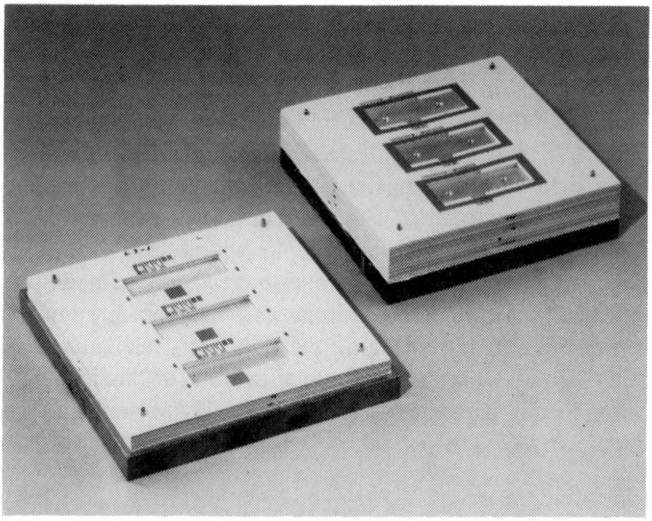

Figure 7 A partial stack (left) and a fully laminated stack (right) for 3 optoelectronic packages

The stacks were then laminated at a temperature of 50°C and a pressure of 2000 PSI. After lamination, these bars were cut into individual packages using a saw. A hole was drilled into the end of the green (unsintered) laminated package to accept the fiber ferrule. This hole was also metallized with tungsten ink. An example of the green package is shown on the left side of Fig. 8.

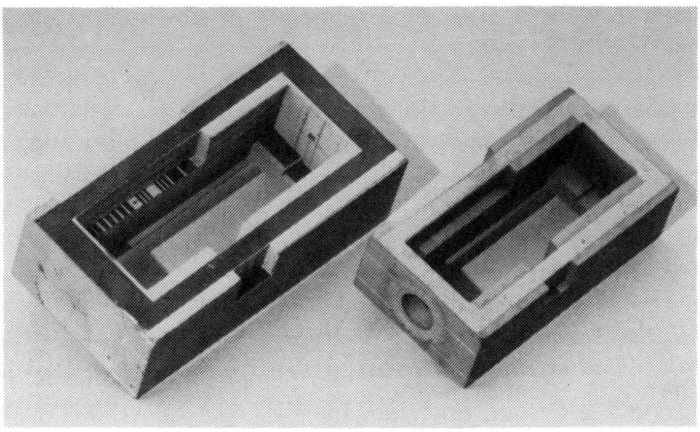

Figure 8 Comparison of the unsintered (left) and sintered (right) optoelectronic package.

The packages were then put into a sintering furnace where the binder was burned out and sintering was carried out at a temperature of 1600°C for 2 hours in a wet hydrogen-nitrogen atmosphere. An example of the sintered package is shown on the right side of Fig. 8. The above described process culminates in the device of Fig. 8 which represents the ceramic package for the laser transmitter, thermoelectric cooler, HIC and optical isolator as outlined earlier. As final steps, nickel was electroplated onto the tungsten to provide a wettable surface for subsequent brazing operations. The base, the seal ring, the lead frames and a ferrule for the fiber to be inserted into the end of the package were then brazed using a Ag/Cu alloy. Finally, the metal parts were electroplated with gold to yield the final package demonstrated in Fig. 4.

SUMMARY

While ceramic packages account for a small segment of the total packaging requirements in telecommunications, they are a very important and visible segment. The intricate design of interconnection available using multilaye ceramic technology, the choices and material properties (dielectric constant, the mal expansion, and thermal conductivity) and the ability to provide hermiticity, a combine to make ceramic packages the choice for highly demanding application Finally, through the example of a large laser transmitter package, it has bee demonstrated that ceramic multilayer packages have application in comple optoelectronic devices.

ACKNOWLEDGEMENTS

The authors wish to thank W. R. Holbrook, P. D. Smeltz, S. Kaufman and J. Lipso for their advice and collaborative efforts in designing and fabricating the optoelec tronic package.

REFERENCES

[1] R. R. Tummala, "Ceramic and Glass-Ceramic Packaging in the 1990s J. Am. Ceram. Soc., 74 [5] 895-908 (1991).

[2] A. Kanda, S. Nozaki and H. Banno "Trends of Ceramic IC Packages an Substrates" J. Ceram. Soc. Jpn. Inter Ed., 97 R39-R50 (1989).

[3] R. R. Tummala, "Ceramic Packaging" Chapter 7, p. 455-522 in Microelec tronics Packaging Handbook, Van Nostrand, Reinhold, NY, 1989.

[4] H. W. Stetson, "Multilayer Ceramic Technology," p. 307-322 in Ceramic and Civilization, Vol. III, High-Technology Ceramics - Past, Present an Future, Am. Ceram. Soc., Inc. Westerville, OH, 1987.

[5] H. C. Bhedwar and H. T. Sawhill, "Ceramic Multilayer Package Fabrica tion," p. 460-469 in Electronic Materials Handbook, Vol. 1, Packagin ASM International, Metals Park, OH, 1989.

[6] S. H. Wen, "Optoelectronic Packaging," J. of Metals, p. 14-18, June, 1988.

[7] D. S. Alles and K. J. Brady, "Packaging Technology for III-V Photoni Devices and Integrated Circuits," AT&T Technical Journal, 68 (1) 83-9 (1989).

[8] J. L. Hokanson, N. R. Dietrich, K. A. Yanushefski, M. J. Yanushefski, C. A Young, W. R. Holbrook, P. D. Smeltz, and T. A. Chynoweth, "A Integrated Laser Transmitter Operating Over a Wide Data-Rate Range i Harsh Environments," ECOC 87: 13th European Conf. in Optical Com munications, Technical Digest, Vol. 3, p. 1-4, 1987.

[9] M. F. Yan and W. W. Rhodes, "Ceramic Optical Packages: Material Requirements and Guidelines for Material Selection," p. 439-453. In Materials Research Society Proceedings, Vol. 108, "Electronic Packaging Materials Science III," R. Jaccodine, K. A. Jackson and R. C. Sundahl, eds., Materials Research Soc., Pittsburgh, PA, 1988.

10] G. M. Chaoui, J. Lipson, S. Barski, N. R. Dietrich, C. E. Gaebe, S. Y. Huang, S. Kaufman, L. C. Upadhyayula and X. L. Yeh. "10 GH_3 Isolated Laser Module," Technical Digest Series 1990 Optical Fiber Communications Conf. OFC 90, p. 109, Optical Soc. of Am., Wash., D.C. 1990.

MULTILAYER MULLITE-GLASS CERAMIC SUBSTRATES FOR MAINFRAME COMPUTER

Nobuyuki Ushifusa, Shosaku Ishihara, and Satoru Ogihara*
Production Engineering Research Laboratory, Hitachi, Ltd.
292 Yoshida, Totsuka, Yokohama, Kanagawa 244 Japan
*Hitachi Research Laboratory, Hitachi, Ltd., 3-1-1 Saiwai, Hitachi, Ibaragi 317 Japan

Abstract

The Hitachi M-880 processor group, the newly developed top-of-the-line processors in HITACHI M Series computer family, feature ultrahigh performance, excellent reliability and eased installation performances. To achieve these design goals, a new set of advanced hardware technologies have been developed in wide areas such as semiconductor, ceramic module board, printed circuit board, electric part, cooling and power feeding. One of them is high performance multilayer ceramic substrates consisting of mullite-glass ceramics and tungsten conductor. Mullite ($3Al_2O_3$-$2SiO_2$) has a low thermal expansion coefficient and a low dielectric constant making it a favorable material for substrate applications. It is difficult, however, to sinter pure mullite ceramics even above 1700°C. Thus, mullite ceramics containing glass additives which could be sintered at about 1600°C were fabricated. Many advantages can be tailored in mullite-glass ceramics by making adjustments of the compounding ratio of mullite and glass phase. Some of them are low dielectric constant

(5.9), high bending strength (210MPa), and low thermal expansion coefficient ($3.5x10^{-6}/°C$). Substrates made of mullite-glass ceramics are suitable for mounting silicon devices of computer processors.

M-880 PACKAGING HIERARCHY

The M-880 processor group consists of one processor board, 730mm by 534mm, capable of mounting up to 20 high density modules through connectors[1]. The high-speed high-density LSI chips, each of which is mounted on a Micro Carrier for Chip (MCC), are mounted and soldered on a square multilayer ceramic substrate (106mm by 106mm) by means of flip chip bonding method using micro solder balls. The square multilayer ceramic substrate is equipped with a cooling mechanism, covered with a module cap, and hermetically sealed with solder. The module cap and the multilayer ceramic substrate are hermetically sealed with solder. On the module cap, a copper-made water-cooling jacket that carries cooling water from water cooler is placed with grease applied between the jacket and the cap. (Figure 1)

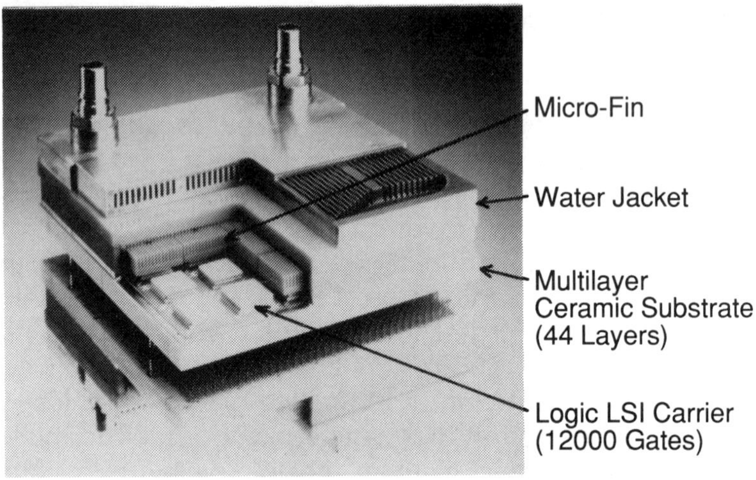

Fig. 1 M-880 packaging hierarchy.

The multilayer ceramic board is made of mullite-glass ceramics which was newly developed and tungsten conductors. The module board consists of a total of 44 layers: 2 outermost layers, 18 signal layers, 11 matching layers, and 13 power and ground layers. The pitch of the lattice on the signal layers is 0.45mm and one line 0.1mm wide is applied in a interval of the lattice. (Figure 2) The matching layers are used to match top and bottom pin pitches to the lattice on the signal layers.

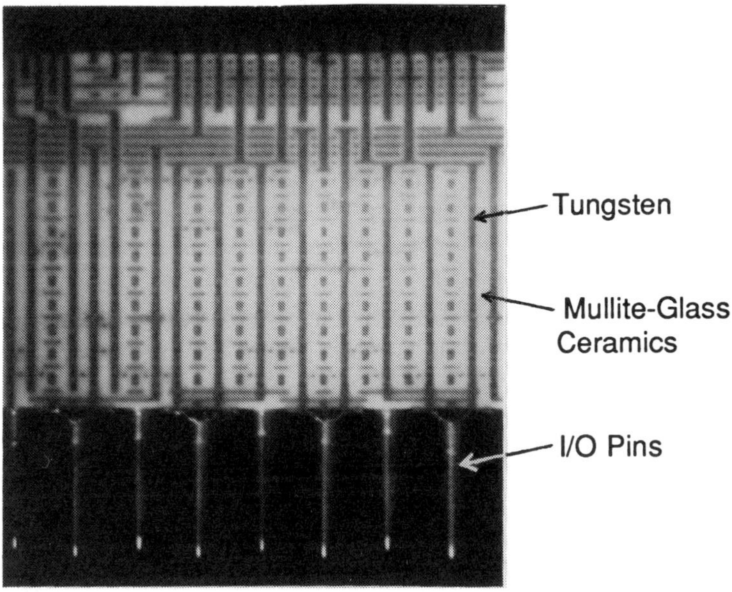

Fig. 2 Cross section of multilayer ceramic substrate.

INTRODUCTION

The required properties for multilayer ceramic substrates are as follows.

(1) Low dielectric constant. This decreases the pitch of the lattice on the signal layers.

(2) High density wiring. This increases the signal transfer speed between LSI chips.

(3) Thermal expansion coefficient close to that of LSI chips: This provides reliability for bonding between substrate and LSI chips.
(4) Thermal expansion coefficient close to that of conductor materials. This provides reliability of packaging multilayer substrates with high density wiring.
(5) High bending strength. This allows cooling units to be fixed on the surface of the ceramic substrates directly, which provides an advantage in the cooling mechanism.

In order to attain ceramics with the above mentioned properties, we selected the combination of mullite-glass ceramics and tungsten conductor.

Mullite is the only binary Al_2O_3-SiO_2 system stable at atmospheric pressure[2]. Because the thermal expansion coefficient and dielectric constant are lower than those of alumina, mullite is expected to have many uses as a substrate material for silicon wafers and chips. However, mullite has poor sinterability and is used extensively as a refractory material. Mullite is difficult to sinter adequately even at 1600°C and, like alumina, needs suitable additives[3,4,5]. It is also necessary to adjust the thermal expansion coefficient of the ceramics closer to that of silicon (3.5×10^{-6}/°C) in order to achieve highly reliable joints with silicon chips. Mullite (5.0×10^{-6}/°C) and glass (1.5×10^{-6}/°C) have low thermal expansion closed to that of Si chip. Mullite-glass ceramics can have different features by adjusting the compounding ratio of mullite and glass. The ceramics must be strong enough to support various connected devices, such as pins for electrical signals and LSI chips. At the same time, it is desirable to reduce the dielectric constant in order to achieve faster transmission of electrical signals. One approach is to combine mullite with a material of lower dielectric constant (dielectric constant of mullite : 7.0). For example, the addition of silica-rich glass will reduce the dielectric constant, because the dielectric constant of silica is the lowest of all inorganic materials. We developed ceramics with a low thermal expansion coefficient, low dielectric constant, and high strength by adding a chemically stable, ternary $MgO-Al_2O_3-SiO_2$ glass to mullite[6].

STARTING POWDERS

Table 1 summarizes the compositions of the starting powders. Cordierite ($2MgO-2Al_2O_3-5SiO_2$), which has a relatively low eutectic temperature and dielectric constant, was selected as the sintering aid for mullite, and the mullite/cordierite ratio was varied (Sample Nos.1-3 and 8-12 in Table 1). The addition of silica to cordierite was also attempted in order to reduce the dielectric constant. Sample Nos.4-7 contained 72wt% of mullite, the stoichiometric cordierite ratio of alumina and magnesia, and silica in excess of the cordierite stoichiometry. Magnesia was added in the form of hydromagnesite ($Mg_5(CO_3)_4(OH)_24H_2O$) because of its high hygroscopicity.

Table 1. Compositions of samples (wt%).

No.	Mullite	Alumina	Silica	Magnesia	(Hydromagnesite)
1	50.0	17.4	25.6	7.0	(16.2)
2	70.0	10.5	15.4	4.1	(9.5)
3	72.0	9.8	14.3	3.9	(9.0)
4	72.0	4.9	21.1	2.0	(4.6)
5	72.0	2.0	25.3	0.7	(1.6)
6	72.0	1.0	26.6	0.4	(0.9)
7	72.0	0.0	28.0	0.0	(0.0)
8	75.0	8.7	12.8	3.5	(8.1)
9	78.0	7.7	11.2	3.1	(7.2)
10	80.0	7.0	10.2	2.8	(6.5)
11	82.0	6.3	9.2	2.5	(5.8)
12	85.0	5.2	7.7	2.1	(4.9)

MULLITE-GLASS CERAMICS

Mullite-Cordierite Ceramics

The mullite ceramics containing a cordierite (14.0wt% MgO, 34.8wt% Al_2O_3, 51.2wt% SiO_2) sintering aid were studied. Figure 3 shows the effect of mullite content on dielectric constant and bending strength.

The dielectric constant increased with mullite content up to 82wt%, and decreased slightly at 85wt%, as a result of inadequate sintering, as anticipated from the relative density and porosity results.

The mullite-cordierite ceramics had a dielectric constant of 6.0 at a mullite content of 50wt%, but the sample deformed during sintering due to the high concentration of the glassy phase, and it was fairly porous.

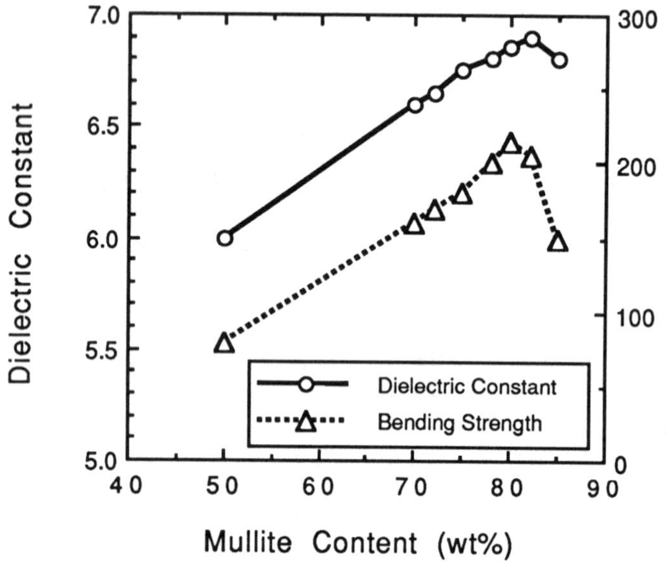

Fig. 3 Dependence of mullite content on dielectric constant and bending strength.
The other composition is cordierite (4.0wt% MgO, 34.8wt% Al_2O_3, and 51.2wt% SiO_2).

The bending strength increased as the mullite content increased up to 80wt%, but started to decrease thereafter. It attained a maximum of about 215MPa at a mullite content of 80wt%. The sample containing 50wt% mullite and having a low dielectric constant had a low bending strength of about 80MPa; this was due to a high concentration of glassy phase which is known to be a relatively low-strength ceramics.

Thus, some of the mullite-cordierite ceramic compositions were found to have a bending strength beyond 200MPa, but such compositions had high dielectric constants. On the other hand, the composition (50wt% mullite) having a dielectric constant of 6.0 had insufficient bending strength (80MPa). These ceramic compositions failed to achieve the properties required of substrates.

Mullite-Glass Ceramics

Attempts were made to add silica to the cordierite sintering aid, because of its low dielectric constant of abut 4, in order to reduce the dielectric constant of the ceramics. Addition of silica to the cordierite composition will raise the melting point of glassy phase, and thus is expected to increase the sintering temperature. The mullite content was varied in a range from 50 to 85wt% to investigate the effects of silica content on the sintering aid (glass). As a result, it was found that samples containing more than 78wt% mullite were sintered insufficiently at 1600°C for 1h, while that containing 50wt% mullite was deformed during sintering, on account of the excess quantity of the glassy phase, and those containing 70 to 75wt% mullite were sintered satisfactorily at the same conditions. The sample containing 72wt% mullite is discussed as an example.

Figure 4 shows the effects of silica content in the glass (consisting of cordierite and silica) on the relative density of the ceramics, where the relative density was calculated using 3.20, 2.50, and 2.22g/cm^3 as the densities of mullite, cordierite and silica, respectively. Silica, although added to the starting powder in the form of quartz, became amorphous during sintering, and thus the data for quartz glass were used.

As the silica content increased, the relative density increased gradually indicating the favorable effects of silica on the densification of the ceramics. The sample could be adequately sintered to a relative density of about 98% by increasing the silica content to 90wt%.

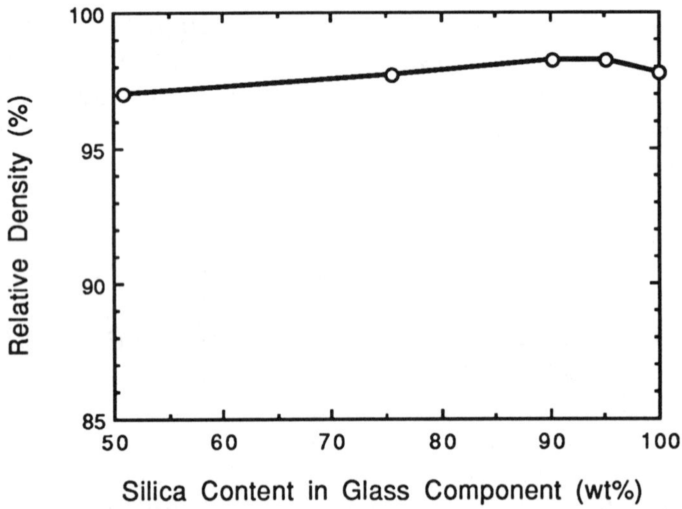

Fig. 4 Dependence of silica content on relative density. Mullite content : 72wt%.

Figure 5 presents the scanning electron microscope (SEM) photographs of the sintered fracture surfaces at different levels of silica content in the sintering aid. The pores decreased, both in size and quantity, as the silica content increased. At the same time, the mullite grains decreased in size as the silica content increased, indicating that silica controlled the growth of the mullite grains.

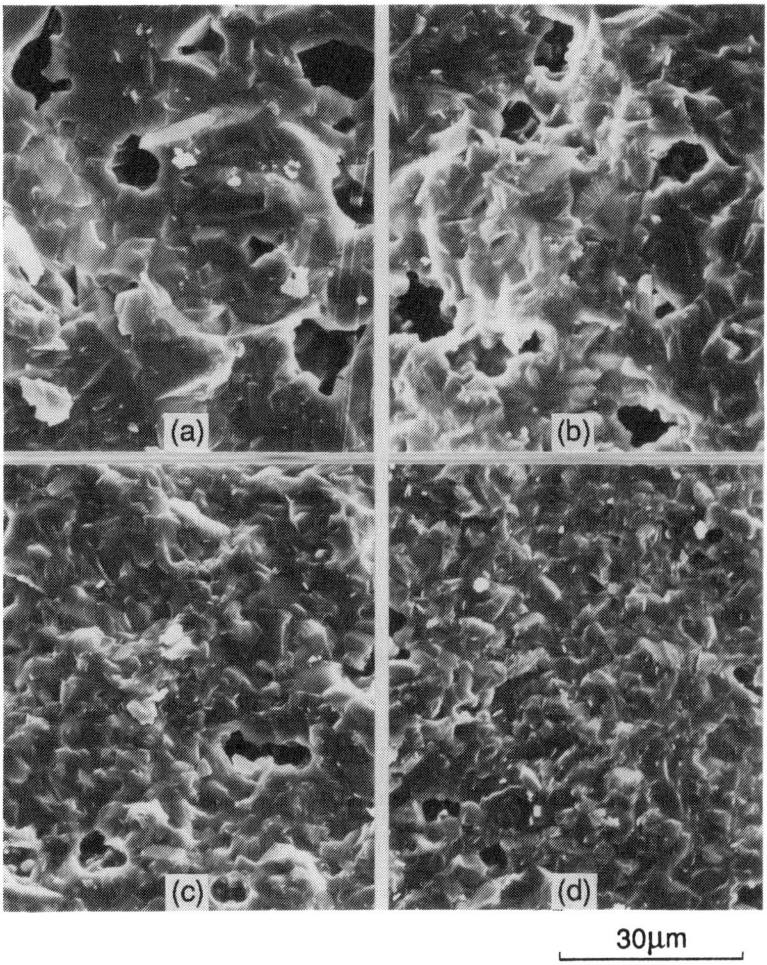

Fig. 5 SEM photographs of fracture surface.
SiO_2 content in glass component (wt%) :
(a) 51.2, (b) 75.5, (c) 90.2, and (d) 95.1.

Figure 6 shows the effects of silica content on the thermal expansion coefficient, dielectric constant, and bending strength of the ceramics.

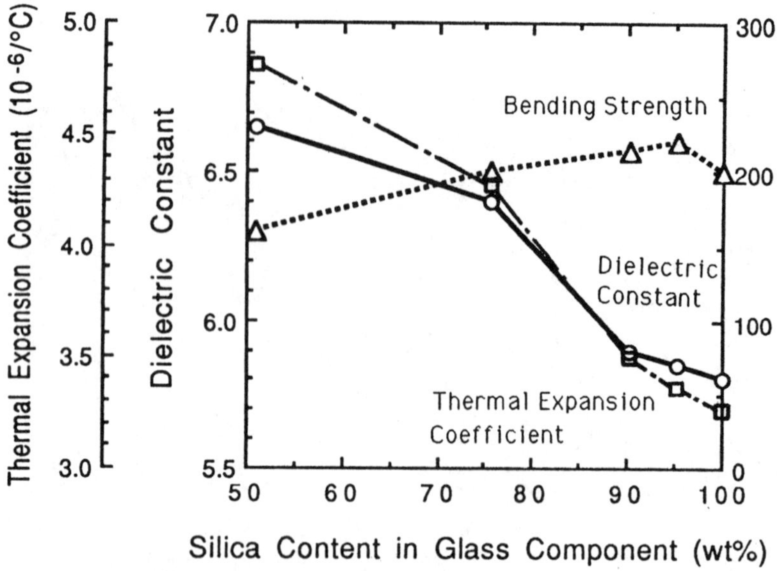

Fig. 6 Dependence of silica content on thermal expansion coefficient, dielectric constant, and bending strength.

The thermal expansion coefficient tended to decrease as the silica content increased, presumably as a result of the transformation of the cordierite and silica in the sintering aid into the amorphous state, to form a SiO_2-rich glassy phase of low thermal expansion coefficient. The thermal expansion coefficient of the ceramics decreased to 3.5×10^{-6}/°C or below when the silica content in the glassy phase was increased to 90wt% or more. Such ceramics, having comparable thermal expansion coefficient to silicon chip (thermal expansion coefficient of the silicon chip = 3.5×10^{-6}/°C), can be used as substrates for silicon chips.

The dielectric constant, like the thermal expansion coefficient, decreased as the silica content increased as a result of the low dielectric constant of silica and was 5.9 or below when the silica content was 90wt% or more. This value is lower than that of an alumina substrate at

9.5, allowing electrical signals to move faster. The propagation delay time of electrical signals is inversely proportional to the square root of the dielectric constant, and hence the delay time of mullite substrates will be about 20% lower than that of alumina substrates.

The bending strength tended to increase as the silica content increased, because silica added to the cordierite sintering aid increased the relative density of the ceramics and decreased the mullite grain size. The bending strength decreased slightly when the silica content was increased to 100%, possibly because of the high melting point of silica which caused inadequate sintering, as shown in Fig. 4. These results indicate the bending strength could be increased to 200MPa or more by increasing the silica content to 75 to 100wt%.

Similar results were observed for the ceramics containing 70 to 75wt% mullite. These ceramics could be sufficiently densified at 1560-1660°C.

Thus, the mullite-glass ceramics with good characteristics as substrates have been developed, containing 70 to 75wt% mullite and 90 to 100wt% silica in the glassy phase.

Microstructure

Figure 7 shows the effects of sintering temperature on the crystalline phase formed. At silica content of 51.0wt%, alumina, silica, and magnesia added as the sintering aid components formed cordierite and a small quantity of spinel at around 1400°C. The cordierite phase disappeared above 1500°C, and an unidentified phase consisting of alumina, silica, and magnesia was formed besides mullite and sillimanite crystals. It was unstable, and the peak positions changed with temperature. At a silica content of 90.2wt%, on the other hand, cordierite and silica in excess of the stoichiometric amount were found at 1500°C and below. At around 1400°C, a small quantity of spinel formed, as was the case with 51.0wt%

silica content. Above 1500°C, cordierite and silica disappeared, and the only crystalline phases found were mullite and sillimanite. Use of silica-rich sintering aid stabilized the crystalline phases.

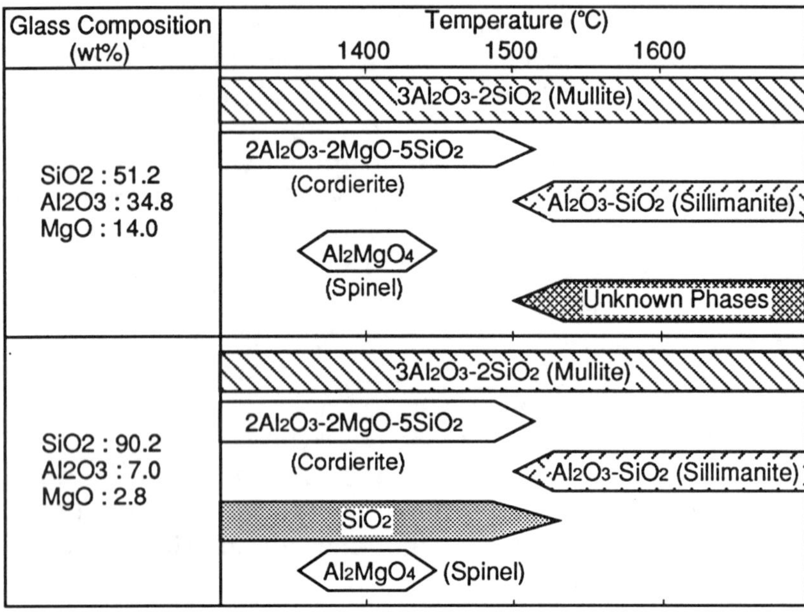

Fig. 7 Crystalline phase change in mullite ceramics at various sintering temperatures.

Figure 8 shows the transmission electron microscope (TEM) photograph and electron diffraction patterns of the mullite-glass ceramics containing 72wt% mullite and 90.2wt% silica, 7.0wt% alumina, and 2.8wt% magnesia in the sintering aid. The grains shown in the TEM photograph are 2-6μm in size, and did not grow much from the mullite grains in the starting powder, presumably because of the action of the SiO_2-rich glass sintering aid.

The electron diffraction results indicated the grains consisted of the mullite and sillimanite phases. An amorphous glass phase was found between the grains. It was a SiO_2-rich phase, as revealed by the qualitative analysis. Magnesia added as the sintering aid component was dissolved by diffusion in grains of mullite and sillimanite and in the glassy phase to form solid solutions. It was not present in the crystalline form.

Sillimanite might be formed (1) by the reactions between mullite and silica added as sintering aid components or (2) from alumina and silica in the sintering aid. In the case of (1), part of the mullite in the starting powder might be dissolved in the glassy phase and reacted with silica to form sillimanite, and silica might be dissolved by diffusion in the mullite grains to change the crystalline structure. In the case of (2), the sintering aid might be transformed into cordierite and spinel during the heating period, leaving behind sillimanite. Sillimanite can be formed by either path.

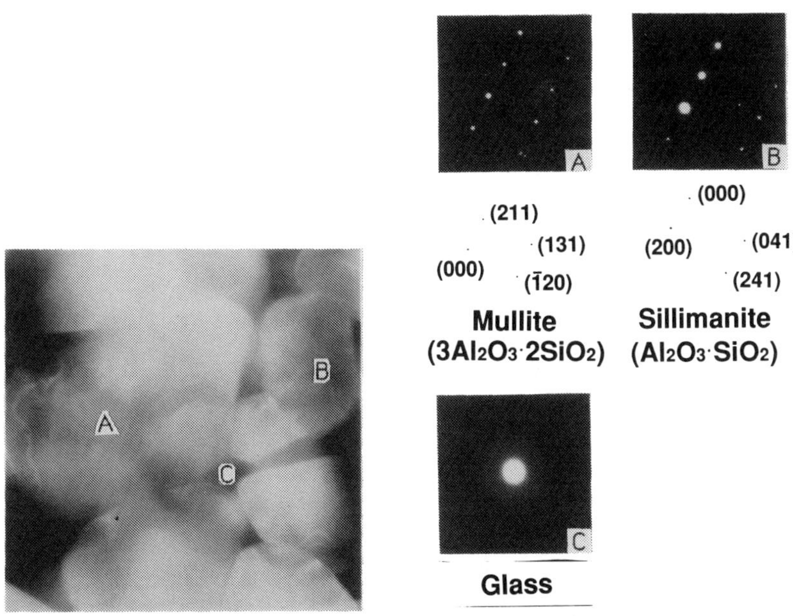

Fig. 8 TEM photograph and electron diffraction patterns of mullite ceramics. Mullite content : 72wt%.

CONCLUSIONS

The glass composition of the sintering aid was studied to develop mullite ceramics which have satisfactory characteristics for substrates. Density, dielectric constant, bending strength, and thermal expansion coefficient of the ceramics were measured.

1) Mullite-glass ceramics having a dielectric constant of 6 or less, a bending strength of 200MPa or more, and a thermal expansion coefficient of about $3.5 \times 10^{-6}/°C$ were formed by increasing the mullite content to 70-75wt% and the silica content in the glassy sintering aid to 90-100wt%. The crystalline phases of these ceramics consisted of stable mullite and sillimanite.

2) Increasing the silica content in the sintering aid decreased the dielectric constant and thermal expansion coefficient, and increased the bending strength of the ceramics.

3) The increased bending strength by the addition of silica resulted from the controlled growth of mullite grains and decreased porosity.

Substrates made of mullite-glass ceramics are suitable for mounting silicon devices of computer processors.

REFERENCES

1) F. Kobayashi, et al., 41st ECTC Proceedings, 693-703 (1991).
2) I.A. Aksay and J.A. Pask, J. Am. Ceram. Soc., 58, 507-12 (1975).
3) N. Kamehara, et al., IMC 1982 Proceedings, Tokyo, 388-93.
4) G.S. Perry, Trans. Brit. Ceram. Bull., 72, 279-83, (1973).
5) S. Kanzaki, et al., Yogyo-Kyokai-Shi, 95, 1213-8 (1987).
6) N. Ushifusa, et al., J. Ceram. Soc. Japan, 98, 377-83 (1990).

LOW TEMPERATURE CO-FIRED MULTILAYER GLASS-CERAMIC PACKAGE

Yuzo Shimada, Yoshinobu Kobayashi* and Kazuaki Utsumi*
1st Computers Division, NEC Corporation
1-10, Nisshincho, Fuchu City, Tokyo 183, Japan

*Material Development Center, NEC corporation
1-1, Miyazaki 4-Chome, Miyamae-ku, Kawasaki, Kanagawa 216, Japan

Glass-ceramic material which can be sintered at 900℃ in air have been found. A multilayer glass-ceramic (MGC) substrate, which has a low dielectric constant and high flexural strength, has been developed by means of green sheet lamination technology. Because of the low firing temperature it is possible to use silver-palladium ($5\mu\Omega\cdot cm$) and gold ($3\mu\Omega\cdot cm$) with low electrical resistivity. The microstructure of the sintered glass-ceramic body was analyzed in order to elucidate high flexural strength. The MGC substrate has applications in high speed computer packages, optical transmission modules and fiber-optic transmitter/receiver modules. It has become possible to achieve high system performance using the MGC package.

INTRODUCTION

Electronic circuit miniaturization is continuing for integrated circuit improvements. The high speed computer systems and the optical communication systems require new packaging technology with high propagation speed and high wiring density.

For high speed computer systems, the use of high density packaging is one of the key factors in reducing machine cycle time, which determines computer system performance[1]. The reduction in the machine cycle time can be achieved by reduction in propagation delays for signals between LSIs. The following formula can be represented as a performance factor Tm for the wiring delay in a given substrate[2].

$$Tm = \tau \cdot \sqrt{S/G} \qquad (1)$$

where τ is the signal delay per unit wiring length, S is the substrate area, and G is the number of gates mounted on the substrate. Taking εr as the dielectric constant for a dielectric material and c as the speed of light, yields the following formula,

$$\tau = \sqrt{\varepsilon r} / c \qquad (2)$$

Lower dielectric constant materials and higher gate density on the substrate become the key points for packaging.

High speed and small size are required for optical communication systems. In order to reduce inductance and capacitance, which might cause electronic crosstalk at high speeds, it is necessary to use new packaging technologies with high wiring density.

The main requirement for high speed computer and communication systems lies in how to reduce delays due to wiring. The following requirements must be met in the packaging substrate in order to improve system capabilities. In terms of construction, multi-chip packaging [MCP] or multi-chip module [MCM] technology with high density wiring and short wiring length are needed, while, in terms of materials,there are two important factors. One is to decrease the dielectric constant for dielectric material near a signal line. The other is to minimize electrical resistance for conductors to prevent undesirable voltage drops.

This paper describes dielectric materials with low dielectric constant and the conductor, the microstructure of glass-ceramic bodies, fabrication processes, and their applications to high speed computer packages and optical communication modules.

FABRICATION,MICROSTRUCTURE,AND PROPERTIES OF MGC SUBSTRATE

To meet the high propagation speed requirement,the authors have developed several low dielectric constant glass-ceramic materials.The relationship between flexural strength and dielectric constant for NEC's glass-ceramic systems is shown in Fig.1.A is the alumina-lead borosilicate glass system [3], B is the cordierite-borosilicate glass system [4], and C is the quartz glass-cordierite-borosilicate glass system [5]. Further,low dielectric permittivity glass-ceramic structure D(εr=3.4) was realized,owing to closed pores in the glass-ceramic body[6].Alumina-lead borosilicate glass system A was applied to packaging substrates for computer and communication systems.

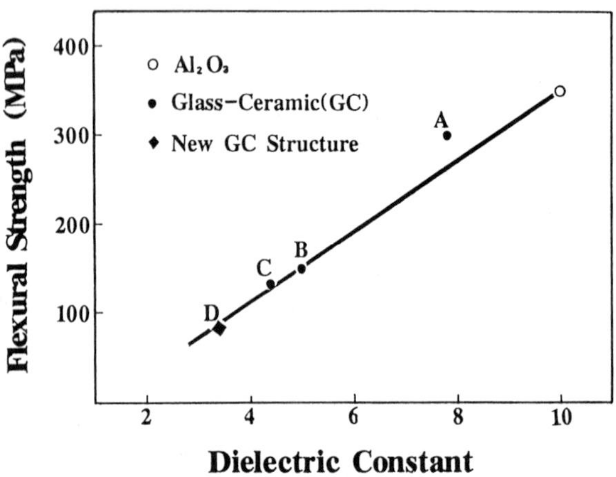

Fig.1 Relation between flexural strength and dielectric constant for dielectric glass-ceramic materials.

A. Fabrication Process

A flowchart for the co-fired multilayer glass-ceramic package fabrication process is shown in Fig.2. The dielectric material is a mixture of alumina and lead borosilicate glass powder, which have controlled particle size, owing to their electrical and mechanical properties and shrinkage control. The composition of the mixture is 55 wt% alumina and 45 wt% lead borosilicate glass. The slurry used consists of mixed dielectric powders and organic liquid vehicles. The slurry is cast into about 200 μ m thick ceramic green sheets by a slip casting process. Via holes are formed through

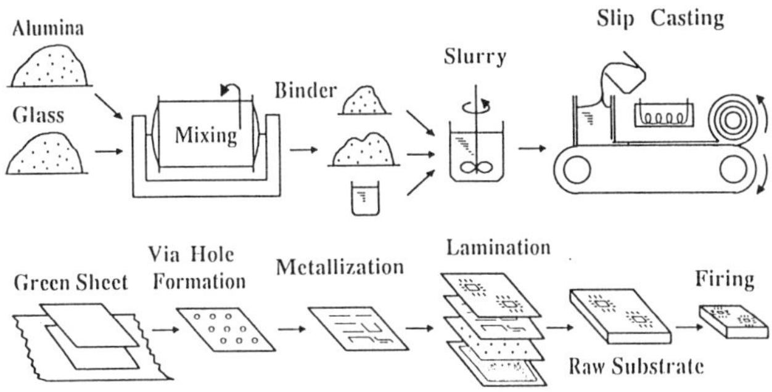

Fig.2 MGC package fabrication process .

the green sheet by a punching machine with punches and dies. A green sheet may have more than 10,000 via holes in a 250mm square area. Low resistivity conductor paste is applied onto the punched sheets. In this process,via holes are filled with paste to form contacts between signal lines. Low electrical resistivity for conductors reduces voltage drop along the line,and prevents a rounded pulse waveform. Since the dielectric material can be sintered at about 900℃, it is possible to use low electrical resistivity conductors with a low melting point, such as silver-palladium, and gold, instead of molybdenum or tungsten refractory metals in the alumina co-fired multilayer substrate. When sintered at 900℃, the resistivities are around 5 $\mu\Omega\cdot cm$ for 95 wt% silver-5wt% palladium, and around 3 $\mu\Omega\cdot cm$ for gold. These conductors with low electrical resistivity reduce voltage drop as well as achieving high wiring density.

The metallized green sheets are precisely stacked in a pressing die in sequence by the stacking machine. These sheets are then laminated together by hot press.Density heterogeneities in the laminated samples influence any shrinkage in the sintered substrate. Therefore, this lamination process should be homogeneously carried out by means of correct dimensional die and punch with flat surfaces. After burn-out processing at about 500℃, sintering is carried out in an electric furnace at about 900℃ in air. Finally, LSI chips, capacitor chips, resistor chips and so on are assembled onto the MGC substrate.

B. Substrate Properties

The typical properties of the MGC substrate, using the alumina - lead borosilicate glass system, are summarized in Table 1.

Table 1 Typical MGC substrate properties.

Substrate:	Material	Alumina 55wt% Lead Borosilicate Glass 45wt%
	Flexural strength	300MPa
	Dielectric constant	7.8 (1MHz)
	Dissipation factor	<0.3% (1MHz)
	Thermal expansion coefficient	$42 \times 10^{-7} deg^{-1}$
	Thermal conductivity	3.6 W/mk
	Linear shrinkage	13.0%
	Shrinkage tolerance	<0.2%

This substrate has high flexural strength, more than 300MPa. Camber is extremely small, though the substrate size is large. The dielectric constant is 7.8, and the dissipation factor is less than 0.3% at 1 MHz. Thermal expansion coefficient is one of the important factors, which must be considered for the packaging substrate. Thermal expansion coefficient (TEC) is 42x $10^{-7}deg^{-1}$ and is close to that of silicon chips. Thermal conductivity is not so high. Substrate linear shrinkage and shrinkage tolerance are exactly controlled by the high level control technologies.

C. Microstructure of Glass-Ceramic Body

When sintered at about 900 ℃, crystallization takes place in the sintered ceramic body on alumina-lead borosilicate glass system A.This reaction leads to high mechanical strength for the MGC substrate. This lead borosilicate glass mainly contains elements of Si,Pb,B,and Ca, and has a softening temperature at around 800℃.

In the case of coarse glass powder,crystallization takes place in the glass, and consequently cristobalite crystal is formed in it, while, in the case of fine glass powder,the microcrystalline structure grows as a result of chemical reaction of the alumina with the lead borosilicate glass. Figure 3 shows a TEM image of the glass-ceramic body in the multilayer substrate. Alumina particles and anorthite crystal phase, which is formed by a reaction between alumina and glass, exist in the vitreous network. This vitreous phase has the element of Al dissolved in the lead-borosilicate glass. Anorthite is formulated as $CaAl_2Si_2O_8$, and is formed because of the reaction of alumina with Si and Ca in the glass.

In the first stage of the sintering process, a glass softening reaction takes place at about 800℃, and alumina dissolves in the glass. Anorthite is then crystallized partially from the glass phase with alumina at more than 800℃.If the crystallization is insufficient,a heterogeneous phase remains in the glass phase. Figure 4 shows the microstructure of a glass-ceramic body sintered under conditions of high speed sintering with short holding time at 900℃. The heterogeneous phase is observed in the sintered body and was identified using energy-dispersive X-ray analysis(EDX). This heterogeneous phase consists of aluminum-calcium-silicon oxide,and changes into anorthite crystal phase when the crystallization is sufficient.

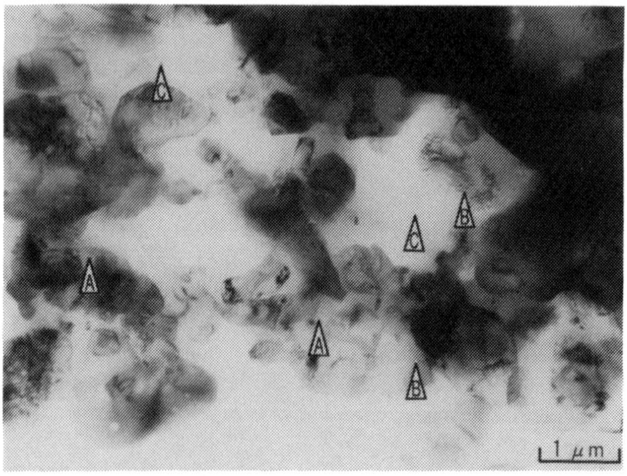

Fig.3 TEM image of the glass-ceramic body in the multilayer substrate.
A:Anorthite B:Alumina C:Glass

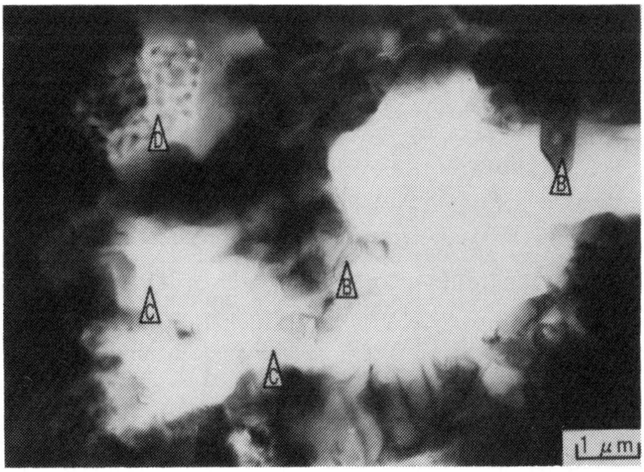

Fig.4 TEM image of the glass-ceramic sintered body at the first stage of sintering process (900℃)
B:Alumina C:Glass D:Separated Phase

Anorthite crystal plays the role of reinforcement in the glass-ceramic body. As a result, the MGC substrate, which is made from the alumina - lead borosilicate glass system, has a high flexural strength of more than 300MPa.

D. Shrinkage control technologies

It is very important to control shrinkage between raw substrate and sintered substrate for fine line formation on the substrate. Since shrinkage is delicately influenced any change in the raw materials, it is necessary to establish shrinkage control technologies. Highly accurate shrinkage can be attained by strict control at every process stage. Figure 5 shows constituent volume proportions in various stages. The shrinkage is greatly influenced by differences between green sheet and raw substrate densities, and between raw substrate and sintered substrate densities.

Shrinkage control factors are mainly packing density and glass-ceramic chemical reaction.The parameters, which influence these forms, are as follows[7].

1) Particle size -powder packing
2) Pressure on lamination -raw substrate density
3) Binder content in slurry
4) Powder composition
5) Additive for crystallization
6) Temperature profile at sintering

Particle size and pressure can be specially altered for shrinkage control, because these parameters can suitably control raw substrate density.

Figure 6 shows the glass powder specific surface area (Brunauer,Emmett,Teller method) effect on the glass ceramic powder volume ratio for the raw substrate and shrinkage. The glass ceramic volume, occupied in the raw substrate, decreases with increasing surface area. In other words, porosity in the raw substrate (see Fig.5) increases as surface area increases. The sintered glass ceramic body is very dense, above 4.0m^2/ g surface area.

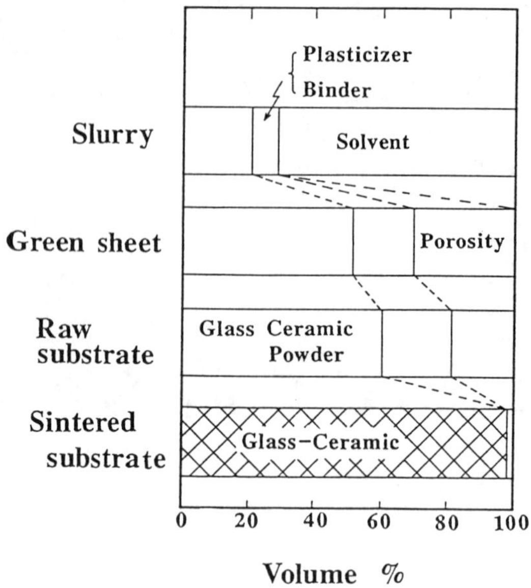

Fig.5 Constituent volume proportions in various stages.

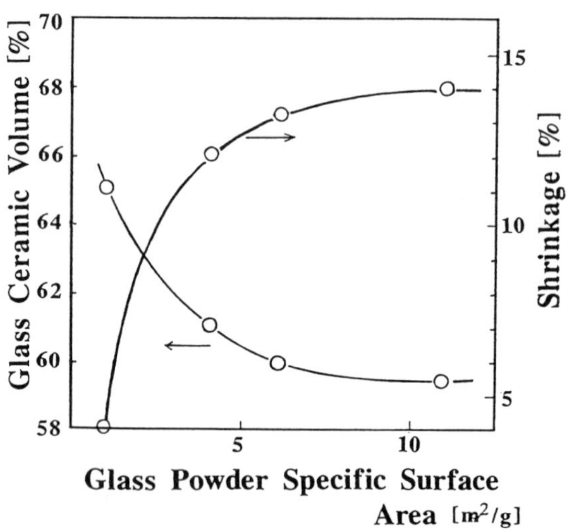

Fig.6 Glass powder specific surface area effect on glass ceramic powder volume ration for raw substrate are shrinkage.

Lamination pressure also be used to control the raw substrate density. The pressure effect on glass ceramic powder volume ratio for the raw substrate and shrinkage is shown in Fig 7. When the pressure is varied, the glass ceramic volume ratio varies from 58 to 62% in the same proportion as the 100 to 250 kg／cm^2 range. Consequently, it becomes possible to precisely control shrinkage by controlling and changing both these factors for the MGC substrate. Figure 8 shows linear shrincage for multilayer substrate samples,which are co-fired with silver-palladium conductors. It is possible to closely control small shrinkage variations during sample production.

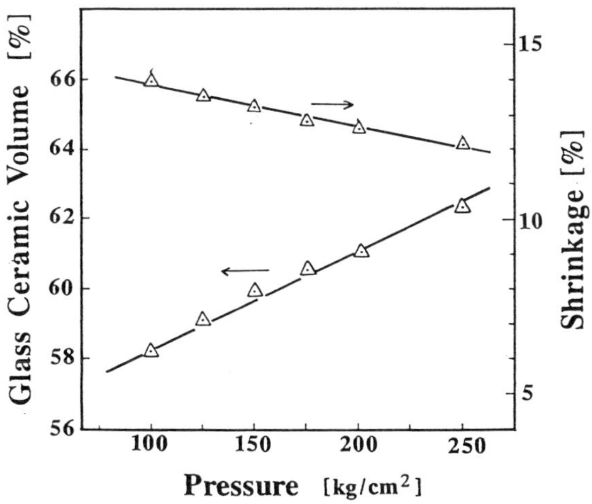

Fig.7 Pressure effect on glass ceramic powder volume ratio for raw substrate and shrinkage.

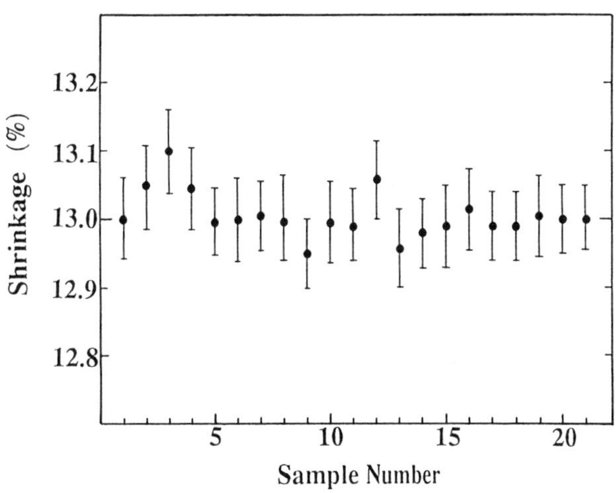

Fig.8 Linear shrinkage for individual samples,co-fires at 900℃ with silver-palladium conductors.

APPLICATION

A. High speed computer packaging

The packaging hierarchy for the high speed computer system consists of 3 levels [8]. The first level is concerned with the connection form for LSI chips. The second level is a multi-chip package level containing high density flipped TAB carriers (FTC), and is most important for a high speed packaging. The third is a board level with ZIF (zero insertion force) connectors. The MGC substrate is applied to the multi-chip package in the second level[9].

Figure 9 shows a cross section of the multilayer substrate (MLS), consisting of MGC base substrate and polyimide thin-film layers,for high speed computer use. There are clock line, power, ground ,and via hole layers in the MGC substrate. Via holes for electrical contact between layers are around 200 μ m in diameter. This substrate has 40 glass-ceramic dielectric layers, of which 14 are conductive layers with 6 clock layers.

Fiqure 10 shows a cut-away view of printed green sheets in the raw MGC substrate. The sintered substrate size is 225mm by 225mm. The substrate thickness is 5.5mm. The MLS, which has polyimide layers on the MGC substrate, is pictured in Fig.11. Polyimide dielectric layers have 8 conductive layers with 4 signal layers. The signal wiring pitch is 75 μ m, and its width is 25 μ m. The cure temperature for polyimide is about 400℃, which is very useful for the combination of fine signal layers with higher pincounts and higher power distribution on the MGC substrate. The adoption of many polyimide layers brings about mechanical stress due to polyimide cure, and differences in thermal expansion between polyimide and MGC substrate. Therefore, a MGC substrate has to maintain high flexural strength.

Figure 12 shows an MGC substrate sectional photograph with input / output (I/O) pins. By using silver 95wt%-palladium 5wt%, the sintered metallization electrical resistivity is 5 $\mu\Omega \cdot$ cm using the thick film printing process, while resistivity for gold is 3 $\mu\Omega \cdot$ cm. I/O pins are prepared to sputtered bonding pads by using 80wt%Au-20wt%Sn solder, with 1.8mm pitch in the lattice. The I/O pin number on the substrate is 11,540. These pins are made of copper alloy, which has low electrical resistivity. The bonding strength hardly changes during a 40 heat cycles test(25℃ → 220℃ → 100℃ → 25℃), and is sufficient, more than 3kg per pin. The multi-chip package can be used to mount up to 100 LSI chips.

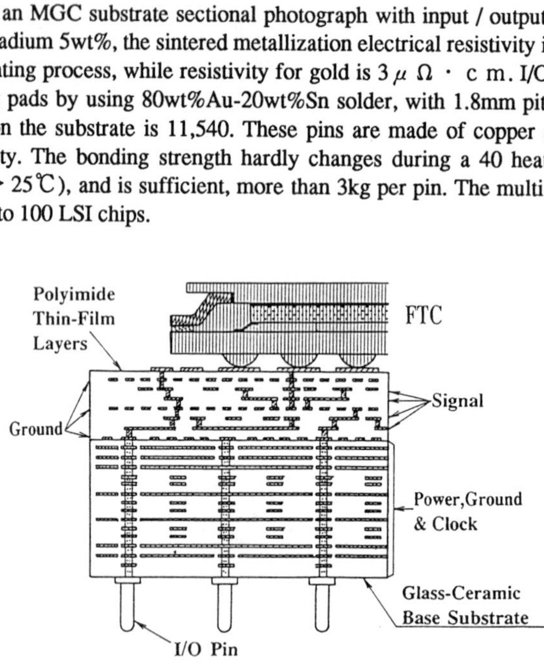

Fig.9 High speed computer use multilayer substrate cross section view.

The reliability tests on electrical migration for the silver-palladium conductor were carried out under high temperature and high humidity conditions. Consequently, high reliability for the MGC package was confirmed. Typical specifications for the MGC substrate, which is used in the MLS for high speed computer systems, are shown in Table 2.

Fig.10 Cut-away view showing printed green sheets in the raw MGC substrate.

Fig.11 MLS which has polyimide layers on the MGC substrate.

Fig.12 MGC substrate sectional photograph.

Table 2 Typical MGC substrate specifications for MLS.

Substrate:	Size	225 X 225mm
	Thickness	5.5mm
	Line width	200 μ m
	Via hole diameter	200 μ m
	Number layers	40
	Conductive layers	14
	Conductor;Material(resistivity)	Ag-Pd(5 $\mu\Omega\cdot$cm)
		Au(3 $\mu\Omega\cdot$cm)
I/O pin:	Number	11540
	Pitch	1.8mm
	Bonding strength	>3kg

B. Optical transmission module

Development of an optical interface based upon the network node interface (NNI) on the CCITT counsel, is being actively carried out. High performance, miniaturization, and low consumption of electric power are required for optical interface modules, accompanied by miniaturization of the communication system. In order to meet these requirements it is necessary to increase packaging density in the module. However, high packaging density brings about problems of chip radiation and increased wiring layers in the substrate. A multi-chip module (MCM) is one of the best techniques which can be used to solve these problems. The MGC substrate is applied to MCM for the 156Mb/s optical interface used in optical communication systems.

Figure 13 shows a cross section of the MCM structure with bare-chip packaging. The MCM contains an MGC substrate with 6 conductor layers, and some ultra small size components (capacitor chips, resister chips, and SAW filter), realize high density packaging. An external view of MCM for a 156Mb/s optical interface is shown in Fig.14. The substrate size is 30mm by 25mm. The substrate has all components on both sides of it except for an LD and a PD. The miniaturization of the optical interface module can be realized by means of bare-chip packaging technique for 3 IC chips, and the dimensions for the new developed module are reduced to 1/7 of the size of the conventional module.

The MGC substrate is advantageous to bare-chip MCM packaging, because of a low dielectric constant (7,8), a 42 x $10^{-7}deg^{-1}$ thermal expansion coefficient, close to that for silicon, and a high thermal conductivity (3.6W/mK)compared with PWB. The conductor material was adopted silver-palladium, in consideration of gold wire bond and soldering to components. The signal wiring pitch is 0.4mm, and its width is 0.2mm. The via hole diameter is 0.2mm, and its pitch is 0.6mm. The I/O pins were adopted lead frame with 1.27mm pitch for high connection density.

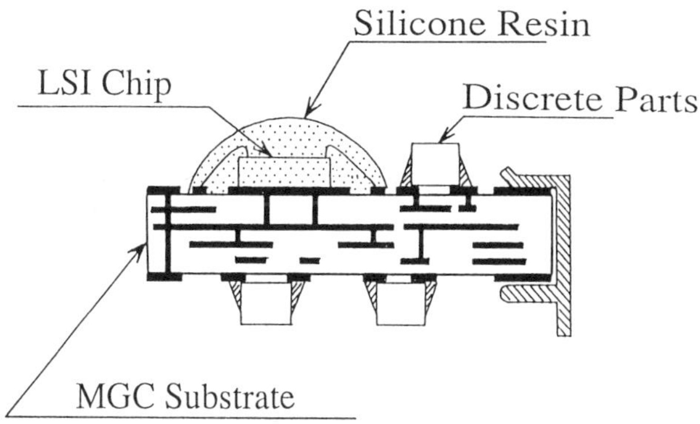

Fig.13 Cross section of MCM structure for optical interface module.

Fig.14 MCM for 156 Mb/s optical interface.

C. Fiber-optic transmitter/receiver

Optical fiber local area network (LAN) systems have been attracting much attention due to the features of optical fiber ,i.e.,wide bandwidth and no electromagnetic noise interference. Recently, there has been an increasing requirement for high speed transmission of various information services, such as voice, data, graphics, database and file transfer. Fiber-optic transmitters/receivers are applied to such LAN systems for electrical-to-optical signal conversion [10]. The MGC substrate is applied to the transmitter/receiver modules to accomplish miniaturization[11]. This part describes the receiver module in particular.

Figure 15 shows an image diagram for a fiber-optic receiver module. Resistor network chip, capacitor network chip, LSI and discrete chips are mounted on the MGC substrate. This substrate has power supply and solid reference layers, to reduce crosstalk coupling noise between the front-end output stages. This substrate has seven conductor layers and six insulator layers. Three of the seven conductor layers were power supply and solid reference planes.

The capacitor network chip is used with a low firing temperature (about 900℃) ceramic material in the ternary system $Pb(Fe_{2/3} \cdot W_{1/3})O_3$-$Pb(Fe_{1/2} \cdot Nb_{1/2})O_3$-$Pb(Zn_{1/3} \cdot Nb_{2/3})O_3$, and has 15 capacitor elements from 12nF to 117nF in it. The dielectric constant for this material is around 14,000 at room temperature. The fabrication process uses green sheet lamination technology, and is basically the same as that for the MGC substrate.

The resistor network chip has 7 resistor elements with a range of from 22 Ω to 3.6 kΩ in the insulation layers. The resistor material is mainly RuO_2, and is co-fired with glass-ceramic insulation material and conductor at about 900℃. The fabrication process is also similar to that for the MGC substrate.

Fiber-optic receivers which use the new multilayer ceramic components (MGC substrate, and capacitor and resistor network chips) and conventional components are shown in Fig.16. The dimensions of the new developed fiber-optic receiver are reduced 1/3 compared with those for a conventional receiver. The MGC substrate size is 21.8mm x 9.8mm x 0.8mm. The capacitor network chip size is 7mm x 5mm x 1.5mm, and the resistor network chip size is 4mm x 5mm x 0.8mm. The multilayer resistor network chip, the LSI and transistors are mounted with solder on one face of the MGC substrate, while the multilayer capacitor network chip is mounted on the other

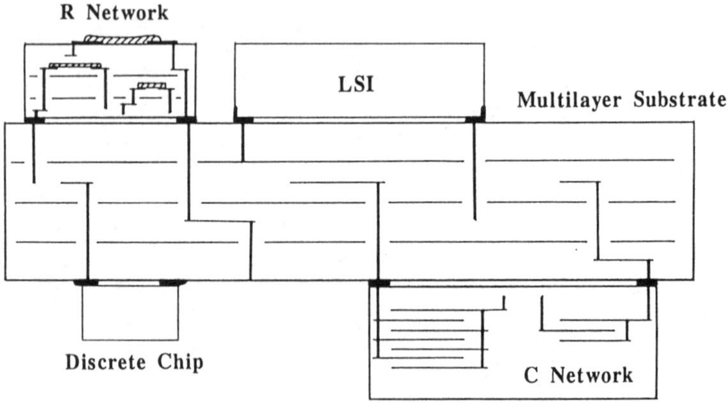

Fig.15 Image diagram for fiber-optic receiver.

face of the MGC substrate. LSI terminals are connected to the multilayer capacitor network chip through via holes in the MGC substrate. The passive component number for the newly developed fiber-optic receiver decreases to 6 chips in comparison to a conventional receiver (26 chips). It has become possible to achieve small size. Bit error rate characteristics of 230 Mbps for the fiber-optic receiver are as good as those for a conventional receiver operating at 200 Mbps.

Fig.16 Fiber-optic receivers with multilayer ceramic components (left) and conventional components (right)

CONCLUSION

VLSI packaging modules for high speed computer systems and optical communication systems require new technology with high propagation speed and high wiring density. The multilayer glass-ceramic (MGC) substrate can meet these requirments, and is advantageous for bare-chip packaging because of its mechanical and thermal characteristics.

This substrate is applied to some modules for computer and communication systems, and it has become possible to achieve high system performance. The low temperature co-fired multilayer glass-ceramic packaging has the following outstanding characteristics:

1) Glass-ceramic materials with low dielectric constant have been developed for high propagation speed.
2) Low electrical resistivity can be realized using Ag-Pd or Au on the MGC package. It is advantageous for achieving a low voltage drop.
3) The MGC substrate has a high flexural strength, because anorthite crystal plays a reinforcing in the glass-ceramic body.
4) The MGC package can support high wiring density in a large scale area through use of highly accurate processes with shrinkage control technology.

5) An MGC multi-chip package with polyimide layers can support high density wiring and high speed signal transmission for high speed computer systems.
6) The MGC substrate is advantageous in bare-chip MCM packaging and miniaturization for optical communication module.

ACKNOWLEDGMENTS

The authors would like to thank the many people who have contributed to this work, especially those members of the Material Development Center, the Transmission Division, and the Computer Division, for their cooperation and discussion. They are very grateful to M.Yonezawa, H.Takamizawa, M.Kuwata, T.Watari, I.Matsusita, M.Endo, T.Makabe for their advice and guidance. They would also like to thank K.Kata and A.Shibuya for their assistance.

REFERENCES

1. T.Watari, "Computer packaging technology for system performance," NEC Res. & Develop., No.98, pp.49-59, 1990.
2. T.Watari,"The NEC SX supercomputer technology, "proc. 1988 IEEE Int. conf. on comput. Design, pp. 54-57, Nov. 1986.
3. Y.Shimada, K.Utsumi,M. Suzuki, and H,Takamizawa, "Low firing temperature multilayer glass-ceramic substrate, "IEEE Trans. Comp, Hybrids, Manuf. Technol., Vol.CHMT-6, pp.382-388, 1983.
4. Y.Shimada, Y.Yamashita and H.Takamizawa, "Low dielectric constant multilayer glass-ceramic substrate with Ag-Pd wiring for VLSI package," IEEE Trans, Comp., Hybrids, Manuf. Technol., Vol.11, No.1, pp163-170, 1988.
5. K.Kata, Y.Shimada and H.Takamizawa, "Low dielectric constant new materials for multilayer ceramic substrate," IEEE Trans Comp, Hybrids, Manuf. Technol., Vol13, No.2 pp448-451, 1990.
6. K.Kata, A.Sasaki, Y.Shimada and K.Ustumi, "New fabrication technology of dielectric permittivity multilayer ceramic substrate," Proc ISHM'90, pp.308-315, 1990.
7. Y,Shimada, Y.Kobayashi, K.Kata, M.Kurano, and H,Takamizawa, "Large scale multilayer glass-ceramic substrate for supercomputer, "IEEE Trans. Comp., Hybrids, Manuf. Technol., Vol.13, No.4, pp751-758, 1990.
8. A.Dohya,T.Wtari,and H.Nishimori, "Packaging technology for the NEC SX- 3/SX-supercomputer, "Proc.40th Electric components & Techno. Conf.pp.525-533, 1990.
9. Y.Shimada, A.Dohya, K.Kata, and J.Inasaka ,"High speed computer packaging with multilayer glass-ceramic substrate, " Proc.ISHM'91, pp.176-182, 1991.
10. Y.Uda,T.Shin'e and K.Takahashi, "High speed optical fiber datalink transmitter/receiver using 1.3 LED and PIN-PD, "Proc.of EFOC-LAN '87, pp.246-250, 1987.
11. Y.Shimada, T.Shin'e, Y.Uda, and R.Nagaoka, "Application of multilayer ceramic components for fiber-optic transmitter/receiver, "NEC Res & Develop.,Vol.32, No.2, pp.195-206, .April, 1991.

ELECTRICAL EVALUATIONS OF MCM-SUBSTRATE FOR HIGH SIGNAL SPEED APPLICATIONS

Rokuro KAMBE, Yukihiro KIMURA, Atsushi KANDA, Ryuji IMAI, Masao KURODA
Research and Development Center, NGK SPARK PLUG CO., LTD.
2808, Oaza Iwasaki, Komaki-Shi, Aichi-Ken, 485 JAPAN

ABSTRACT

As the ceramic substrate for Multichip Module (MCM), MCM-C, which applies multilayer cofired ceramic technology, and MCM-D, which applies thin film multilayer technology can be offered.

In this report, the characteristics of several kinds of new ceramic materials, which is applicable to MCM-C are described, and the status of MCM-D, which is the highest density packaging substrate is also described. And the comparisons of their signal transmission characteristics are investigated and discussed.

As the results, MCM-C, which applies lower dielectric constant material, higher conductive metallization and higher density wiring can show better signal transmission characteristics. And furthermore MCM-D, which applies copper-polyimide thin film multilayer on the cofired alumina and aluminum nitride substrate shows the best characteristics for high speed signal transmission.

1. INTRODUCTION

At this time, there is a strong demand to achieve higher speed signal transmission in electronic equipment such as computers, telecommunication systems, etc. A very effective solution that complies with these demands is achieved by applying the concept of the Multichip Module (MCM), which strives to mount a number of Large Scale Integrated Circuits (LSIs) as close together as possible on a single substrate.

The MCM is capable of realizing very high speed signal transmission and the reduction of interchip wiring lengths for the effective downsizing of sophisticated electronic systems. The classification of these MCMs with the appropriate materials and processes involved are listed in Table 1.

Table 1. Types of MCM substrates

	Definition
MCM–L	High density laminated printed circuit board
MCM–C	Cofired multilayer ceramic substrate
MCM–D	Thin film multilayer on ceramic, silicon, metal, substrates

2. THE REQUIRED CHARACTERISTICS OF MCM SUBSTRATES

2.1 The electrical characteristics

For the MCM substrates, it is very important to maintain precious control over signal line characteristic impedance (Z_O) and to reduce crosstalk noise (K_B). Table 2 shows how to realize the downsizing and high speed signal transmission in MCM substrates.

Table 2. Required transmission characteristics of MCM substrates

Issues	Approach
Propagation delay time	Lower dielectric constant materials
Voltage drop	Higher conductivity metallization
Power and ground noise	Decoupling capacitors
Signal noise	Termination resistors
Wiring method	Fine line screen printing and Photolithographic patterning

2.2 The characteristics of reliability

A critical feature of MCM design is substrate reliability. It is well known that in order to achieve high level reliability in the packaging system substantially, all of the characteristics listed in Table 3 must be met.

Table 3. Required reliability characteristics

Issues	Items
Thermal resistance	Low thermal resistance design High thermal conductivity ceramics
Thermal stress	Thermal expansion matching to Si chip
Mechanical strength	High substrate strength
Bondability, Mountability	Metallization adhesion strength
Sealing	Hermeticity

3. MCM-C

3.1 Required characteristics of MCM-C ceramic materials

Alumina ceramics has been used most widely for MCM-C cofired substrates because of their high reliability and mass productivity. Recently, several new ceramic materials have been developed and brought into practical use in order to exploit their particular characteristics.

Table 4 shows electrical and thermal properties of these advanced materials. Polyimide, which is used primarily as an MCM-D dielectric, silicon and GaAs are listed for reference. Alumina is the most conventional material and its properties are well known. AlN has a high thermal conductivity and low temperature coefficient of expansion (T.C.E.), close to that of silicon. Mullite has a low dielectric constant and low T.C.E.. Glass ceramics have not only a low dielectric constant and low T.C.E. but also highly conductive metallization traces with low sheet resistance to be used by virtue of its low firing temperature. However, mullite and glass ceramics have poor thermal conductivity and therefore necessitate thermal management by other means.

Fig.1 shows thermal expansion curves for these ceramics and for silicon. It can be seen that the thermal expansion behavior of AlN, mullite, and glass ceramics is similar to that of silicon.

Fig.2 illustrates a comparison of thermal resistance between two similarly configured alumina and aluminum nitride pin grid array IC packages. The AlN package improves thermal resistance down to a value half that of the alumina package[1].

Table 4. Properties of ceramic materials

MATERIAL	T.C.E. (×10⁻⁶/℃)	DIELECTRIC CONSTANT 1MHz	DIELECTRIC LOSS 1MHz (%)	THERMAL CONDUCTIVITY (W/m·k)	SPECIFIC RESISTIVITY (μΩ·cm)
ALUMINA	6.8	9.8	0.03	18	20
ALN	4.3	8.9	0.03	170	20
MULLITE	4.0	6.6	0.2	5	20
GLASS CERAMICS	3.0	5.1	0.3	3	5
POLY-IMIDE	20~70	3.2	0.2	0.2	2
SILICON	3.4	12.0	---	125	---
Ga/As	6.0	13.1	---	54	---

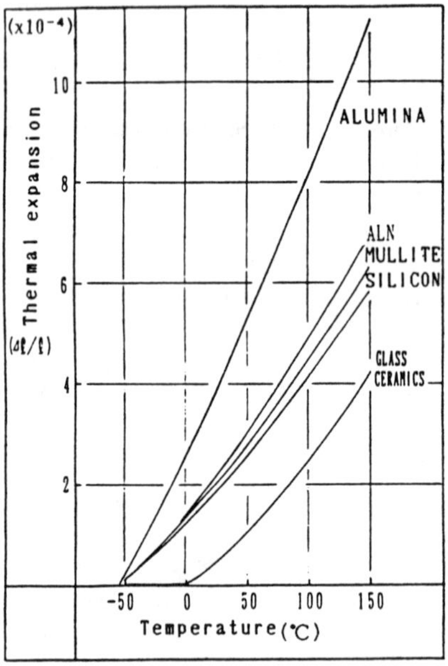

Figure 1. Thermal expansion of ceramic materials

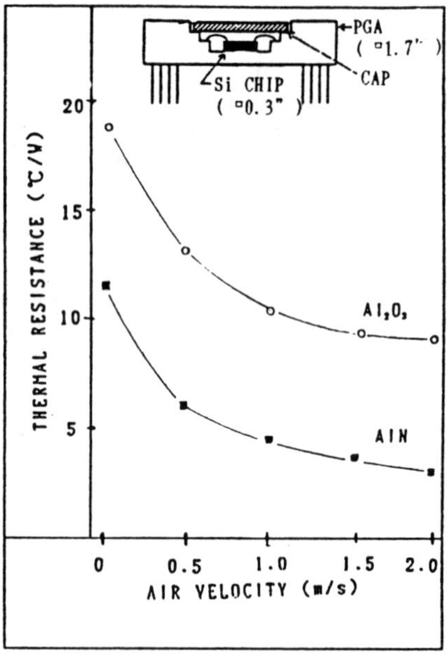

Figure 2. Thermal resistance of packages

3.2 The co-firing process of various ceramics

All of ceramic materials used for MCM-C substrates follow the same process flow; mixing, tape casting, punching, printing, laminating, cofiring, brazing, and plating. But each ceramic has a unique and very specific set of cofiring process conditions (see Table 5). In the case of alumina and mullite, the ceramic and metal is cofired at 1500 ℃ ~ 1600 ℃ in hydrogen gas in the liquid phase.

On the other hand, aluminum nitride is sintered at a higher temperature, 1700 ℃ ~ 1800 ℃. In the solid phase sintering of aluminum nitride, the yttrium aluminate compounds which were formed by the reaction of oxygen with the impurities of the raw material promote sintering. At that time, the distortion in the AlN crystal structure gradually disappears. And then, the transmission of the phonon grows up and the higher thermal conductive ceramics is made. Adhesion between the tungsten metallization ink and the aluminum nitride is accomplished by mechanical interlocking.

In the case of glass ceramics, sintering is accomplished by melting the SiO_2 amorphous glass phase. The sintering temperature is 800 ~ 1000 ℃ which is lower than the other ceramics. Therefore, glass ceramics can be sintered in combination with low resistance copper metallization. However precise control must be maintained over the sintering gas, since copper is easily oxidized(2).

One of the key point in the cofiring operation is binder removal, or the debinder process. If the green tape binder is not adequately removed prior to the actual onset of sintering, excess levels of carbon will be retained. As a result the shrinkage of the sintering will have dispersion and the characteristics of the conductor metallization will be inferior(3).

Table 5. Cofiring conditions of various ceramics

	Al_2O_3	Mullite	AlN	Glass Ceramics (Cordierite)
Additive	SiO_2 MgO CaO	SiO_2 Al_2O_3 MgO	Y_2O_3 CaO	———
Metallization	W, Mo	W	W	Cu, Au
Cofiring Temperature / GAS Atmosphere	1500 ~ 1600℃ / H_2 GAS	1500 ~ 1600℃ / H_2 GAS	1700 ~ 1800℃ / N_2 GAS	800 ~ 1000℃ / H_2 GAS

4. MCM-D

4.1 Comparisons of MCM substrates

MCM-D substrate which consists of deposited (sputtered, CVD, etc.) metallization systems which are patterned using photolithographic techniques, and insulation layers of low dielectric constant organic materials are attractive for high precision, high density MCM substrates. Several kinds of ceramic substrate, silicon wafers, and metal substrates are commonly used as the platform to construct thin film, multilayer configurations. Approximate design rules for MCM substrates; MCM-L, MCM-C, and MCM-D are shown in Table 6[3].

The refinement of screen printing technology is being promoted in order to densify the MCM-C, however at the present it seems that for all practical purposes the line width limit is around 100 μ m at present. On the other hand, because of the use of photographic processes, the line widths for MCM-D substrates is around 25 μ m. It is clear from this table that MCM-D substrates are in many ways superior to other two types of MCM substrates.

Table 6. Comparisons of MCM-L, MCM-C, and MCM-D

Items	MCM-L (PC board)	MCM-C (cofired ceramic)	MCM-D (thin film multilayer)
manufacturing process	copper foil photoetching	green sheet screen printing	sputtering, plating photolithography
line width	100 μ m	100 μ m	25 μ m
line pitch	250 μ m	200 μ m	75 μ m
via size	300 μ m	100 μ m	50 μ m
via pitch	1000 μ m	250 μ m	150 μ m

4.2 Outline of MCM-D

(1) Metallization

Thin film metallization deposited by sputtering or other physical vapor deposition (PVD) methods is commonly used for the conductive patterns on MCM-D substrates. Thin film metal is followed by an electrical plating process to increase electrical conductivity of the traces.

The materials for thin film metallization are selected from copper, aluminum, gold, and so on. Similarly, copper, gold and other metals are selected for electrical plating. Copper is mainly adopted because of its advantages such as high conductivity, low cost, and easy platability.

Metal pattern definition is accomplished by means of photolithography, that is coating photoresist, exposure and development, and etching.

(2) Insulation layer

Various materials such as polyimide, Benzocyclobutene (BCB), or SiO_2 are used as the insulation material between metallization layers. Polyimide which has low dielectric constant (3.0 ~ 3.4) and high temperature resistance (~ 450 ℃) organic material is most commonly used for insulation material.

The MCM-D which consists of the above copper conductor and polyimide is referred to as copper-polyimide multilayer substrate.

4.3 Combination of MCM-D and cofired ceramic substrate

Figure 3 shows an example of an actual MCM substrate which is a combination of MCM-D with cofired alumina substrate, and it is called as MCM-D/C. The configuration of the MCM-D part has five metallization layers insulated by polyimide.

The five metallization layers consist of two signal layers sandwiched by two ground and/or power layers and a top layer. Signal layers are in the dual strip line configuration and have controlled characteristic impedance.

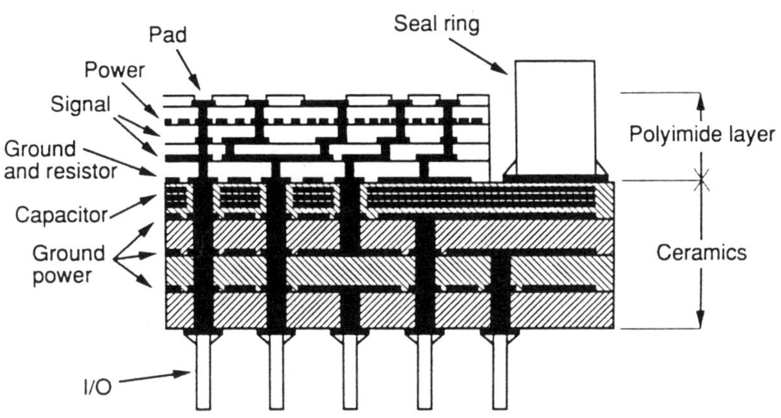

Figure 3. MCM-D/C, thin film multilayer combined with cofired ceramic substrate

Furthermore, this example also has thin film termination resistors on the ceramic substrate and cofired decoupling capacitor using thin ceramic layers.

The advantages of this MCM-D/C are as follows;

a. The configuration of MCM-D/C is compatible with MCM-D placed in the proper package, so that module assembly is simplified.

b. Low resistance and large current power supply of various potentials is possible through the power and ground planes formed in the cofired ceramic base substrate.

c. Signal I/O and power supplying pins can be placed in optimum locations, that is, at short distances, so that it is advantageous for high speed and high power applications.

5. SIGNAL TRANSMISSION CHARACTERISTICS

5.1 Electrical evaluation of various MCM substrates

As the clock frequencies of LSIs increase , various signal and ground noises become severe in the transmission lines of high speed MCM substrates. So the evaluation of these electrical characteristics in the high frequency region is very important.

For this reason ,a cofired alumina test substrate and copper-polyimide one have been designed and fabricated for electrical evaluation. The sizes, pictures, and cross-sectional schematics of these substrates are shown in Figure 4.

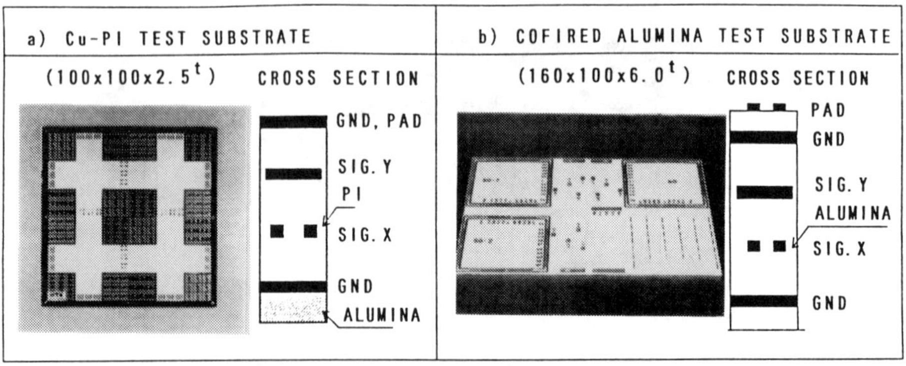

Figure 4. Pictures and cross-sectional schematics of the test substrates

They are modeled for use as large MCM-C and MCM-D substrates respectively. These are in the dual stripline configuration and each configuration size is designed to be 50 Ω matching by simulation. The most important electrical characteristics of these substrates in high frequency region ,such as characteristic impedance, propagation delay time, and crosstalk noise are evaluated by Time Domain Reflectmetry(TDR)[4].

Moreover in the case of other new ceramic substrates , such as aluminum nitride, mullite, and glass ceramics, electrical characteristics are also simulated and compared[5].

As for the basic line width, as shown in Table 6, it is 100 μ m in cofired ceramics made by screen printing, and 25 μ m in copper-polyimide made by photolithography. Furthermore the influence of manufacturing methods and dielectric constant of insulating materials on electrical performance is also investigated.

5.2 Impedance matching structure

As is well known, impedance matching is very important to reduce signal reflection noise. The measured line impedance of these test substrates is 52 Ω in copper-polyimide and 51 Ω in cofired alumina, which are approximately the same as the designed value of 50 Ω. As measured and simulated results, in the 50 Ω matching structure, the relationship between each insulation thickness and dielectric constant of various ceramic substrates including copper-polyimide is shown in Figure 5.

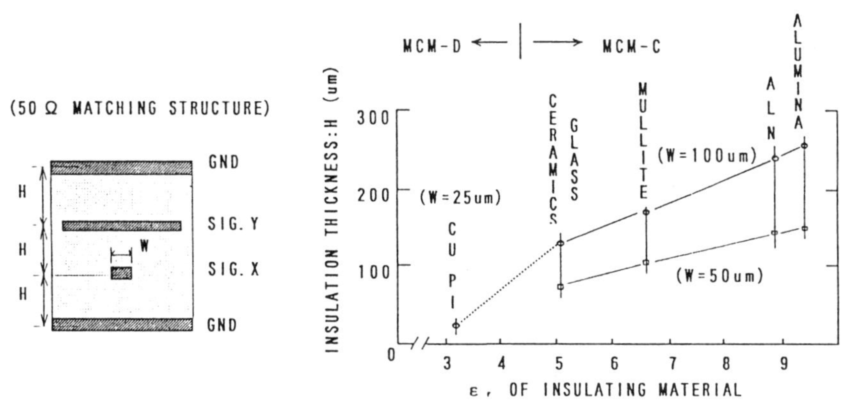

Figure 5. Relationship between ε_r of ceramics and impedance matching structure

From these results, in lower dielectric constant ceramic material, the impedance matching structure is much thinner. Furthermore, if the line width of ceramic substrates could be finer from 100 μ m to 50 μ m, impedance matching structure could be also much smaller and thinner, that would enable high density multilayer wiring and packaging. Especially in copper—polyimide substrate, because of its lower dielectric constant of 3.2 and its finer line width of 25 μ m, this advantage is most outstanding.

5.3 propagation delay time

The propagation delay time is very important for high speed signal transmission to prevent the timing error of clock rate. The propagation delay time (t_{pd}) of the unit line length is determined by the dielectric constant of insulating material as shown in equation (1)

$$t_{pd}=\sqrt{(\varepsilon_r)}/(c_o) \quad (1) \quad (c_o=\text{light velocity})$$

The calculated and measured value of t_{pd} is shown in Figure 6.

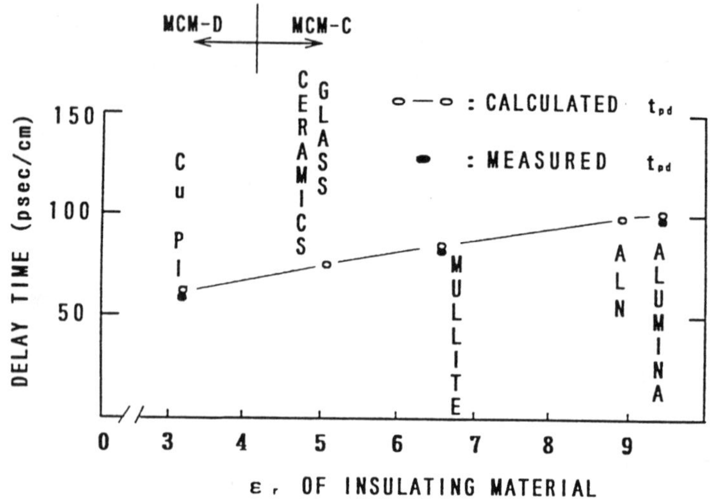

Figure. 6 Relationship between ε r of ceramics and propagation delay time

So lower dielectric constant material is more suitable for high speed signal transmission, because of its shorter t_{pd}.

5.4 Backward crosstalk noise

In the case of high density and fine pitch wiring, backward crosstalk noise (K_B) becomes one of the most important problems, especially in high speed signal transmission. The relationship between the K_B level and signal line spacing in various ceramic substrates is shown in Figure 7.

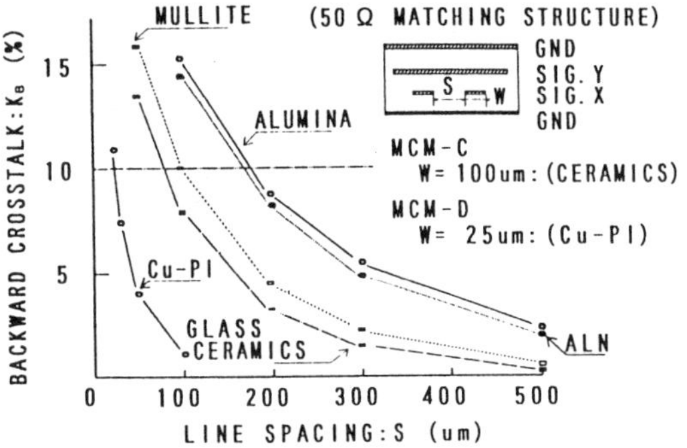

Figure 7. Relationship between crosstalk noise and various ceramic materials

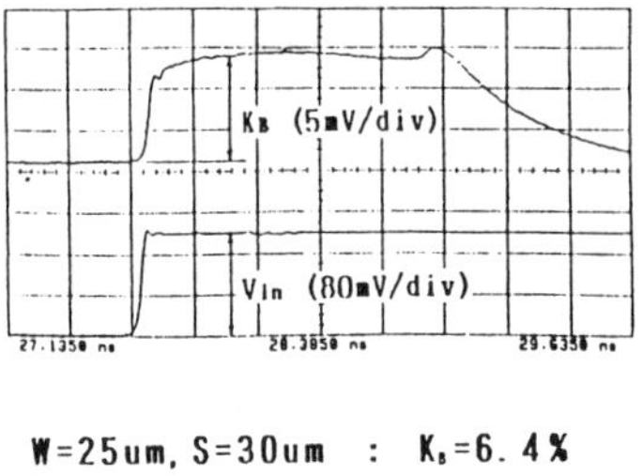

Figure 8. Crosstalk noise waveform of Cu–PI substrate

From these results, in case of MCM–C, as lower dielectric constant ceramic substrates, such as glass ceramics are used, K_B becomes lower, which makes finer line pitch and down sizing possible. Especially,in copper--polyimide substrates, K_B is significantly decreased. For example, as shown in Figure 8, a copper–polyimide substrate can keep K_B less than 10% in case of only 30 μ m line spacing[3]. On the other hand, cofired alumina needs more than 200 μ m line spacing to keep K_B less the 10%.This is one of the greatpst advantages of MCM–D.

5.5 Companies of each electrical performance

From these measured results, the electrical design rules required to satisfy 50 Ω matching and low K_B(less than 10%) are decided, on condition that line width is 100 μ m and 50 μ m in cofired ceramic substrates and 25 μ m in copper–polyimide substrate, respectively.

Based on these design rules, the total signal line length per cm^2 with one pair of dual stripline configuration ,as shown in Figure 9, can be roughly calculated for each substrate. The calculated lengths is shown in Figure 10.

It means the wiring density ratio of signal line. The higher this value is, the more possible is the down sizing of the substrate and reduction of signal line length, which can realize high speed signal transmission. These line lengths are the maximum value, which are only decided by electrical design rules, and by the limitation of manufacturing processes, the real values will be a little bit lower.

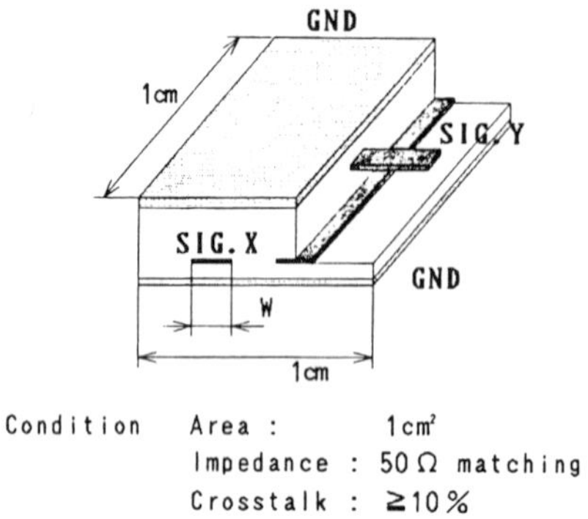

Figure 9. Dual strip line model to calculate the wiring density ratio

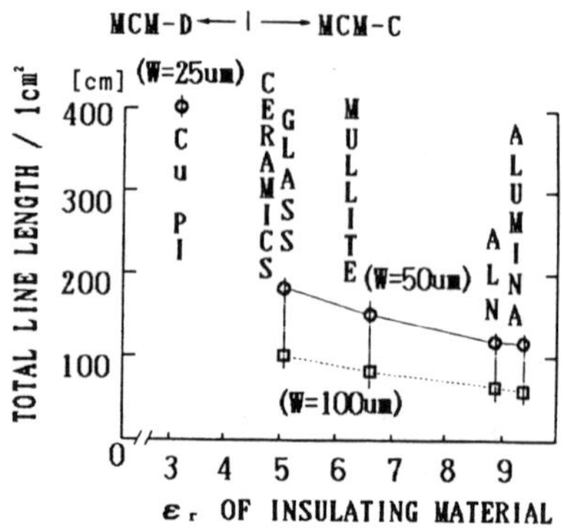

Figure 10. Wiring density ratio of various MCM substrate

So this Figure 10 shows that copper–polyimide used for MCM–D has the highest electrical performance. On the other hand in case of cofired ceramic substrate used for MCM–C, by using a lower dielectric constant ceramic material and developing higher density wiring methods, it will be possible to improve electrical performance near to that of copper–polyimide .

6 CONCLUSION

The characteristics of insulating materials and the electrical designs of substrates for MCMs have been investigated. As for electrical performance, by using low dielectric constant materials and high density wiring methods, the downsizing of substrates and the reduction of signal line lengths become possible, which can achieve high speed signal transmission.

In the case of cofired ceramic substrates used for MCM–C , glass ceramics is most suitable from the standpoint of electrical performance, because of its low dielectric constant and high conductive metallization such as Cu and Au. But the task to overcome low mechanical strength is needed.

On the other hand, copper–polyimide substrate used for MCM–D has the most excellent electrical performance. So, it can be applied to the most high end applications. Especially, in the configuration called MCM D/C shown in Figure 3, as a base substrate of copper–polyimide , cofired alumina with decoupling capacitors of thinner tapes, which can reduce ground noises, is usually used , because of its higher dielectric constant and higher reliability. So, this substrate is one of the best solutions for high speed signal transmission.

In the near future, not only conventional alumina , but also new advanced ceramic materials, with such as low T.C.E, high thermal conductivity, and low dielectric constants, will be used suitably, utilizing their own merits, in order to meet various demands , which are not only high end technology but also low end applications. Furthermore the application of these materials in the higher frequency region such as micro wave bandwidth, will need to be investigated, corresponding to the rapid development of LSIs.

7. ACKNOWLEDGEMENT

The authors wish to express many thanks to Mr.H.Banno and Mr.R.KRAUSE for their advice and encouragement . The authors also would like to thank their co-workers, especially T.Sukegawa, M.Nakata in the R&D Center, NTK Technical Ceramics Division, NGK SPARK PLUG CO., LTD. for their contribution to this study.

8. REFERENCES

[1]A. Kanda, S. Nozaki, and H. Banno "Trends of Ceramic IC Packages and Substrates" J. Ceram. Soc. Jpn. Inter. Ed. Vol.97 (pp39-50), 1989.

[2]K. Kondo, M. Okuyama, and Y. Shibata, "Low Firing Temperature Ceramic Material for Multilayer Ceramic Substrates", The Am.Ceram.Soc.,Inc.Advances in Ceramics, Vol.19, Multilayer Ceramic Devices. Edited by J.B. Blum, and W.R. Cannon, pp.77-87, 1986.

[3]R. Kambe, M. Kuroda, R. Imai, and Y. Kimura, "Copper Polyimide Multilayer Substrate for High Speed Signal Transmission", 41st ECTC Proceedings, pp.14-19, 1991.

[4]R. Kambe, "High-Density Multilayer Wiring Substrate for High-Speed Signal Transmission", Ceramic Bulletin, Vol.71, No.6, pp,962-968, 1992.

[5]A.Feller, H.R.Kaupp, "Crosstalk and Reflections in High-Speed Digital Systems", AFIPS Conference Proceedings Fall JT Computer Conference, vol.27, pp.511-525, 1965.

SENSOR PACKAGING FOR THE 90's

Joseph M. Giachino
Ford Motor Company
17000 Rotunda, ETC-C375
Dearborn, MI 48212

ABSTRACT

Sensor packaging is the key to a viable sensor. The purpose of various sensor packages, the characteristics to be considered when designing a package (with specific examples) and the future direction of sensor packaging is discussed.

INTRODUCTION

Engineers realize that sensors are one of the keys to the effective utilization of control systems. Many do not realize that sensors themselves are systems. Figure 1 details the components that make up the system called a sensor. The main functional blocks that go into the system are the transduction or sensing element, the electronics and the package. Some sensors do not have any electronics but all have transduction elements and packages. For many years packaging was regarded as a mundane task needed only to protect the element and electronics from the environment.

During the last decade, with the drive for increased performance and reduced cost, engineers have realized that packaging is a crucial part of sensor design.[1,2] Not the least in influencing this attitude is the realization that a significant cost of a sensor is packaging (on the order of 50%). Packaging also effects final performance and durability of the sensor.

Modern packaging[3] is relying heavily on new materials technology (e.g., ceramics, plastics, thick film) and computer modelling techniques (e.g., FEA) to meet the customers demands.

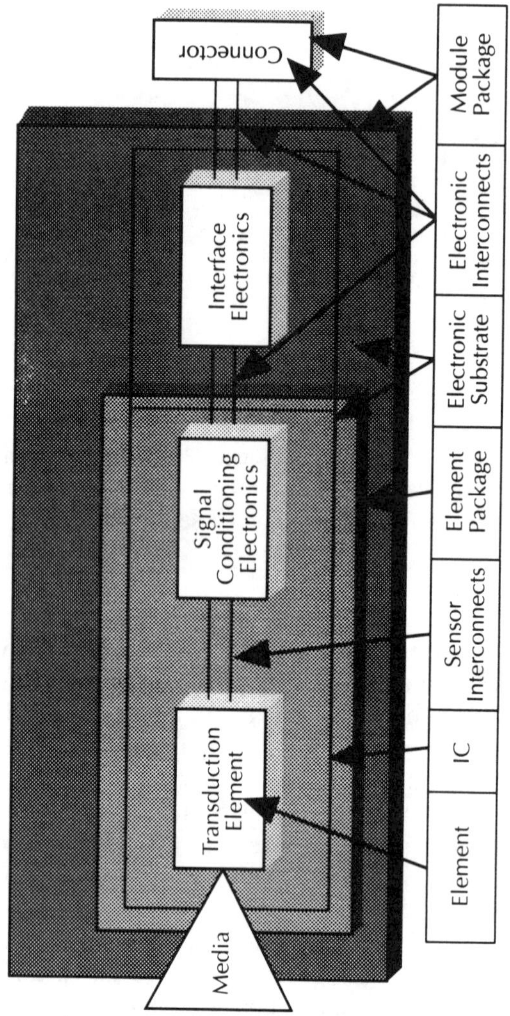

FIGURE 1. GENERIC SENSOR SYSTEM

PACKAGE TYPES

Sensor packages can be divided into two basic types: first level packages and final packages.

First level packages are subassemblies usually required to protect the element during subsequent final packaging and for isolation from unwanted signals during calibration.

An example of a very simple first level package is the coating (e.g., epoxy, glass) given to thermistors to protect them from the handling and processing that goes into making a temperature sensor.

Final packages are used to protect the element and the electronics from the environment and to accommodate installation of the sensor in the system. Figure 2 shows a first level package assembly and the final package for a pressure sensor used in Ford vehicles.

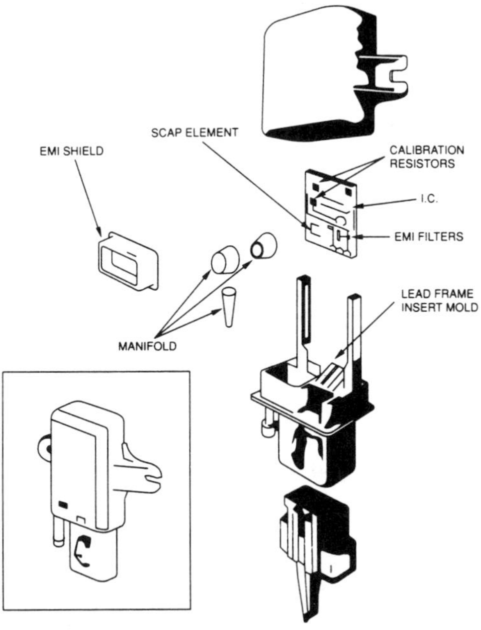

FIGURE 2. FORD SILICON CAPACITANCE ABSOLUTE PRESSURE (SCAP) SENSOR

In some cases, the first and final package can be one. This is being done with monolithic integrated silicon sensors.

KEYS TO SENSOR PACKAGING

All sensors are required to meet the specific needs of the particular market they are addressing. One of the most demanding is the automotive sensor market. Table 1 lists the parameters sensed and Table 2 indicates the breadth of the transduction element technologies used to produce these sensors.[4] Table 3 gives a synopsis of the powertrain environment.[4] All these sensors must be

TABLE: PARAMETERS SENSED	
• Manifold Pressure	• Engine Knock
• Manifold Charge Temperature	• Vehicle Speed
• Engine Coolant Temperature	• Vehicle Height
• Throttle Position	• Vehicle Direction
• Exhaust Gas Valve Position	• Wheel Speed
• Piston Position	• Fluid Level
• Exhaust Gas Oxygen	• Oil Pressure
• Outside Temperature	• Fuel Level
• Air Flow	

TABLE 2: TECHNOLOGIES UTILIZED
• Thick Film
• Thin Film
• Ceramics
• Variable Reluctance
• Hall Effect
• Optics
• Silicon Processing

TABLE 3: AUTOMOTIVE ENVIRONMENT	
Temperature: -40° to 150°C	
Mechanical Shock: up to 50g	
Mechanical Vibration: up to 15g	
Exposure to:	Fuel
	Brake Fluid
	Oil
	Transmission Fluid
	Ethylene Glycol
	Salt Spray
EMI: 200 Volts/Meter	

packaged for their specific application to last for the life of the vehicle, be producible in the millions of units per year, and be low cost. The following are the key concerns that must be addressed to produce a cost-effective sensor. The sensor package MUST be designed concurrently with the element, electronics and manufacturing processes to address these concerns. Many technical and market place failures occur because the package is designed independently and as the final task. The key characteristics to be considered in sensor packaging are:

- Transduction element connections
- Media compatibility
- Measurand application
- Isolation from unwanted inputs
- Ease of assembly
- Ease of test/calibration

TRANSDUCTION ELEMENT CONNECTIONS

There are two basic transduction element connections: electrical and mechanical. Electrical connections are usually welded, soldered, or wire bonds. In silicon-based sensors, notably, pressure sensors and accelerometers, wire bonds are common. In non-silicon based sensors, soldering and welding are more common.

Mechanical connections vary more than the electrical. In silicon based sensors anodic bonding the silicon element to a glass (e.g., Corning 7740) substrate is widely used. Anodic bonding[5] requires the two pieces be at an elevated temperature at or above 300°C and subjected to an electric field at or above 600V. The silicon is biased negatively and while the exact mechanism is not fully understood, it has been demonstrated that a hermetic seal is formed.[6] The time, temperature, and field can be varied to minimize the stress levels at room temperature. A feature of this process is that one can form a hermetic seal around an inclusion. This has been used commercially to form a hermetic electrical contact when joining silicon and glass.

Another technique being used to mechanically connect silicon elements to substrates is fusion bonding.[7,8] Fusion bonding requires placing the silicon element in intimate contact with a silicon substrate and placing them in an oven at approximately 1000°C. The two units will fuse together. An advantage of this method over anodic bonding is that since both parts are silicon stress levels in the final assembly are very low.

Both these techniques are usually done at the silicon wafer level making them commercially attractive. These techniques are also being explored for first level packaging of elements.

Another technique used for connecting elements to substrates involves glass seals. One method developed involves using low temperature melting boron-doped glass to join silicon elements to substrates.[9] An advantage of this process is that it involves low temperatures (as low as 450°C) and no electrical fields.

Other techniques used for both silicon and non-silicon elements include epoxies, solders, and glass frits.

Some designs have combined the electrical and mechanical connections.[10] This is shown in Figure 3. Here flip chip solder bumps were used to make electrical and mechanical connections to the elements.

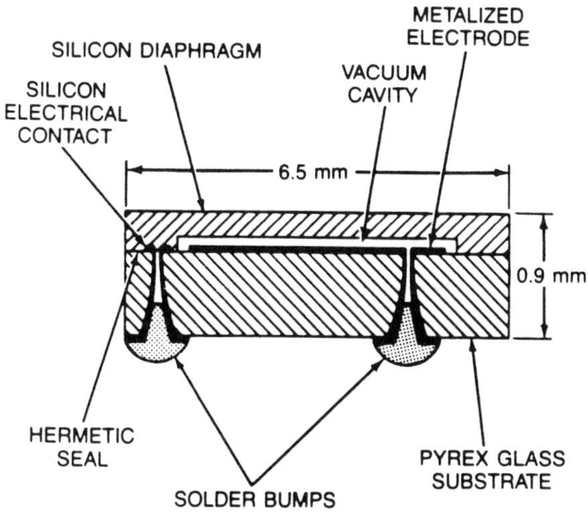

FIGURE 3. THE SCAP SENSING ELEMENT (Cross Section)

This step in packaging, usually the first, is the most critical in determining both the ultimate functional performance and durability of the final sensor.

MEDIA COMPATIBILITY

Media compatibility is one area the sensor industry has emphasized for without this the sensor is useless. The techniques vary from the simple one of encasing the entire sensor in metal or plastic as is common with thermistor based temperature sensors to complex ones such as pressure sensors utilizing metal diaphragms with oil filled cavities for very harsh media.

The drive for low cost sensors, especially for automotive applications, have resulted in new materials being developed for media compatibility. This has varied from ceramic and glass coatings used for oxygen sensors and mass air flow sensors to fluorosilicones used in pressure sensors. Many of these materials and their application techniques have been developed for specific companies and therefore are not available in general. Two that are available for general use are polyimide and paralyene. Both of these can be applied as conformal coatings and offer media compatibility with a large variety of materials.

The most simple yet often overlooked guide to media compatibility is to ensure that one thoroughly understands the environment in which the sensor will be used. This includes the isolated occasions when the environment contains an "unusual" element. An example is additives that are added occasionally by some motorists to gasoline and transmission fluids.

MEASURAND APPLICATION

Easily interfacing with the measurand and not causing the sensor to give false readings or completely fail is obviously crucial to sensor package design. To do this successfully requires that the sensor design team understand how and where the sensor will be physically mounted and how the data will be used. An example is that while pressure sensors used to determine air intake in an engine do not care if the air is in laminar or turbulent flow, a thermal mass air meter can be strongly influenced by flow conditions. Therefore, in many mass air meters the package includes an air flow conditioning structure as part of the package design.

Another example are accelerometers where it must be ensured that the package can be rigidly mounted to the surface and transmit the acceleration to the element. A package that flexes would obviously cause problems.

ISOLATION FROM UNWANTED INPUTS

One of the most vexing problems for sensor designers is to obtain the measurand and not other signals that effect the transduction element. The most common unwanted signal is temperature. One common way of negating the temperature signal is to use electronic compensation. To use electronics effectively, the unwanted signal must be consistent with temperature and preferably as small as possible. Correct design of the package can reduce the magnitude of the unwanted signal and make electronic compensation unnecessary.

One common packaging method of isolating unwanted signals is to use a compliant seal between the transduction element and the package. This in effect acts as a spring absorbing the stress. While this is effective for pressure sensors, it is not viable for accelerometers where a rigid mount is necessary. To effectively design a package that includes rigid mounts requires computer based mechanical design techniques.

CAD/CAM tools have allowed mechanical modeling to accurately describe stress, deflection, resonant frequencies, and other mechanical parameters so that rigid transduction element mounting can be designed instead of empirically determined. Modeling has also made possible the prediction of final package mounting conditions on the performance of the sensor.

EASE OF ASSEMBLY

Packaging must be designed to allow simple assembly of various final package configuation required for different applications and for automated assembly where high volume is required.

An example of a packaging system where both these goals can be met is with plastic injection molding. Temperature sensors with different dimensions, with "O" ring seals and threads are produced in million quantities using plastic injection molding and one common manufacturing process.

Another example is shown in Figure 4 where silicon pressure sensors and accelerometers are first level packaged (assembly, wire band, and test) on large ceramic plates containing up to 100 sensors and then singulated at the end of the process.[11] This allows for automated and efficient assembly.

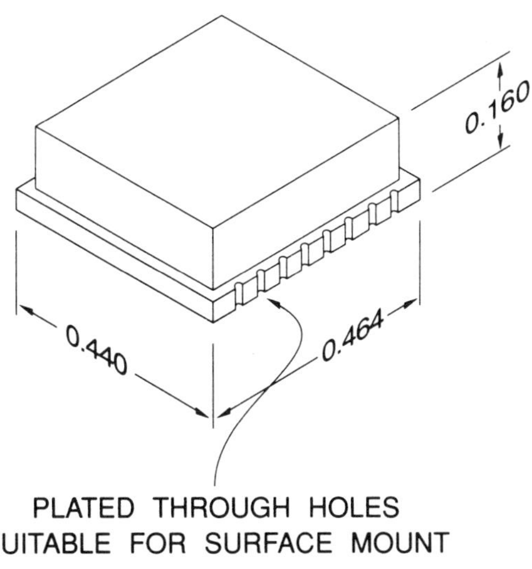

FIGURE 4. CERAMIC FIRST LEVEL PACKAGE
(Courtesy of Lucas Nova Sensor)

Figure 5 shows another assembly method to achieve a variety of first level package configurations with minimum manufacturing process changes.[12] The conductive seal interconnect system eliminates wire and tab bonding connectors. The final package can be easily and quickly modified to use whatever type part is required.

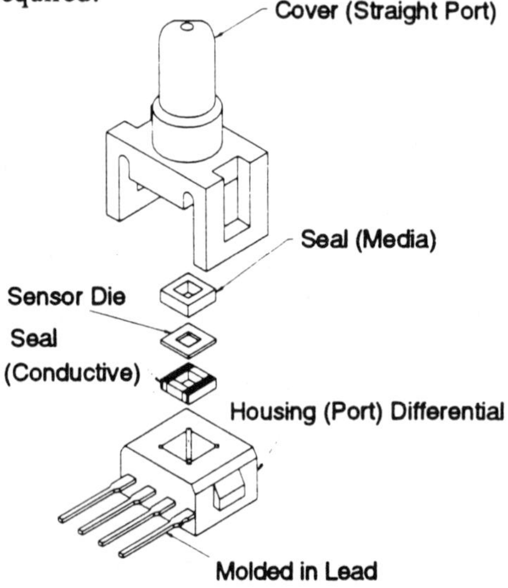

FIGURE 5. PLASTIC FIRST LEVEL I PACKAGE (Courtesy Honeywell)

For developing a family of low-cost sensors, ease of assembly must be addressed at the start of the design process.

TEST/CALIBRATION

This is one of the most expensive manufacturing step in many sensors. It can be both labor and capital intensive as the physical phenomena (e.g., pressure, flow) must be controlled and applied to the sensor while the electrical circuit is adjusted. Proper design of the package can reduce both the labor and equipment required.

The first level package is the key to reducing test/calibration costs. First level packages must be designed for isolation from unwanted environmental inputs and final package effects. The first level package should also be designed for ease of adjustment. The adjustments are usually laser trimming, fuse flowing, or potentiometer adjustment.

An example is SCAP, (Figure 2) where the first level package was designed to be insensitive to temperature stress to the transduction element, eliminate stray capacitance effects and to be planar to allow for automated handling during laser trimming of the electronic circuit. This allowed for rapid calibration of the sensor at the first level package level. The final package provides environmental protection.

FUTURE TRENDS

A trend in sensor packaging is to design the first level package as part of the transduction element. This is particularly attractive with silicon based sensors where electronics can be incorporated on the periphery of the transduction element and wafer bonding techniques used to produce low-cost individual devices. This trend will increase as new materials and fabrication techniques are commercially exploited. Diamond films are an area that shows promise for allowing more compact structures combining transduction element, electronics, and package as well as providing environmental protection. MCM packages being developed for IC's offer new opportunities for silicon based sensors. New materials, such as electrically conductive polymers[12], offer opportunities for new package designs and assembly techniques.

CONCLUSION

The area of sensor packaging will continue to diversify and grow to meet the demands of the market place. Computer modelling techniques and new materials will offer engineers packaging options that are not presently economically viable. The key to success will as it always has depend on the designers ability to blend new materials, processes, and computer tools to meet the customer's needs.

REFERENCES

1. J. Paulsen and J. Giachino, "Powertrain Sensors & Actuators Driving Toward Optimized Vehicle Performance," 39th IEEE Vehicular Conf., Vol. II, 1989, p. 574.

2. R. E. Suloff, "Silicon Sensors for Automotive Application," Transducers 91 International Conference on Solid State Sensors and Actuators, IEEE 91, ch. 2817-5, p. 170-176.

3. P. Chapman, "Sensors," Vol. 9, March 1992, p. 3-4.

4. J. Giachino, "Solid State Sensor and Actuator Workshop," Tech. Digest IEEE, Hilton Head Island, SC, June 1986.

5. G. Wallis and D. I. Pomerantz, "Filed-Assisted Glass-Metal Sealing," J. Appl. Phys., Vol. 40, 1969, p. 3946.

6. K. B. Albaugh et al., "Mechanisms of Anodic Bonding of Silicon to Pyrex Glass," Tech. Digest IEEE Solid State Sensor Workshop, Hilton Head Island, SC, June 6-9, 1988, p. 109-110.

7. K. Petersen et al., "Silicon Fusion Bonding For Pressure Sensors," Tech. Digest IEEE Solid State Sensor Workshop, Hilton Head Island, SC, June 6-9, 1988, pp. 143-147.

8. P. Barth, "Silicon Fusion Bonding for Fabrication of Sensors, Actuators and Microstructures," Transducers '89, The 5th International Conference on Solid-State Sensors & Actuators, June 25-30, 1989, Montreux, Switzerland, COMSAT, Lausanne I, Switzerland, P. 263.

9. L. A. Fields and R. S. Muller, "Fusing Silicon Wafers With Low-Melting Temperature Glass," Transducers '89, The 5th International Conference on Solid-State Sensors & Actuators, June 25-30, 1989, Montreux, Switzerland, COMSAT, Lausanne I, Switzerland, p. 365.

10. M. E. Behr, C. F. Bauer, and J. M. Giachino, "Miniature Silicon Capacitance Absolute Pressure Sensor," 3rd International Conference on Automotive Electronics, 1981, pp. 255-260.

11. J. Bryzek, K. Petersen, L. Christel, and F. Pourahmadi, "New Technologies for Silicon Accelerometers Enable Automotive Applications," SAE International Congress, Feb. 24-28, 1992, Paper 920474.

12. F. W. Kules and R. H. Estes, "Solderless Flip Chip Technology," Hybrid Technology, February 1992, pp. 24-27.

LOW TEMPERATURE CERAMIC FILMS FOR MICROELECTRONIC APPLICATIONS

Grish Chandra, Robert C. Camilletti, Mark J. Loboda
Dow Corning Corporation, Midland, Michigan 48686

ABSTRACT

Low temperature ceramic film technology from designed metastable organometallic and inorganic precursors is relatively new. These films have suitable physical, chemical, morphological, microstructural, electrical, and mechanical properties for electronic applications. The films can be deposited by conventional electronic industry processing methods at temperatures from 175 to 400°C. Applications of primary interest include inter-level dielectric planarization and coatings for protecting integrated circuits from severe environments.

INTRODUCTION

Many of the major advances of the last few decades in communications, consumer electronics, transportation and information processing resulted from applications of integrated circuits. The increasing demand for greater performance (speed) and higher density (functionality) for complex integrated circuits is being met by continuing development of devices, circuit design and enhanced processes for further shrinkage of the circuit geometry. Essential for manufacturers to meet these integrated circuit challenges will be advancements in the preparation of the raw materials and the technology required to form and package the integrated circuit final product.

The industry requires dielectric materials as films for diffusion and ion implantation masks, for capping doped films to prevent the loss of dopants, for gettering impurities, for insulation between layers of circuit conductors, and for passivation to protect devices from impurities, moisture and scratches. These films must meet many requirements: be free of flaws, have uniform thickness over the large number of wafers processed at one time, be planar over the entire circuit surface and fill gaps between adjacent conductors, and have controlled and reproducible structure, composition, and purity. Normally selections of dielectric materials are based on the film characteristics: electrical, optical, thermal, mechanical, chemical, and morphological/microstructural.

The properties of thin ceramic films largely depend upon the deposition process. Thin film deposition technologies are either solely physical, such as evaporative methods, or solely chemical, such as gas and liquid phase chemical processes. There are also many processes that are based on glow discharge and reactive sputtering which combine both physical and chemical reactions [1].

Dielectric materials being used in electronic circuits include: silicon dioxide, borosilicate glass (BSG), phosphosilicate glass (PSG), borophosphosilicate glass (BPSG); silicon nitride [plasma nitride (SiN_xH_y) and "UV nitride"]; silicon oxynitride (SiO_xN_y); and silicon carbide, a-SiC:H. However, processes normally used by the semiconductor industry to deposit these dielectrics as thin films frequently cause particulate formation, and are hazardous (when silane is used). These processes generally yield films with residual hydrogen and carbon, crack and pinhole defects, and the films exhibit poor step coverage.

This paper focuses on two dielectric materials, silicon dioxide and silicon carbide, their new precursors, and improved processes for deposition as thin ceramic films. These thin films insulate integrated circuits and protect devices from environmental stresses.

PREPARATION AND PROPERTIES OF SILICA AND SILICON CARBIDE FILMS

Dow Corning is pioneering the development of designed precursors and deposition processes for in situ formed thin film ceramics on integrated circuits. This approach converts metastable chemical precursors to ceramic films at relatively low temperatures:

$$\begin{matrix}\text{Organometallic} \\ \text{or} \\ \text{Inorganic species}\end{matrix} \xrightarrow{175\ -\ 400^{\circ}C} \text{Ceramic films} \qquad (1)$$

Organometallic or inorganic precursors are converted to ceramic films by pyrolysis, reactive gas chemical conversion, chemical vapor deposition, or plasma enhanced chemical vapor deposition [2]. Conversion of designed organometallic or inorganic polymeric precursors to a ceramic is particularly interesting since some conversion processes need only moderate temperatures, as low as 175°C, and a high level of purity can be maintained.

The precursors may be prepared with very high purity by distillation and other purification processes. This purity affords consistent film formation processing, uniform film characteristics, and functional stability of the electronic circuits. The precursors may be applied to surfaces by chemical vapor deposition (CVD) or plasma-enhanced chemical vapor deposition (PECVD) processing, or others from a solution, such as by conventional spinning or spraying, before conversion to a ceramic film.

Challenges being pursued are development of higher purity precursors, lower temperature and cleaner (no particulates) processes, and simpler application techniques while ensuring processing safety and maintaining low overall cost.

Silica Film and its Precursor

Pure silica films are frequently the optimum ceramic for deposition directly on circuit surfaces. These films, when properly prepared, have suitable mechanical, dielectric, thermal and film integrity characteristics and also can planarize the surface for deposition of any subsequent layer. The subsequent layer may be required for photolithography processing, circuit conductor metallization, or additional environmental protection.

A relatively high density pure silica film can be prepared by the oxidation of hydrogen silsesquioxane, $(HSiO_{3/2})_n$. Commonly known as "H-resin," this carbon-free silica precursor is synthesized by the "scarce water" hydrolysis of trichlorosilane in a benzene/SO_3 fortified sulfuric acid reaction medium [3]:

$$C_6H_6 + H_2SO_4 \text{ --------> } C_6H_5OSO_2H + H_2O \quad (2)$$

$$nHSiCl_3 + 3/2nH_2O \text{ -----> } [HSiO_{3/2}]_n + 3nHCl \quad (3)$$

This process is followed by an extensive work-up of the organic layer to remove the residual acid; the yield is typically 90-95%. A modified synthesis replaces benzene with toluene for greater process control and reduced hazards [4].

H-resin is a stable, white solid, soluble in common organic solvents, and unique as a silica precursor: it softens and flows at ~150 - 200°C [5]. It consists of two components. One is a volatile component (sublimes below 100°C at 0.01 Torr), $(HSiO_{3/2})_n$, with cage structures with n=8, 10, 12, 14 and 16. These low molecular weight structures are suitable CVD precursors [6]. The other component is soluble non-volatile amorphous $HSiO_{3/2}$ resin. It consists of higher molecular weight cages (n>16) and various cages strung together in a more-or-less random bead-chain fashion [3]. This resin can be fractionated, using super-critical fluid, to reduce the molecular weight distribution [7].

H-resin can be prepared with high purity; the concentration of critical mobile cations (Na^+, K^+, Ca^+) can typically be limited to <20.0 ppb and anions (Cl^-) to <10 ppm. This silica precursor can be dissolved in organic solvents for application by conventional spinning, spraying, dipping or flow coating processes.

The viscous flow of H-resin over the surface topography prior to conversion to silica substantially aids in planarizing and gap filling between adjacent metallized areas.

For conversion to silica, the H-resin film is oxidized:

$$[HSiO_{3/2}] \xrightarrow{O_2} [HOSiO_{3/2}] \text{ -----> } SiO_2 + H_2O \quad (4)$$

This oxidation takes 0.5 hour at 700-800°C or 1-2 hours at 400°C in a furnace. The oxidation time can be reduced to only 5-10 minutes at 400°C by rapid thermal processing, such as with the use of an arc or tungsten-halogen lamp [8]. The oxidation temperature can be further reduced by processing in a wet ammonia atmosphere; then at 250°C the oxidation time is 2 hours or at 175-200°C, 3 to 4 hours. The silica film density decreases somewhat as the oxidation temperature decreases. The infrared spectral features of the H-resin derived silica is similar to those of thermal oxide grown on silicon surfaces as shown by the FTIR spectra in Figure 1; virtually no residual SiH or SiOH is normally detected in H-resin derived silica.

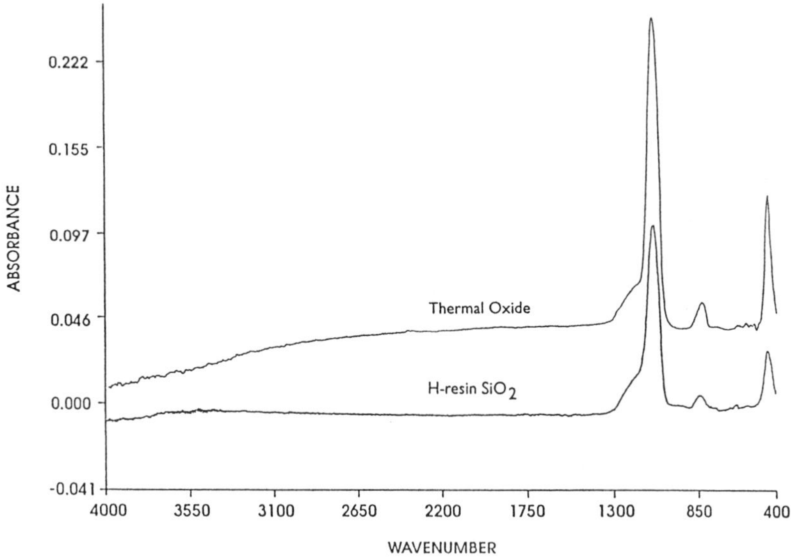

Figure 1. FTIR spectra comparing H-resin derived silica (oxidized in wet NH_3 at 250°C for 2 hours) to thermally grown oxide.

Oxidation of all the precursor species would result in a theoretical weight gain of about 11% which reduces the volume contraction. Weight gains of 7-8% are achieved, as shown in Figure 2, by the addition of 10 - 20 ppm Pt catalyst to retard the volatilization of the low molecular weight species. Silica layers of high quality with a thickness from 50 to 500 nm have been routinely prepared from a single coating of the precursor solution on silicon and gallium arsenide surfaces.

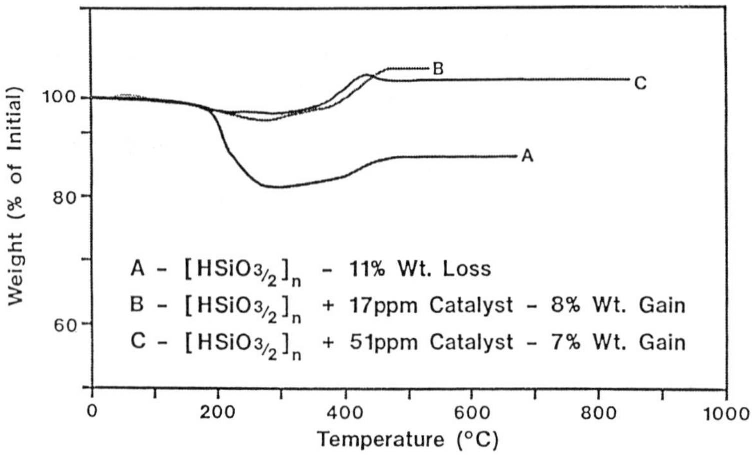

Figure 2. Thermogravimetric analysis of H-resin.

The principal properties of 150-300 nm thick silica films prepared by 300-450°C furnace oxidation of H-resin are listed in Table 1. These silica films, when properly applied in situ on silicon integrated circuits under clean conditions, reliably smooth and planarize the circuit surfaces and they seal the bond pads [9].

Table 1. Properties of Silica Films

Films deposited on Si by 300-450°C furnace oxidation

Si:O ratio	~1:2 (RBS)
Dielectric constant	3.7-4.1 (0.1 MHz - 7 GHz)
Dielectric loss factor	0.02 (0.1 MHz), 0.04 (7 GHz)
Volume resistivity	5-9 x 10^{10} ohm-cm
Dielectric strength	>75 V/µm
Refractive index	1.44-1.45
Density	1.6-2.0 g/cm^3
Water diffusion constant	6 x 10^{-14} cm^2/s
Coating uniformity	$\pm$2%
Freedom from pinholes	passes MIL-STD glassivation test [10]
Elastic modulus	40-70 GPa
Hardness	3.5-7.0 GPa
Stress	1.0-3.0 GPa (tension)
Coefficient of thermal expansion	1.5 ppm/°C

Silicon Carbide Film and its Precursor

Amorphous hydrogenated silicon carbide (a-SiC:H) films are also valued as protective barrier coatings on integrated circuits. Silacyclobutane (SCB), $H_2\overline{Si(CH_2)_3}$, is a suitable precursor for forming these films. It is synthesized in three steps by initially adding trichlorosilane to an allyl halide (Eq. 5) followed by the cyclization (Eq. 6) and reduction to silacyclobutane (Eq. 7) [11]:

$$HSiCl_3 + CH_2{=}CHCH_2X \longrightarrow Cl_3SiCH_2CH_2CH_2X \qquad (5)$$

$$Cl_3SiCH_2CH_2CH_2X \longrightarrow \overline{Cl_2SiCH_2CH_2CH_2} \qquad (6)$$

$$\overline{Cl_2SiCH_2CH_2CH_2} \longrightarrow \overline{H_2SiCH_2CH_2CH_2} \qquad (7)$$

An overall yield of 50 to 60% has been obtained. Silacyclobutane is a colorless liquid which boils at 43°C, has a density of 0.779 g/cm^3 and a refractive index of 1.442.

Silacyclobutane is decomposed by thermal processes at ~500°C to form amorphous a-SiC:H. More rapid film deposition, ~100 nm/min, is obtained by using PECVD at temperatures from 400°C down to 150°C. A wide range of film properties are obtained by varying the deposition temperature, plasma frequency and power, pressure, and concentrations of the precursor gas in the PECVD process [12].

The reactive intermediates in the decomposition of silacyclobutane are believed to be silene, $>Si{=}CH_2$ and/or silylene, $:SiHCH_3$. These coordinatively unsaturated intermediates, with the desired 1:1 ratio of Si:C, adhere to the surface of heated substrates to form a film of a-SiC:H. Adherent, scratch resistant films have been deposited on silicon, silica, GaAs, and metal surfaces.

The composition, porosity, mechanical, thermal, dielectric and optical properties of silacyclobutane-derived a-SiC:H films were determined for 200 to 500 nm thick films deposited on planar structures. These a-SiC:H films were formed at 175-250°C using a 10% silacyclobutane-argon mixture. Results of Auger, RBS and other analyses are presented in Table 2.

Table 2. Characteristics of Thin a-SiC:H Films

Si:C atomic ratio	0.8 - 1.2
Composition (typical)	67 wt% silicon, 28 wt% carbon, 5 wt% hydrogen (bound, trapped)
Dielectric constant	4-6.5 (100 Hz-1.0 MHz)
Volume resistivity	10^{12} - 10^{14} ohm-cm
Optical bandgap	2.2-2.6 eV
Refractive index	1.9-2.3
Film density	1.7-2.0 g/cm^3
Elastic modulus	~100 GPa (nanoindenter)
Hardness	15-17 GPa (nanoindenter)
Film stress	0.1 - 2.0 GPa (compression)
Coefficient of thermal expansion	~3.0 ppm/°C (similar to that of silicon)

MIL-STD Glassivation test [10] of a-SiC:H coated interdigitated aluminum patterns on silicon substrates revealed no pinholes in the films.

INTERLAYER DIELECTRIC APPLICATION

Interlayer dielectric formation between closely spaced circuit conductors is recognized as the critical process for manufacturing sub-micron interconnected integrated circuits. As device geometries are reduced to gain higher integrated circuit functional performance, the requirements for dielectric interlayer planarization and gap filling become more stringent. Conventional spin-on glass (SOG) materials [13] are no longer adequate as trench width (distance between parallel conductors) shrinks to 0.5 µm and less. Solving reliability problems from dielectric cracking, delamination, moisture absorption and poisoned vias (non-conductive paths between conductors) requires new precursor materials.

Silica interlayers formed from electronic grade H-resin solutions (now commercially available as Dow Corning® Flowable Oxide) overcome these limitations [14,15]. The improvements are due to the unique combination of characteristics:

- Precursor stable as a solid and in solution
- Carbon-free precursor composition
- Precursor melts and flows prior to conversion to silica; helps to fill gaps and planarize surface
- No acidic stabilization - eliminates potential corrosion problems
- Easy to clean up - prevents particle generation.

A thermomechanical analysis (TMA) of H-resin is shown in Figure 3. The large drop around 200°C indicates plastic deformation associated with material flow [15]. At this temperature, the precursor flows over the circuit surface. The high solids stability of the precursor solution permits thick coatings to be applied as a single layer. The absence of carbon in the precursor helps prevent poisoned vias from forming between circuit conductors.

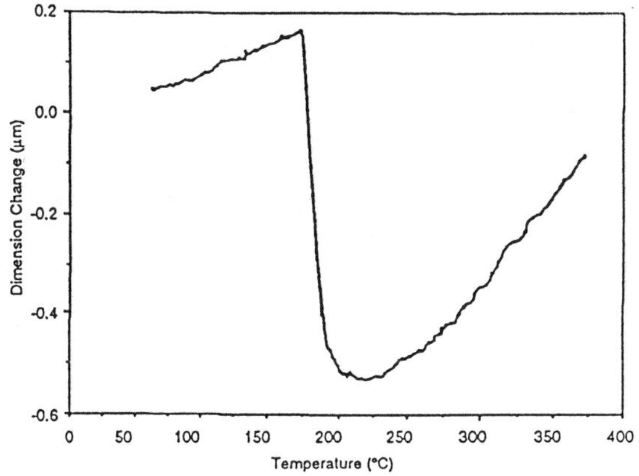

Figure 3. Thermomechanical analysis of hydrogen silsesquioxane.

The process for forming the silica interlayer typically involves spin coating of a high solids H-resin solution, heating to 200°C on the spin coater for 60 seconds for solvent removal and flowout of the precursor, and conversion to silica in an oxidizing ambient (with no Pt catalyst) at 425°C for 1-2 hours.

The degree of planarization and gap filling is revealed by cross sectional SEM. Figure 4 shows an example of complete gap fill in sub-micron (600 nm), high-aspect ratio (2:1) trenches. Almost complete planarity was achieved on 1.1 µm aluminum steps without cracking. Figure 5 demonstrates planarization of a sub-micron structure with a 120 nm gap width and step height of 500 nm (4:1 aspect ratio).

With these advancements in interlayer dielectric processing, this flowable oxide silica precursor is expected to extend the capabilities of integrated circuit processing to geometries smaller than 0.35 micrometers [14].

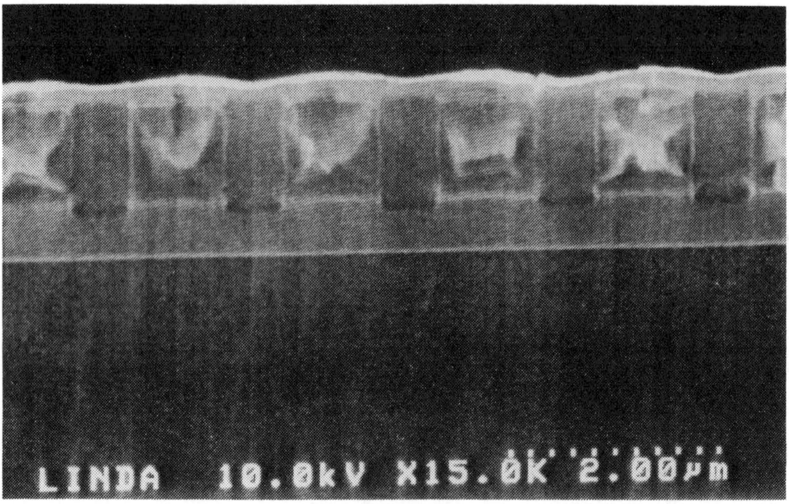

Figure 4. Gap fill and local planarity of H-resin derived silica coating on 1.1 μm conductor step height with 2:1 aspect ratio. 15000X.

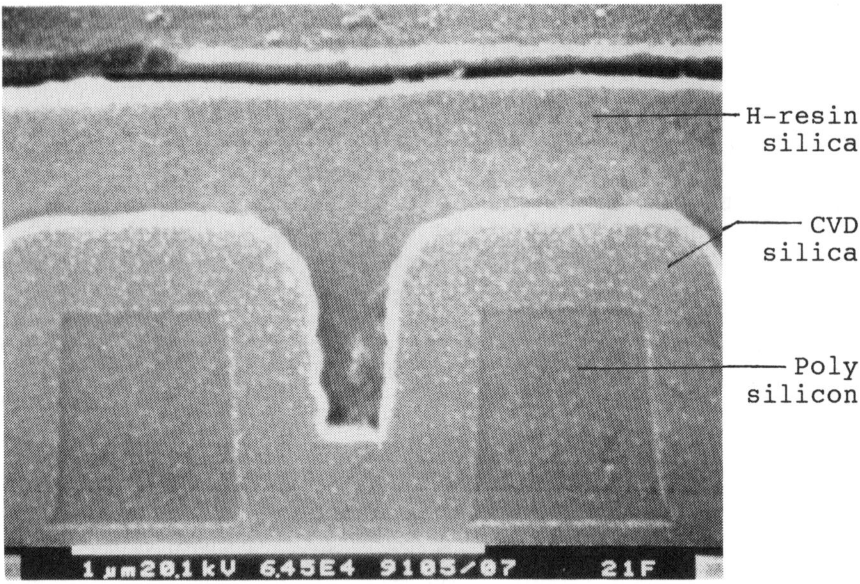

Figure 5. Planarization over 0.12 μm wide gap with step height of 0.50 μm (4:1 aspect ratio).

PROTECTION OF INTEGRATED CIRCUITS

In situ formed ceramic coatings are a novel environmental protection concept of Dow Corning for integrated circuits. Ceramic coatings are less permeable to ions, moisture and organic vapors than polymeric coatings [16] and thin ceramic coatings are less prone to cracking than thicker coatings. Thus, ceramic coatings of sufficient density and optimum thickness, processed by procedures which are compatible with integrated circuit fabrication, can protect integrated circuits if the films are continuous (no surface connected porosity, pinholes, cracks, loss of adhesion) [17]. Multiple studies have shown that a two-layer ceramic coating provides excellent protection: a silica layer for smoothing the circuit surface and filling crevices, and a separate dense plasma silicon carbide barrier layer as shown in Figure 6.

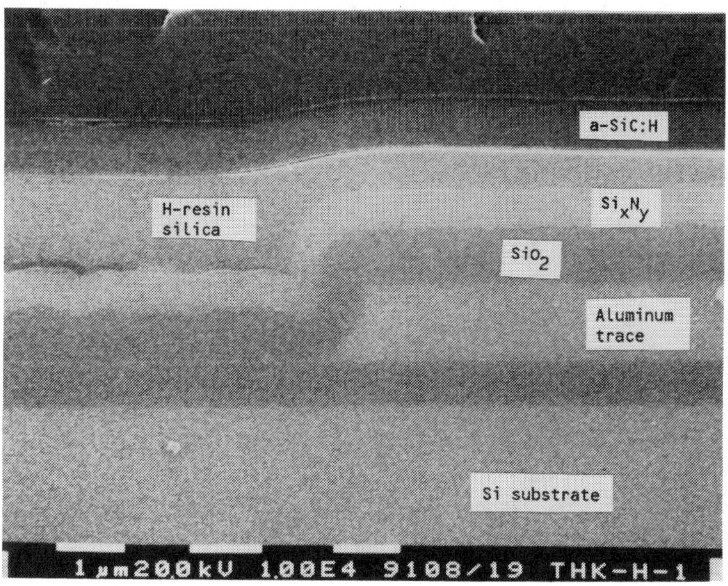

Figure 6. Cross section of two-layer silica/a-SiC:H coating on silicon CMOS integrated circuit.

The protection of silicon integrated circuits was demonstrated by dual-layer ceramic coatings on Motorola 14011B silicon CMOS chips. The circuit chip, with a primary passivation coating of silica/silicon nitride applied by the manufacturer, was mounted in a standard 14-pin side-brazed alumina package without a lid. The circuits had aluminum inner bond pads and gold surfaced external bond pads.

Intensive studies were made of these integrated circuits with a two-layer ceramic coating: an initial silica coating ~200 nm thick (derived from H-resin processed at 175-250°C) and a ~500 nm thick top coating of SCB-derived a-SiC:H processed by PECVD at 175-250°C. All electrically active surfaces, including the bond wires, were coated; see Figure 7. Complete cleaning of the circuits prior to film deposition is essential for producing suitable ceramic coatings.

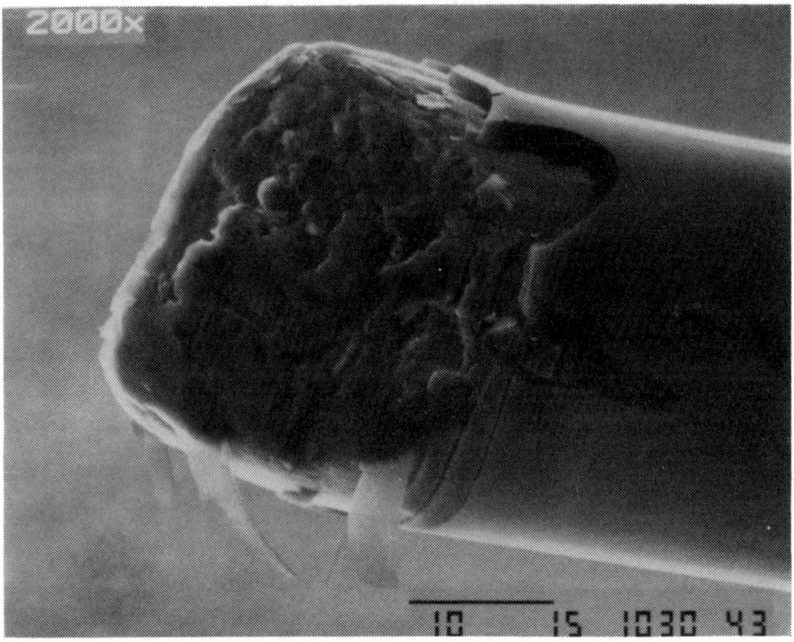

Figure 7. Silica coating on aluminum leadwire.

These coated circuits were exposed to autoclave conditions (water saturated air) at 121°C and 100 kPa pressure, in a separate study to HAST (Highly Accelerated Stress Testing) conditions (159°C, 400 kPa pressure, 10 volt circuit bias, 85% RH), and in a third study to -65°C to +150°C thermal cycling at a rate of one complete cycle/hour. Circuits which exhibited any significant change in operating (functional) characteristics were classified as failed; the results are listed in Table 3. These results show that the in situ formed ceramic coatings provided nearly complete protection of these sensitive circuits.

Whereas all unprotected circuits failed within 100 hours of either autoclave or HAST exposure, only one of 20 ceramic-coated circuit failed after 500 hours of exposure to any of these accelerated environmental stress conditions. An examination of an uncoated circuit after 100 hours of autoclave exposure revealed extensive corrosion of the aluminum bond pads.

Other studies showed that the protection provided by thin film ceramic coatings is further enhanced by applying a thin silicone gel topcoat [17]. This topcoat also mechanically protects the circuit from stresses during subsequent equipment assembly and use.

Table 3. Environmental Stress Exposure of Two-Layer Silica/a-SiC:H Coated Silicon CMOS Integrated Circuits

Exposure	Number of Circuits	Cumulative Failures		
		100 h	300 h	500 h
Autoclave	20	0	0	0
HAST	20	1	1	1
Thermal cycling	20	0	0	0

Multilayer ceramic film protection has the potential of eliminating costly ceramic packaging of integrated circuits, improving circuit performance and reliability, and reducing circuit size and weight. The level of protection afforded by the ceramic coatings in these studies has led to a new concept for circuit packaging. Combining these ceramic coatings with conventional low cost plastic molding of integrated circuits should provide high reliability without the conventional hermetic sealing in large ceramic packages. The coating technology may be applicable to sealing integrated circuit chips at the wafer stage, prior to chip separation for "chip-on-board" system assemblies [17].

CONCLUSIONS

Metastable organometallic and inorganic precursors can be processed at low temperatures to yield ceramic films suitable for interlayer dielectrics and for protecting integrated circuits. Hydrogen silsesquioxane and silacyclobutane were demonstrated to produce quality silica and a-SiC:H films, respectively.

The silica films can be used to form interlayer dielectrics to insulate adjacent circuit conductors in complex integrated circuits. Unique features of this silica precursor are its softening and melting at low temperatures prior to conversion to silica and its carbon-free composition. The resulting films completely fill gaps between the circuit conductors and planarize the surface for subsequent processing steps.

Another application for ceramic thin films is to protect integrated circuits from environmental stresses. A novel circuit packaging concept of Dow Corning uses designed precursors for in situ formation of protective multilayer ceramic film coatings. The protective performance of these coatings on silicon CMOS circuits was demonstrated under exposure to three industry recognized stress conditions. This concept may eliminate costly, bulky and imported ceramic packages for integrated circuits and thus reduce the circuit size and weight, lead to improvements in circuit speed and reliability, and reduce circuit cost. This concept can be applied to single chips and to multichip packages.

REFERENCES

1. K. Schuegraf, ed. Handbook of Thin-Film Deposition Processes and Techniques, Noyes Publications, New Jersey (1988).

2. R. Baney and G. Chandra, "Preceramic Polymers," Mark-Bikales-Overberger-Menges, Encyclopedia of Polymer Science and Engineering 13, 2nd edition, John Wiley and Sons, New York, 312-344 (1988).

3. C. Frye, W. Collins, J. Amer Chem Soc, 92, (19), 5586 (1970).

4. H. Bank, T. Martin, M. Cifuentes, US Patent 5010159 (23 April 1991).

5. D. Ballance, R. Camilletti, T. Gentle, US Patent 5,145,123 (8 Sept 1992).

6. T. Gentle, US Patent 5,169,955 (24 Nov 1992).

7. T. Gentle, L. Hanneman, K. Sharp, US Patent 5,063,267 (5 Nov 1991).

8. G. Chandra, T. Gentle, US Patent 5059448 (22 Oct 1991).

9. G. Chandra, "Low Temperature Ceramic Coatings for Environmental Protection of Integrated Circuits," Mat Res Soc Symp Proc 203 (1991).

10. MIL-STD 883C, Method 2021.3, Procedure B.

11. G. Cawthon, J. D'Errico, W. Schulz, Jr., US Patent 4,973,723 (27 Nov 1990).

12. M. Loboda, J. Seifferly, F. Dall, "PECVD of a-SiC:H Films Deposited from Organosilicon Precursors," to be published J. Vac. Soc. Technol. A. (June/July 1993).

13. S. Gupta, "Spin-On Glass for Dielectric Planarization," Microelectronic Manufacturing and Testing, 1 (April 1989).

14. Pei-Lin Pai, C. Konitzer, "Manufacturing Feasibility of Planarization Process," VMIC Conference (June 9-10, 1992).

15. D. Ballance, K. Scheibert, J. Tietz, "Low Temperature Reflow Planarization Using a Novel Spin-on Interlevel Dielectric," VMIC Conference (June 9-10, 1992).

16. R. Traeger, "Hermeticity of Polymeric Lid Sealants," Proc 25th Annual Elect Components Convention, 361 (1976).

17. R. Camilletti, G. Chandra, "Low Temperature Ceramic Coatings for Chip-On-Board Assemblies," Proc 2nd Intl Surface Mount Conference (1992).

Section III. Processing, Structure, and Properties

STRUCTURE-PROPERTY RELATIONS IN LOW PERMITTIVITY DIELECTRICS

Kelley Markowski, David P. Cann, Robert E. Newnham
Materials Research Laboratory, The Pennsylvania State University,University Park, PA 16802

ABSTRACT

The minimization of the dielectric constant in substrate materials in packaging applications is of great importance to the overall performance and design. By manipulation of the structure, the dielectric properties can be tailored to specifications. This paper presents a review of the existing low permittivity (K) materials from the viewpoint of structure-property relationships and basic dielectric theory.

INTRODUCTION

The field of microelectronics is a rapidly growing industry that is currently grossing 500 billion dollars worldwide. Packaging products range from consumer goods such as compact disc players, to supercomputers able to preform four trillion complex calculations a second. This diversified market encompasses a variety of unique materials and technologies.[1] Of these technologies, low K packaging materials hold a position of priority.

Increasing signal processing speed, signal processing capacity and decreasing crosstalk have become causes of concern in VLSI (Very Large Scale Integration) packages.[2] As new technologies evolve, the underlying property of greatest importance is the permittivity since it is the limiting material property for all three concerns. Recent research on composite materials has given dielectric constants as low as 1.6, which is significantly lower than the standard alumina substrates, (K=10).

The role of the microelectronic package has greatly expanded from merely serving as housing for semiconductor circuits.[3] The dielectric constant of the substrate has a direct influence on the signal distribution characteristics, since the signal propagation speed, the impedance of signal traces, and crosstalk between signal traces are all directly influenced by the dielectric properties of the substrate. Given these conditions, manufacturers of high speed integrated circuits require materials with very low dielectric permittivity, K~3. The glass-bonded alumina substrates used at present have a relatively high K~9.5 which imposes severe restrictions on the design of high density packages.

The propagation speed (V) is related to the inverse square root of the dielectric constant by

$$V = \frac{c}{\sqrt{K}} \tag{1}$$

where c = speed of light in a vacuum. Through this relationship it can be seen that low K materials with dielectric constants of about 2 have twice the speed of alumina. A comparison of low permittivity packaging materials is shown in Fig. 1. Based on this graph, it is clear that polymers and silica-based glass-ceramics have an advantage over the standard ceramics used in packaging.

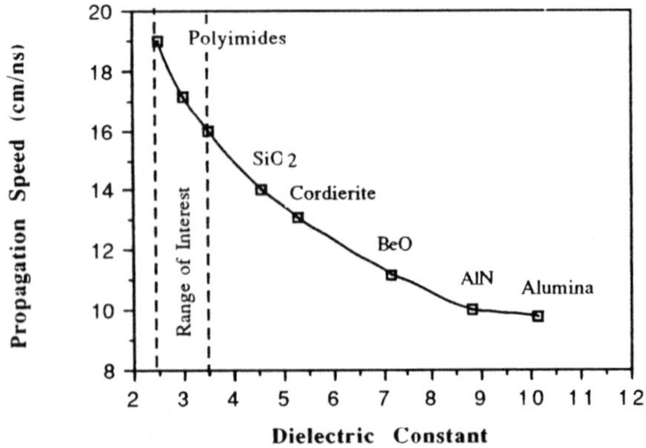

Fig. 1. Propagation speed vs. dielectric constant for several packaging materials.

Interconnections in a package are typically constructed in either a microstrip or stripline configuration. In a microstrip, the signal traces are on the surface of the dielectric. Conversely, in a stripline configuration the signal traces are embedded in the dielectric itself. The processing of the the stripline configuration is carried out with the three-dimensional multilayer interconnect processing required for high density components. The impedance of the signal trace is proportional to the ratio of the inductance per unit length to the capacitance per unit length. Thus, by lowering the dielectric constant, the thickness of the layers can be reduced for a given impedance level..

In microelectronic packages neighboring signal traces couple through mutual inductance and capacitance, introducing noise into the circuit. Such noise is referred to as crosstalk. Crosstalk is common in high density designs because the signal traces are very close to each other, and becomes increasingly troublesome at high frequencies. For a permittivity of 9.5 and a signal rise time of 200 ps, crosstalk ratios as high as -8 dB have been estimated for adjacent signal traces with characteristic impedances of 50 ohms.[4] To control this problem, ground planes must be moved in closer relative to spacings between adjacent signal traces, in order to maintain the line impedance of 50 ohms. This can only be accomplished by

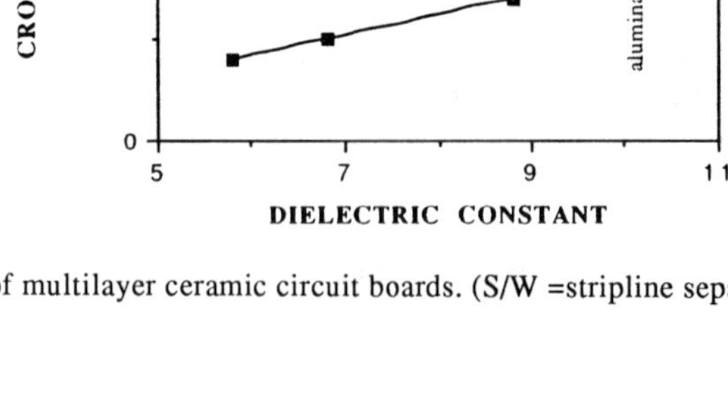

Fig.2. Crosstalk of multilayer ceramic circuit boards. (S/W =stripline separation/width ratio)[6]

reducing the dielectric permittivity, thus allowing a higher signal trace density.[5] The influence of the separation and width of the conductors, or signal traces, in a stripline configuration is illustrated in Fig. 2 [6]. As the ratio of separation to width increases, the crosstalk between signal traces decreases. With a low dielectric constant material, greater wiring density can be achieved by decreasing the separation distance while still maintaining a manageable level of crosstalk.

BASIC DIELECTRIC THEORY

Polarization Contributions (Ionic & Electronic)

There are four basic polarization mechanisms: electronic, ionic, dipolar, and space charge. The total polarizability, α_T can be written as the sum of four terms representing the most important contributions to the polarization,

$$\alpha_T = \alpha_{el} + \alpha_{ion} + \alpha_{dip} + \alpha_{sp\ ch} . \qquad (2)$$

For low K materials, electronic and ionic polarizations are the only contributions that need be considered. In these materials there are no permanent dipoles and only a few free charge carriers.

The electronic contribution responds to AC fields up to approximately 10^{15} Hz. Electronic polarization is the chief contribution to K at all frequencies in covalently-bonded materials such as diamond in which ionic and dipolar effects are absent.[7] Ionic polarizability responds to fields up to approximately 10^{13} Hz. At low frequencies, the dielectric constant, K, of the alkali-halides is about 5. The refractive index for visible light is approximately 1.5, so that the high frequency dielectric constant ($K=n^2$) for these materials is about 2.2. The difference (5 - 2.2 = 2.8) shows that for alkali-halides there exists both an electronic and ionic contribution to the dielectric constant. For MgO, the alkaline earth oxides, K ~ 10 and n^2~3.[8] A comparison of the electronic and ionic contributions for other materials can be seen in Table 1 and Fig. 3. As shown, the more open, low density structures have lower K and n^2 values than the close packed, high density structures.

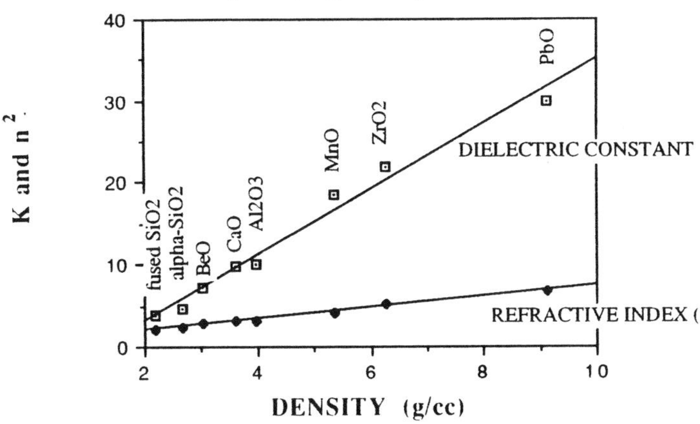

Fig. 3. Density vs. K and n^2

Table 1. High and Low Frequency Dielectric Constant

Material	n^2	K	Material	n^2	K
NaCl	2.25	5.6	KCl	2.13	4.7
NaBr	2.62	6.0	KBr	2.33	4.8
Al_2O_3	3.10	10.1	MgO	2.95	9.7
SiO_2	2.37	4.5	BeO	2.95	7.4
CaO	3.28	11.8	Diamond	5.76	5.8

As stated earlier, only the electronic and ionic contributions need be considered for the dielectric behavior of low K materials. An approximation of the dielectric constant can be obtained with the Clausius-Mossotti equation,

$$\alpha_d = \frac{3}{4\pi} V_m \frac{(K-1)}{(K+2)} \tag{3}$$

where α_d is the dielectric polarizability and V_m is the molar volume. Using the dielectric polarizabilities of the molecules and ions (Tables 2 and 3), it is possible to approximate the dielectric constant.[9]

R.D. Shannon successfully modeled the dielectric constants of 63 oxides and four fluorides with approximately 0.5% accuracy.[9] The molecular polarizability of a compound (A_2BX_4) can be represented in two ways. It can be divided into the molecular polarizabilities of the simpler compounds-[$\alpha_d(A_2BX_4) = 2\alpha_d(AX) + \alpha_d(BX_2)$] ,where α_d, the dielectric polarizability, includes both the electronic and ionic polarization contribution. Alternatively, the complex substance can also be represented by its ionic polarizabilities: $\alpha_d(A_2BX_4) = 2\alpha_d(A^{2+}) + \alpha_d(B^{4+}) + 4\alpha_d(X^{2-})$.

Table 2. α_d Values for Common Ions

Anions	α_d	Cations	α_d	Cations	α_d
F^-	1.63 Å^3	B^{3+}	0.05 Å^3	Mg^{2+}	1.31
O^{2-}	2.01	Be^{2+}	0.30	Fe^{2+}	2.22
OH^-	2.18	Al^{3+}	0.79	Ca^{2+}	3.15
		Si^{4+}	0.85		

Table 3. α_d Values for Common Oxides

Oxide	α_d	Oxide	α_d
BeO	2.15 Å^3	CaO	5.20
MgO	3.32	TiO_2	7.18
FeO	3.89	Al_2O_3	7.70
ZnO	4.47	Fe_2O_3	10.67
SiO_2	4.84		

The correlation between density and refractive index can be illustrated with polymorphic structures of silica (Fig. 4).

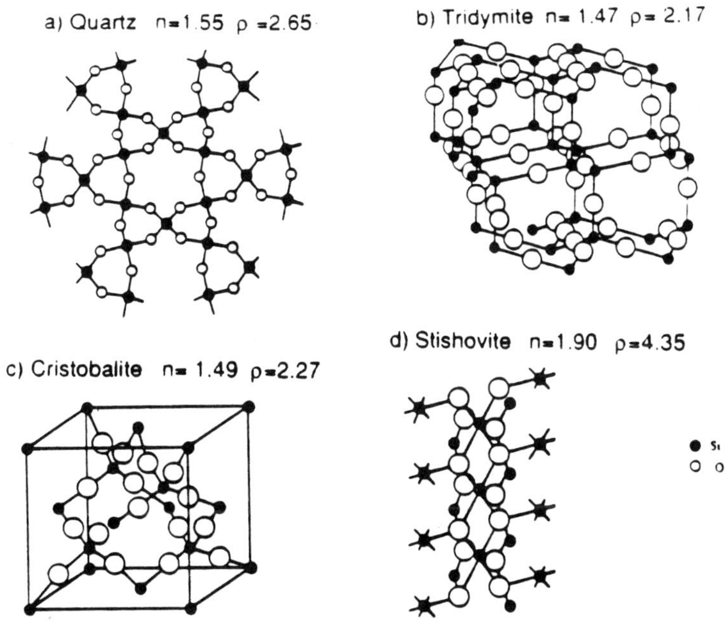

Fig. 4. Silica Structures; a) quartz b)tridymite c)cristobalite and d)stishovite.

Using the unit cell dimensions and ionic radii, the percent "porosity" or open space inherent in a material's structure can be computed. The correlation between the percent "porosity" and dielectric constant in common packaging materials can be seen in Fig. 5.

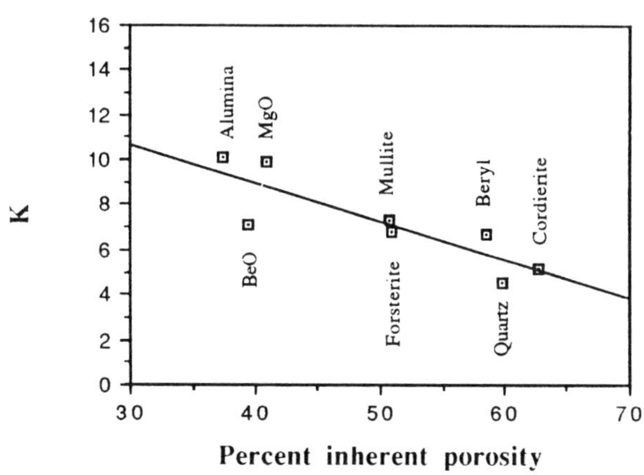

Fig. 5. Percent inherent porosity vs. K.

A prediction for the dielectric constant of low K materials can be determined by graphing an empirical coefficient (referred to in this paper as dielectric coefficient) against K. A dielectric coefficient, d, for simple oxides can be calculated using,

$$d = \frac{|\text{Difference in Electronegativity, cation-anion}|}{S_m} \qquad (4)$$

where S_m is the number of oxygen atoms divided by the molecular weight of the oxide.The dielectric coefficient of Al_2O_3 can be calculated as follows, $S_m = 3/101.96 = 0.029$ and $d= |1.5-3.5| / 0.029 = 68.97$. S_m has previously been used in the calculation of density and optical constants of glasses.[10]

Table 4. Calculated dielectric coefficients for some common oxides.

Oxide	Dielectric coefficients (d)	K
SiO_2	37.47	4.56
BeO	50.02	7.16
MgO	92.71	9.90
Al_2O_3	68.97	10.13
CaO	140.2	12.0

Density alone proved insufficient to account for the various contributions to the dielectric constant in predicting K. In a study of a wide range of silicates the dielectric constant was investigated as a function of density. However, plots of K vs. ρ for a variety of materials showed considerable scatter.[11] A list of dielectric coefficients is given in Table 4. A least squares fit of the dielectric coefficients vs. K for simple oxides is,

$$K=3.49 + 0.066\,d \qquad (5)$$

To illustrate its use for complex oxides we calculate the predicted dielectric constant for beryl, $Be_3Al_2Si_6O_{18}$ using oxide additivity.. The molecular weight of beryl is 537.5. The constituent oxides are $3BeO+Al_2O_3+6SiO_2$, for which the weight fractions are, respectively, 3 x 25 / 537.5, 102/537.5, and 6 x 60 / 537.5. The dielectric coefficient for the complex oxide can then be calculated, using the values given in Table 4. {(3 x 25/537.5 x 50.02)+(102/537.5 x 68.97)+(6 x 60.09/537.5 x 51.52)}=54.63. Substituting this value into the equation 12 predicts K = 7.10, which is in good agreement with the observed value of 6.75. The equation generally correlates within 10% for a wide range of oxides, as shown in Table 5.

Table 5. Prediction of the dielectric constant for common oxides.

Oxide	K(predicted)	K(actual)	% difference
$BeAl_2O_4$ chrysoberyl	7.80	8.93	12.7
$MgAl_2O_4$ spinel	8.47	8.18	3.5
Be_2SiO_4 phenakite	6.85	6.21	10.3
Al_2SiO_5 kyanite	7.61	7.80	2.4
$MgCaSi_2O_6$ diopside	8.90	8.60	3.5
$Be_3Al_2Si_6O_{18}$ beryl	7.10	6.75	5.2
$Al_6Si_2O_{13}$ mullite	7.53	7.3	3.2

<u>Composites</u>

All matter can be viewed as composites, even "solid" single crystals. Regarding the dielectric constant there are four components in the composite: open space K=1; covalent matter K~5; ionic matter K~10; and active ferroelectric matter K~ 10^2-10^4. In ceramics both the ionic and electronic contributions are important. Low K packaging materials show a correlation between percent ionic character, density and dielectric constant. Polymers, which are ~10% ionic and have an open structure, have a dielectric constant ~2-3, while silicates, which are ~50% ionic with an open structure, have a dielectric constant ~4-5.

Two models for predicting the dielectric constant of multicomponent systems are the series and parallel mixing rules. Series mixing significantly influences K. For a two phase material with V_1, V_2, K_1 and K_2 being the volume fractions and dielectric constants of phase one and phase two, respectively. The equation representing the series mixing model is given in Table 6. Parallel mixing on the other hand does not affect the dielectric constant as significantly as the series model. The equation representing the parallel mixing model is also given in Table 6.

It has been suggested that as higher operating speeds of GaAs are achieved, substrates with K<3 will be needed for these systems. When considering inorganic systems no single phase material has a dielectric permittivity in this range. This requirement suggests that a dense single-phase material will not be possible and composites must be considered. The basic advantage of composites is the ability to lower the dielectric constant by introducing a second phase with a lower K. Various mixture relations are available, four common relations are listed in Table 6.[12] A graphical representation of these mixture rules is shown in Fig. 6. using SiO_2 glass and air with K=3.8 and K=1, respectively.

Table 6. Dielectric Mixing Rules for Two Phase Composites.

Rule	Equation
Series Mixing	$\frac{1}{K} = \frac{V_1}{K_1} + \frac{V_2}{K_2}$
Parallel Mixing	$K = K_1V_1 + K_2V_2$
Logarithmic Mixing	$\ln K = V_1 \ln K_1 + V_2 \ln K_2$
Maxwell's Mixing	$K = \dfrac{V_2K_2\left(\frac{2}{3} + \frac{K_1}{3K_2}\right) + V_1K_1}{V_2\left(\frac{2}{3} + \frac{K_1}{3K_2}\right) + V_1}$

Utilizing the line of thinking discussed earlier, the "perfect" low K material, in regards to dielectric constant, would be a mostly covalent material with an open structure. One class of materials which fits this description is the aerogels. Various processing techniques involving SiO_2 have provided some very promising low K packaging alternatives. These materials are two phase composites of SiO_2 and a gas or vacuum phase with K=1 (pores). The types of SiO_2-porosity composites which will be discussed are colloidal SiO_2 gels (aerogels), sputter deposited SiO_2 films, thick film coatings of porous SiO_2, and microporous glass structures. Aerogels will be discussed in greater detail in the following section on materials.

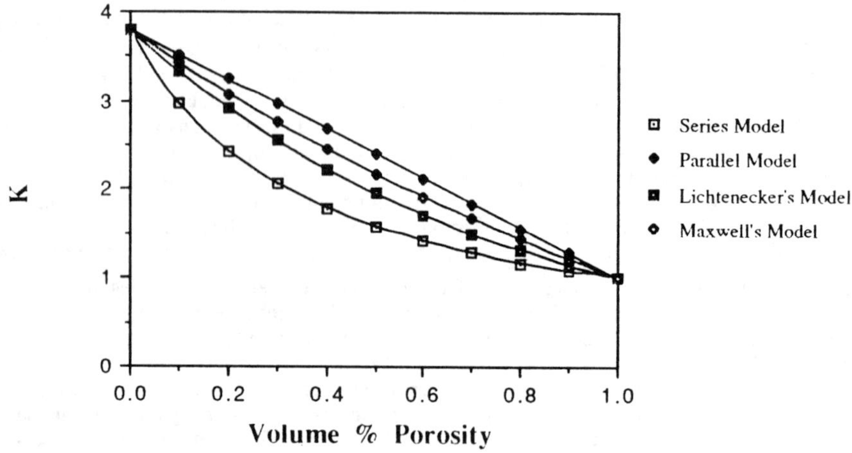

Fig. 6. Mixture relations for a two phase material, SiO_2 and air (porosity).

MATERIALS

A listing of the important properties relevant to packaging applications for a variety of materials is given at the end of this paper in Table 9.

Ceramics

Alumina α-Al_2O_3, the traditional packaging material has many deficiencies as a packaging material when compared to the ideal insulator. One of the deficiencies is a relatively high K=10.1. Others include relatively poor thermal conductivity and a coefficient of thermal expansion (CTE) mismatch with Si. Nevertheless, it remains the material of choice, and is probably used in more than 90% of the ceramic packages produced today.[1]

The comparatively high dielectric constant of alumina can be attributed to its high density ρ=4.0, as well as its large ionic and electronic contributions to its dielectric constant. The percent open space (37%) inherent in the alumina structure is also quite low in comparison to the lower K, more open structures, such as silica (60%) and cordierite (63%). Al_2O_3 has a hexagonal close packed structure with two thirds of the octahedral sites occupied (Fig. 7).[13]

Another widely used packaging material is beryllia, BeO. The dielectric constant of BeO (7.16) is lower than that of Al_2O_3, and it also has good thermal conductivity k=260 W/mK. While BeO also has ionic and electronic contributions to its dielectric constant and has approximately the same percent openness as Al_2O_3 (BeO=39.5%) it has a lower density ρ=3 g/cc which is partly responsible for its lower dielectric constant. BeO has the wurtzite (ZnS) structure with both the Be and O tetrahedrally coordinated (Fig. 8).

Cordierite ($Mg_2Al_4Si_5O_{18}$) has both a very low density ρ=2.51 g/cc and a very high percent of inherent openness 63%. These structural characteristics can be seen in its very low dielectric constant K=5.0. Cordierite has an orthorhombic unit cell with a=9.72Å, b=17.06Å, and c=9.34Å.

Forsterite (Mg_2SiO_4) is another example of a common packaging material. Like cordierite it also has an orthorhombic unit cell with a=4.76Å, b=10.21Å and c=5.98Å. The structure of forsterite is shown below. Its density is 3.2 g/cm^3 and its percent inherent porosity is 51%. Its dielectric constant K=6.2, however is higher than that of cordierite K=5.0, which can be attributed to its higher density and less open structure.

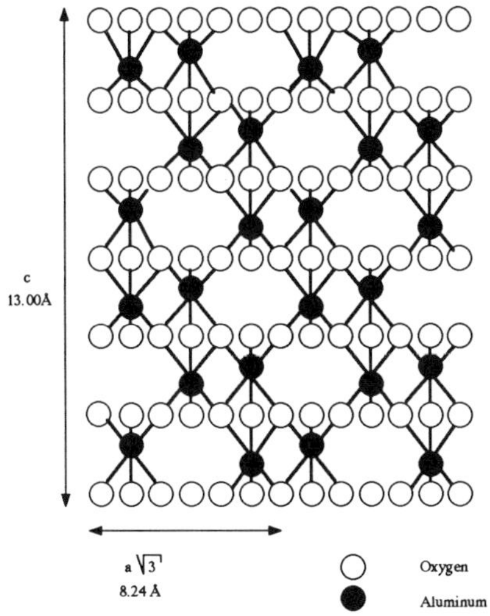

Fig 7. α-Al_2O_3 structure.

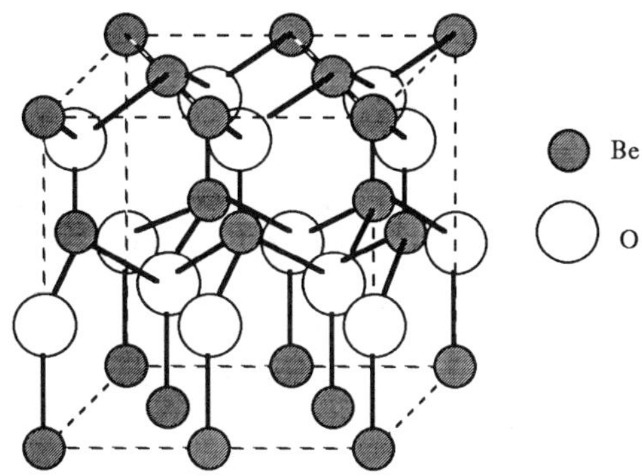

Fig. 8. BeO Structure.

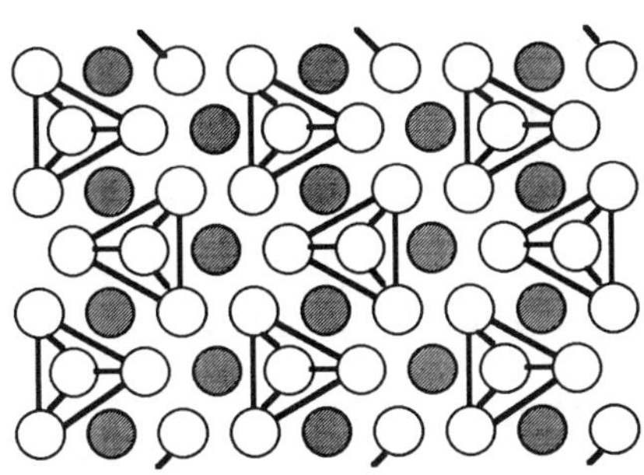

Fig. 9. Forsterite Structure, showing both octahedral Mg^{2+} (dark circles) and separated SiO_4 tetrahedra.

Polymers

Thin film polymer systems, such as polyimide are important in the design of thin film multi-chip modules (TFMCM). When used on a ceramic or silicon substrate to handle the signal interconnects, they have the advantage of a very low dielectric constant.[14] Polymers, in general, offer very low dielectric permittivity, but their thermal stability and expansion properties are not compatible for many packaging uses. For example polyimides ($nC_{23}H_{12}N_2O_4$) have a dielectric constant of 3.5, but a thermal expansion of 35 x

10^{-6}/°C and a thermal conductivity of only 0.2 W/mK. In comparison to traditionally used Al_2O_3 (K=10.1, α=7.8 x 10^{-6}/°C, k=50 W/mK), the thermal properties are problematic, while the dielectric properties offer the possibility of higher signal speed and reduced crosstalk.[15] Two of the better systems appear to be a new, low stress polyimide[16] and benzocyclobutene[17], both of which have dielectric constants less than 3.

Aerogels

Silica can be processed into an aerogel with 87-97 percent porosity. The down side to this method is the severe degradation of the mechanical properties of the material when large scale porosity or moisture is introduced. A common reaction producing silica aerogels is shown below:

Tetramethyloxysilane + water → Orthosilicic Acid + Methanol

$Si(OCH_3)_4 + 4H_2O \rightarrow Si(OH)_4 + 4CH_3OH$

After drying,

$nSi(OH)_4 \rightarrow n(SiO_2) + 2nH_2O\uparrow$

Colloidal Silica

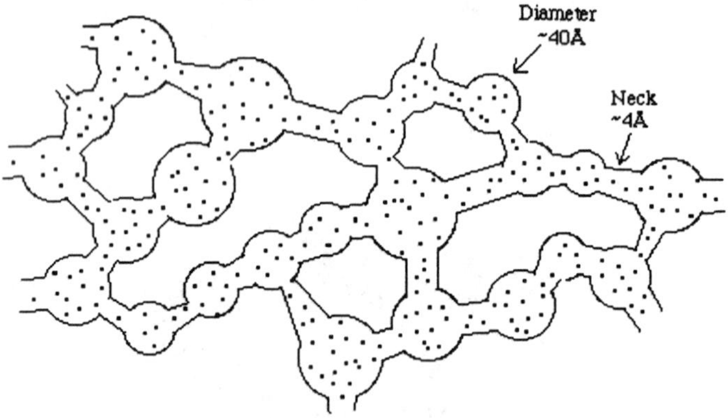

Fig. 10. Microporous SiO_2 Aerogel.

Because of their high porosity, aerogels offer the possibility to form materials with very low dielectric constants. A typical dielectric constant of an aerogel measured at 10 KHz was K=1.5.[18] In this example the SiO_2 aerogels were prepared under neutral conditions with an apparent density of ρ_a=0.35 g/cm^3 and an overall porosity of 82%. In a similar study,[5] aerogels prepared with 55-65% porosity were found to have relative permittivities in the range of 1.6 to 2.0.

Porosity can be introduced to a lesser scale into ceramics such as silica by a variety of methods. One such method is to introduce polymer spheres into the ceramic which can be burned out. The result is a material with spherical cavities and a lower net dielectric constant. The same effect can be achieved by using sol-gel methods, chemical leaching (i.e. vycor), or microspheres. Another disadvantage to the introduction of porosity into the material is the lowering of its breakdown strength. Also, the resistivity of the material will tend to decrease as the porosity increases.

Thick Film SiO_2

Thick film coatings of porous SiO_2 have been processed using a commercial colloidal silica (Ludox As-40) sol mixed with 3 weight percent PVA solution in water.[5] The viscosity of this mixture was allowed to increase through evaporation of the water until the viscosity reached 10-100 poise. The solution was then applied to a low resistivity silicon substrate and spun at 2000-3000 rpm for 5 minutes. The film is finally dried in air for 12 hours and fired for one hour at 500°C. The thickness of the films ranged form 1 to 25 µm, depending on the viscosity of the solution and the films were approximately 40% porous. The relative permittivity of these samples was approximately 2.2.

Sputter Deposited SiO_2 Films

Reactive RF Sputtering was used to generate porous SiO_2 films. The dependence of deposition rate and porosity of the films on the partial pressure of O_2 and argon was used to tailor the morphology of SiO_2 films.[19] For reactive sputtering a five inch Si wafer was used as the target, and three different gas pressures were selected: 1) $P_{Ar} = 27$ mtorr and $P_{O2} = 3$ mtorr (10%), 2) $P_{Ar} = 37$ mtorr and $P_{O2} = 3$ mtorr (7.5%) and 3) $P_{Ar} = 4.4$ mtorr and $P_{O2} =$ 1 mtorr (2.2%). The dielectric constants of the material produced under each condition are demonstrated below. In all cases the RF power was maintained at 120 watts and the substrate target distance was 3.5 cm. The films were deposited on 2.54 cm diameter low resistivity (0.001-0.0006 ohm-cm) p-type Si wafers. As shown in Fig. 15, as the total pressure in the system is increased, more porosity is introduced thus reducing the permittivity[5,19].

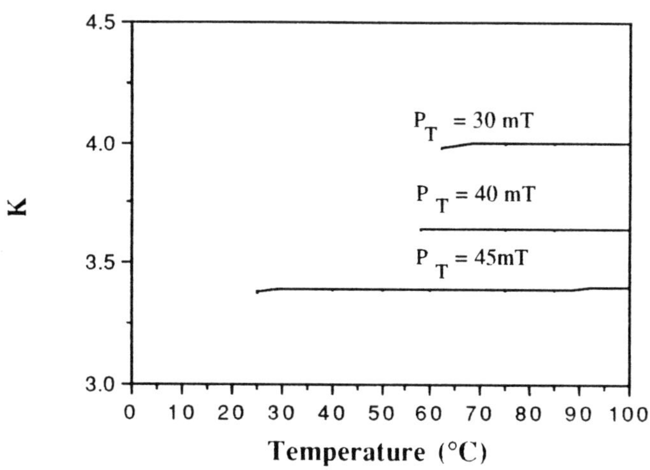

Fig. 11. K vs. temperature of reactively sputtered SiO_2 films (P_T = total gas pressure).

Microporous Glass Structures

Commercially available porous glass of high silica content (Corning 7930) has also been investigated for its low K packaging applications.[20] Dielectric properties of both unaltered samples as well as samples leached with hydrofluoric acid and cleaned with hydrochloric acid were studied. The dielectric behavior of these samples is shown in Fig. 12.

The HF leaching of Porous Vycor, decreased the low frequency dielectric constant. The decrease of K indicated may be attributed to the increase in porosity due to the HF leaching. The 10% HF leaching did not show any reduction in K over the untreated glass which may be caused by the collapse of the silica structure.

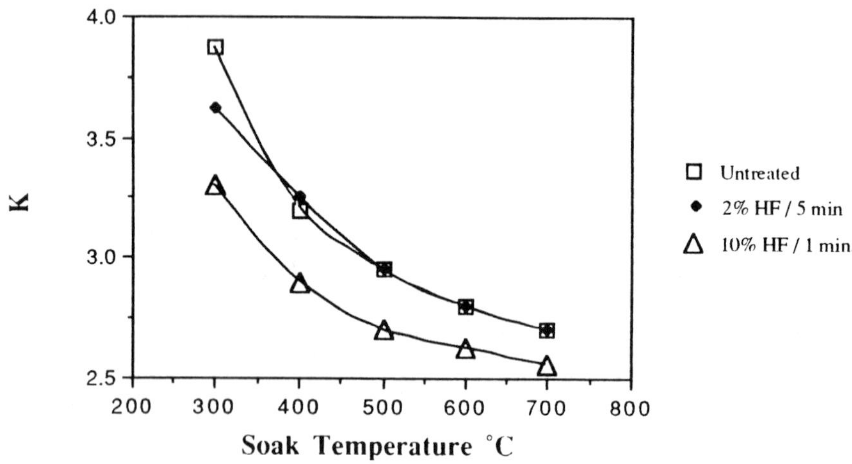

Fig. 12. Dielectric Constant vs. Temperature of Hydrofluoric Acid Soak.

Macro-Defect-Free Cements

The incorporation of hollow glass microspheres into calcium aluminate cements was examined as a technique used to reduce the dielectric constant and loss[21]. Additions of up to 40 vol.% microspheres into SECAR 71 cement paste resulted in a lowering of the dielectric constant from ~4.5 to 2.7. The dielectric constant of the SECAR 71/PVA cement pastes containing various quantities of microspheres was modeled as a combination of Maxwell's and Lichtenecker's mixing rules. The relationship between the volume percent microspheres and dielectric constant is shown below.[21] As the density of the composite decreased, due to additions of microspheres, the dielectric constant decreased.

Table 7. Density, Dielectric Constant of SECAR 71/PVA Composites containing 0,20,30, and 40 Volume Percent Hollow Glass Microspheres.

Volume percent Microspheres	Average Density (g/cm^3)	Dielectric Constant (1 MHz)
0	2.01	4.36
20	1.67	3.42
30	1.54	3.25
40	1.32	2.71

Ceramic / Polymer Composites

Polyimide-silica nanocomposite thick films[22] and planarized coatings[23] have been studied for their potential packaging applications. Thick film polyimide-silica nanocomposites were processed by mixing silica with tetraglyme to wet the powder. The fluorinated polyimide was also dissolved in tetraglyme and heated to 80°C, to increase the dissolution rate. The two components were combined by adding small amounts of the silica-tetraglyme slurry, which was constantly mixed. Resultant compositions after solvent removal ranged from 0 to 30 weight % silica and 0 to 21 weight % silica for both micrometer and nanometer size fillers, respectively. The slurries were then three roll milled and cast onto glass sheets using an aluminum stencil and screen printer and cured.

The dielectric constant of the polyimide increased slightly with the silica addition, K=2.66 to K(after silica addition)~ 2.9. The 30 weight % silica addition showed a dramatic decrease in the dielectric constant K=2.0, due to the presence of porosity introduced into the samples during processing. The silica filler gave approximately a 30% reduction in thermal expansion.

In a similar study[23] conventional 96% alumina substrates were spin coated with polyimide and polyimide / silica films in an attempt to study a composite substrate with a planarized surface texture, a dielectric constant below 3, and a thermal expansion coefficient close to 5.4 ppm/°C. The dielectric constant of the substrate material was lowered by planarizing a conventional 96% Al_2O_3 substrate with a low dielectric constant polyimide and polyimide / silica composite films. The surface roughness was also decreased, which is important for good adhesion of the substrate metallization.

The 91 and 100 volume % films produced smoother surfaces than the original alumina substrate surface. The 94 and 97 volume % films had larger surface roughness because they contained polyimide rich regions surrounding silica island. The 100 volume % film had K=2.4, and the 91 volume % film had K=3.0.

Table 8. Surface Roughness of Planarized Coatings.

Material	Average Surface Roughness
96% Alumina Substrate (uncoated)	1.0 μm
100 vol. % polyimide coating	0.1 μm
97 vol. % polyimide coating	0.6 μm
94 vol. % polyimide coating	0.8 μm
91 vol % polyimide coating	0.2 μm

Ceramic-Fiber Composites

Al_2O_3 continuous-fiber laminates with a polyimide matrix, called Avimid K composites have been fabricated from fiber prepegs made of aligned fibers in a binder resin.[24] The prepegs, or plys, are oriented either unidirectionally for an aligned composite, or staggered to yield quasi-isotropic behavior in plane. The laminate is then cured at an elevated temperature.

The orientation of the fibers in the matrix can be used to anisotropically alter the dielectric properties of the composite. The dielectric constant as a function of volume % fiber loading in both parallel and transverse orientations is shown in Fig. 13[24]. For unidirectionally oriented fibers the out-of-plane dielectric constant, which determines the signal propagation delay, is depressed due to the lack of connectivity in the z-direction. The same behavior is observed in-plane, perpendicular to the fibers. This anisotropy in the composite's properties, arising from the morphology of the composite provides an opportunity to the design composites with desired behavior.

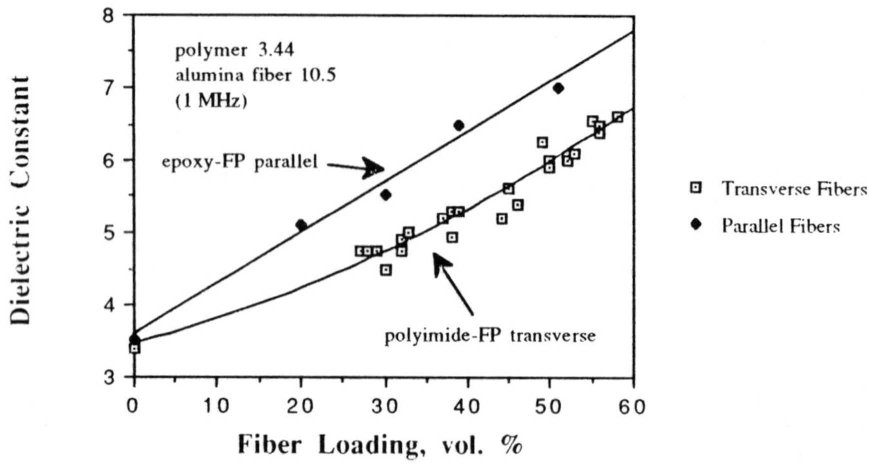

Fig.13.K vs. vol % fibers for parallel and transverse fibers in ceramic-fiber composites.[24]

SUMMARY

A variety of methods of lowering the dielectric constant of various materials have been discussed. Density and the percent inherent porosity have a direct correlation to the dielectric constant. The polarizability, type of bonding, and bond character also have an influential role in single phase materials. The dielectric constant was significantly lowered in porous silica composites by lowering the density. The incorporation of glass microspheres into cements also lowered K. New developments in ceramic fiber composites prove very instrumental in tailoring the properties of dielectrics through various orientations of the fibers in the matrix. A list of the properties of common packaging materials is shown in Table 9[20].

Table 9. Properties of Common Substrate Materials.

Material	Thermal * Expansion (10^{-6}/°C)	Thermal Conductivity (cal/cm sec °C)	Density (gm/cm^3)	Dielectric Constant
96%Al_2O_3	6.4	0.50	3.7	10
BeO	6.1	0.50	2.9	7.16
Sapphire	5.8	0.098	3.98	9.4-11.5
Cordierite	2.3	0.002	2.5	4.1-5.4
Si_3N_4	3.0	0.08		6.0
Zircon		0.012		7.8
AlN	2.7	0.37	3.26	10
MgO	12.8	0.09	3.6	8.2
$MgAl_2O_4$	6.6	0.02	2.8	7.5
Steatite	7.8-10.4	0.008	2.8	5.9-6.1
Mullite	4.3-5.0	0.006	2.5	6.2-6.8
Forsterite	10.6	0.004-0.01	2.8	5.8-6.7
Porcelain	3.5-5.5	0.003-0.017	2.3-2.6	6
Fotoceram	10.4	0.004	2.46	5.6
Fused SiO_2	5.6	0.0034	2.20	3.9
SiC (α)	5.1			6.7
Polyimide/quartz	6-8	0.0005		3.4
Epoxy/quartz	5.0	0.0004		3.6
Polyimide/Kevlar	5.6-5.8	0.0003		3.5-3.6
Teflon/Fiberglass	8	0.0006		2.5-3.5
Thermoplastics	20	0.0004		2.8-3.2

* Si: $\alpha = 3 \times 10^{-6}$; GaAs: $\alpha = 6 \times 10^{-6}$

References

[1] R.R. Tummala, "Ceramic and Glass -Ceramic Packaging in the 1990's,"*J. Am. Ceram. Soc.*, **74** [5] 895-908 (1991).

[2] K. Kata, Y. Shimada and H Takamizawa, "Low Dielectric Constant New Materials for Multilayer Ceramic Substrate,"IEEE/CHMT 1989 Japan IEMT Symposium pp.24-27.

[3] R.R. Tummala and E.J. Rymaszewski, *Microelectronic Packaging Handbook*: pp.25-63. Van Nostrand-Reinhold, New York, 1989.

[4] B.K. Gilbert, D.J. Schwab, and M.L. Samson, "Packaging and Interconnection Problems and Solutions in High Clock Rate Digital Processors," Proc. 1985, VHSIC Packaging Workshop, Naval Surface Weapons Center, Silver Springs, MD, June 5-6, 1985.

[5] L.E. Cross and T.R. Guraraja, "Ultra-Low Dielectric Permittivity Ceramics and Composites for Packaging Applications," *Mater. Res. Soc. Symp. Proc.*, **72**, 53-65 (1986).

[6] N. Kamehara-Fujitsu Lab Ltd.; private communication.

[7] R.E. Newnham, *Structure-Property Relations* ; pp. 78-79. Springer-Verlag, New York, 1975.

[8] N.F. Mott and R.W. Gurney, *Electronic Processes in Ionic Crystals* ; p. 11-20. Oxford Press, London, 1940.

[9] R.D. Shannon," Factors Determining the Dielectric Constants of Oxides and Fluorides," National Institute of Standards and Technology Special Publication 804, Chemistry of Electronic Ceramic Materials, Proceedings of the International Conference held in Jackson, WY, August 17-22, 1990, issued January 1991.

[10] M.L. Huggins and K.H. Sun,"Calculation of Density and Optical Constants of a Glass from its Composition in Weight Percentage,"*J. Am. Ceram. Soc.*, **26** [1] 4-11 (1943).

[11] G.R. Olhoeft, *Physical Properties of Rocks and Minerals* ; pp. 257-329, McGraw-Hill, New York, 1981.

[12] A.R. Von Hippel, *Dielectric Materials and Applications* ; p.5. The Technology Press of M.I.T. and J. Wiley & Sons, Inc., New York, 1954.

[13] R. Newnham, and Y. DeHaan, Z. Krist **117**, 235 (1962).

[14] J.L. Sprague ,"Multilayer Ceramic Packaging Alternatives," IEEE Transactions on Components, Hybrids, and Manufacturing Technology, Vol. 13, no.2, June 1990.

[15] U. Chowdhry & A. Sleight, *Ann. Rev. Mat. Sci.* **17**,pp. 323-340, 1987

[16] B.T. Merriman, J.D. Craig, A.E. Nader, D.L. Hoff, M.T. Pottiger, and W.L. Lautenberger, "New low CTE polyimide for inorganic substrates," 39th ECC 1989 Proc., 1989, pp. 155-159.

[17] T.G. Tessier, G.M. Adena, and I. Turlik, "Polymer dielectric options for thin film packaging applications," in 39th ECC 1989 Proc., 1989, pp. 148-154.

[18] A.A. da Silva, D.I. dos Santos and M.A. Aegerter, "Dielectric Response of Silica Aerogels," *J. Non-Cryst. Solids* 95-96 [2], 1987, pp. 1159-1166.

[19] A.P. Giri, "Sputter Deposited Silica Films as Substrates for Microelectronic Packaging Applications;" Ph.D. Thesis. The Pennsylvania State University (1984).

[20] J. K. Yamamoto,"Dielectric Properties of Porous Glass in the Microwave Region;" M.S. Thesis. The Pennsylvania State University (1986).

[21] P. Sliva, "The Development and Processing of Calcium Aluminate Cement as a Low Relative Dielectric Permittivity Material;" Ph.D. Thesis. The Pennsylvania State University (1988).

[22] D. Reed,"Silica-Polyimide Composites for Microelectronic Packaging;" M.S. Thesis. The Pennsylvania State University (1991).

[23] M. Mulvihill,"Low Dielectric Constant Planarization Coatings for Microelectronic Packaging;" M.S. Thesis. The Pennsylvania State University (1991).

[24] J.D. Bolt and R.F. French,"Polymer/Ceramic Composites: New Properties For Electronics",*Adv. Mat. & Proc.* 7, 1988, pp. 32-35.

PROCESSING AND PROPERTIES OF GLASS/CERAMIC COMPOSITES WITH LOW DIELECTRIC CONSTANT PREPARED FROM MICROCOMPOSITE PARTICLES

M.D. Sacks, R. Raghunathan, I.Y. Park, and G.W. Scheiffele, Department of Materials Science and Engineering, University of Florida, Gainesville, FL 32611

ABSTRACT

Glass/ceramic composites with low dielectric constant and low loss tangent were fabricated using microcomposite particles which consisted of inner cores of a crystalline phase (α-Si_3N_4 or α-Al_2O_3) and outer coatings of a glassy phase (SiO_2 or B_2O_3-SiO_2). Powder compacts prepared with microcomposite particles (SiO_2/Si_3N_4) showed enhanced densification compared to conventionally-processed compacts prepared with a mixture of glass and crystalline particles (SiO_2 + Si_3N_4). The sintering temperatures required for densification of compacts prepared with microcomposite particles decreased significantly as the concentration of B_2O_3 in the siliceous coating increased. This was attributed to the decreased viscosity of the coating as the B_2O_3 content increased. SiO_2/Si_3N_4 composites with ~9-10 vol% closed, isolated, spheroidal pores (~2 μm dia.) were prepared by using a mixture of microcomposite particles and polystyrene particles. The incorporation of microporosity resulted in a lower dielectric constant (~4.4 at 1 MHz) compared to a fully dense sample (~4.9 at 1 MHz) with the same SiO_2/Si_3N_4 volume ratio (~60/40). The samples had the same loss tangent (2×10^{-4}) and average flexural strength (183 MPa) values. SEM observations of fracture surfaces suggested that the strength values for these samples could be increased by reducing the concentration of large processing-related flaws (e.g., powder agglomerates).

INTRODUCTION

The ceramic materials used in microelectronics packaging applications must have (i) excellent electrical properties (low dielectric constant, low loss tangent, high resistivity), (ii) good thermal properties (e.g., coefficient of thermal expansion, CTE, match with chip and/or metallizing materials), (iii) resistance to corrosion (e.g., due to moisture), and (iv) adequate mechanical properties (strength, fracture toughness) to withstand stresses imposed during fabrication and handling of the package. In recent years, many investigations have been directed toward the development of ceramic materials with lower dielectric constant (in order to increase signal propagation speed) and co-firing compatibility with lower resistivity metals such as copper, silver, etc. (in order to produce packages with higher circuit density). In particular, glass/ceramic compositions have been utilized to achieve both lower dielectric constant and lower sintering temperatures.[1-14] This includes (i) "glass-ceramic" materials in which glass powder compacts are sintered to high relative density (low porosity) and can be subsequently crystallized to nearly glass-free materials and (ii) "glass + ceramic" composites in which powder compacts prepared with mixtures of glass and crystalline ceramic particles are sintered to high relative density. The crystalline components usually have higher dielectric constant than the glassy phases, but they also improve mechanical and thermal properties of the material. A difficulty in processing "glass + ceramic" composites is that powder compact densification rates usually decrease as the volume fraction of crystalline ceramic particles increases.[14,15] Composite densification may be severely impeded when the concentration of "non-sinterable" inclusions is sufficient to form a connected particulate network (i.e., percolation path). Thus, it may be necessary to (i) limit the amount of crystalline phase in the powder mixture and/or (ii) modify the glass composition so that the melt viscosity is lower at the sintering temperature. These approaches have been effective in producing hermetic composites with high relative density, but they impose compositional constraints which limit the range of physical properties that can be achieved.

Another approach for producing "glass + ceramic" composites is based on the processing of glass/ceramic microcomposite particles.[14,16] These particles consist of crystalline cores and outer coatings of amorphous SiO_2. As schematically illustrated in Fig. 1, the controlled spatial distribution of the two phases prevents the formation of a connected network of the non-sinterable

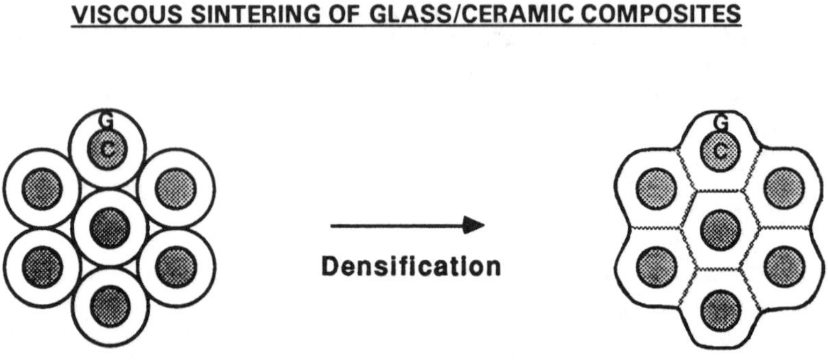

FIG. 1. Schematic illustration of viscous sintering of glass (G)/crystalline ceramic (C) composites using powder compacts prepared with microcomposite particles. The particles have an outer layer of glassy phase and a core of crystalline phase.

crystalline particles in the powder compact. Furthermore, the distribution of the amorphous phase facilitates densification via the viscous flow mechanism. Thus, lower sintering temperatures can be achieved and composites with higher volume fraction of crystalline phase can be produced. In some systems, it is also possible to react the glass and ceramic phases (after densification) to produce a fully crystalline material. The latter process is known as transient viscous sintering (TVS).[16,17]

The present study focuses on the development of glass + ceramic composites using microcomposite particles which consist of (i) crystalline Si_3N_4 or Al_2O_3 cores and (ii) amorphous SiO_2 or borosilicate coatings. A technique for introducing controlled microporosity into the samples is also utilized in order to produce composites with lower dielectric constant.[14]

EXPERIMENTAL

Microcomposite Particles

The preparation of silica-coated microcomposite particles has been described previously.[17] SiO_2 coatings were deposited on core particles using tetraethyl orthosilicate (TEOS)▲ as the starting material. The core particles used in this study were α-Al_2O_3♦ and α-Si_3N_4.◂ The as-received powders were first fractionated using gravitational and centrifugal fluid classification techniques[18] in order to remove hard agglomerates. The fractionated Al_2O_3 and Si_3N_4 powders had median Stokes' diameters in the range ~0.2-0.35 μm, as determined by a centrifugal photosedimentation technique.† The microcomposite particles were prepared by mixing core particle/ethanol suspensions with TEOS/ethanol solutions. SiO_2 was precipitated onto the core particles by adding ammoniated water to the suspension. The final SiO_2/core particle ratio in the microcomposite particles can be varied by adjusting the concentrations of the components in the initial coating suspension.[17] In this study, the composite particles were prepared with approximately 40 vol% crystalline phase and 60 vol% SiO_2.* After precipitation, the suspension of coated particles was filtered and the collected

---

▲ Fisher Scientific, Fair Lawn, NJ.

♦ AKP-50, Sumitomo Chemical America, New York, NY.

◂ SN-E-10, Ube Industries (America), Inc., New York, NY.

† Model CAPA-700, Horiba Instruments, Inc., Irvine, CA.

* This composition is approximate because it is calculated from the ratio of TEOS/crystalline core particles used in the coating suspensions. However, the precipitation of SiO_2 onto the core particles does not occur with 100% yield, i.e., some siliceous species remain in solution. Thus, the crystalline phase concentration in the composite particle may be slightly higher than the calculated value.

powder was washed with deionized water and subsequently dried. Figures 2A and 2B show transmission electron micrographs (TEM)▼ for typical SiO_2/Si_3N_4 and SiO_2/Al_2O_3 composite particles, respectively. It is evident that the SiO_2 coatings are uniformly deposited onto the core particles. X-ray photoelectron spectroscopy§ was used to confirm that the external layer of the particles was SiO_2.[19] X-ray diffraction¶ and high resolution transmission electron microscopy‡ showed that the SiO_2 coatings were amorphous.[19]

Microcomposite particles with borosilicate coatings were formed in the same manner as described above except that precipitation was carried out using mixtures of tributyl borate (TBB)▸ and TEOS. The ratio of B_2O_3/SiO_2 in the coating was altered by varying the TBB/TEOS ratio in the starting solution. Borosilicate/Si_3N_4 particles with ~23 wt% B_2O_3 in the silicate phase and borosilicate/Al_2O_3 particles with ~10 and ~17 wt% B_2O_3 in the silicate phase were prepared.* Since the amount of core particles was kept constant during the coating experiments, the silicate coating/core particle ratio varied with the amount of B_2O_3 added. The volume ratio was ~65/35 for the borosilicate/Si_3N_4 particles, while the ratios were ~62/38 and ~64/36 for the borosilicate/Al_2O_3 particles with ~10 and ~17 wt% B_2O_3, respectively, in the silicate phase.** Figures 2C and 2D show TEM micrographs of borosilicate-coated Al_2O_3 particles.

Green Compact Preparation and Sintering

SiO_2/Si_3N_4 composites were prepared using both silica-coated microcomposite particles and a mixture of SiO_2 and Si_3N_4 particles. In the latter case, the SiO_2 powder was prepared by solution-precipitation using TEOS as the precursor and the Si_3N_4 was the same fractionated powder used in preparation of the silica-coated microcomposite particles. These powders were mixed in aqueous suspension in a ratio of ~60 vol% SiO_2/~40 vol% Si_3N_4 (i.e., the same ratio as

---

▼ Model 200CX, Japan Electron Optics Co. Ltd., Tokyo, Japan.
§ Model XSAM800, Kratos Analytical, Manchester, United Kingdom
¶ Model APD 3720, Philips Electronic Instruments Co., Mt. Vernon, NY.
‡ Model 4000-FX, Japan Electron Optics Co. Ltd., Tokyo, Japan.
▸ Johnson Matthey Alfa Products, Danvers, MA.

* The B_2O_3 concentrations in the silicate phases were determined by inductively coupled plasma (ICP) spectroscopic analysis (Plasma II Model 5800, Perkin Elmer Corp., Norwalk, CT).

** The silicate coating/core particle ratios are approximations calculated from the ratios of components in the coating solutions and estimated true densities of the borosilicate phase.

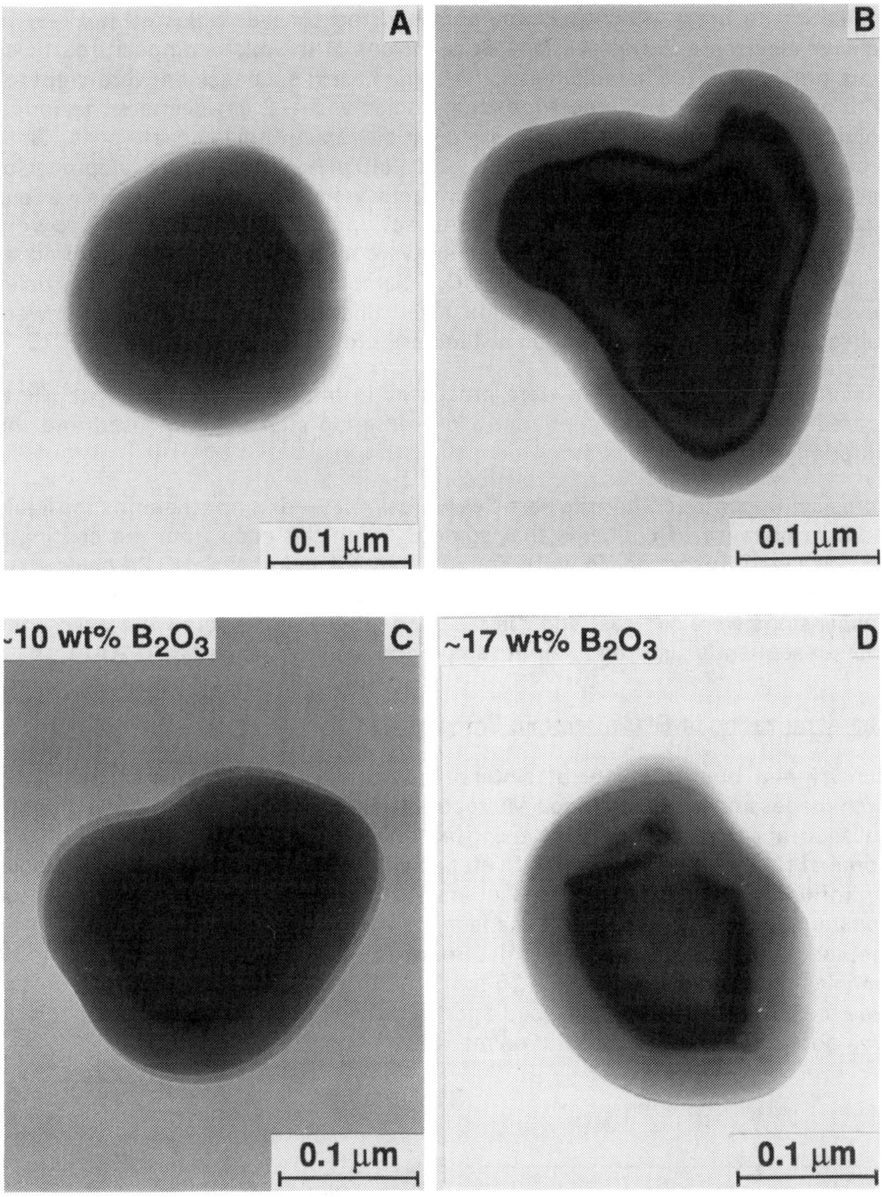

FIG. 2. TEM micrographs of microcomposite particles: (A) SiO_2/Si_3N_4, (B) SiO_2/Al_2O_3, (C) borosilicate (~10 wt% B_2O_3)/Al_2O_3, and (D) borosilicate (~17 wt% B_2O_3)/Al_2O_3.

in the microcomposite particles). The suspension was subjected to ultrasonication to break down agglomerates and the pH was adjusted (to ~9) to achieve electrostatic stabilization. Suspensions of the microcomposite particles were prepared in the same manner. In some cases, suspensions were prepared with mixtures of the microcomposite particles and ~2 μm diameter, spherical polystyrene (PS) particles (synthesized by dispersion polymerization[20]). Both types of particles dispersed well in water (at pH≈9) due to development of negative surface charges. Green compacts were prepared from the various suspensions by slip casting into preforms set upon plaster blocks. The powder compacts (disks and rectangular bars) were oven-dried and pre-sintered at temperatures in the range 600-800°C. Samples prepared with PS particles developed ~2 μm spheroidal voids after this treatment. Compacts were subsequently sintered in nitrogen at temperatures in the range 1150-1300°C.

Silica-coated Al_2O_3 particles were processed in the same manner as the silica-coated Si_3N_4 particles except that a pre-sintering step was not used and the sintering atmosphere was air.

Borosilicate-coated composite particles were processed using ethanol as the liquid medium because of concerns that some B_2O_3 might leach from the coating if particles were suspended in water for an extended period of time. Particles were mixed with polymeric dispersants and ultrasonicated to break down agglomerates. Suspensions were slip cast and the resulting powder compacts were oven-dried and subsequently sintered in air at temperatures in the range 600-1200°C.

Characterization of Glass/Ceramic Composites

Density and open porosity of sintered compacts were determined using the Archimedes immersion method. Microstructure observations were made on both polished and fracture surfaces using scanning electron microscopy (SEM).▲▲ Some of the polished samples were etched with a dilute (~0.5 - 2 wt%) aqueous hydrofluoric acid solution. Measurements of dielectric properties (dielectric constant and loss tangent) were made at 1 MHz using a non-contacting electrode method.◆◆ Average strength was determined by four-point bending tests on 8-10 samples having dimensions of ~25 mm x ~5 mm x ~2 mm. Vickers hardness was measured using a 500 g load and 20 s dwell time.◄◄ Fracture toughness was determined by the indentation method of Anstis et al.[21]

▲▲ Model JSM-6400, Japan Electron Optics Co., Ltd., Tokyo, Japan.

◆◆ HP 4192 Impedance Analyzer and HP 16451B Dielectric Test Fixture, Hewlett-Packard Co., Palo Alto, CA.

◄◄ Micromet 3, Buehler, Ltd., Lake Bluff, IL.

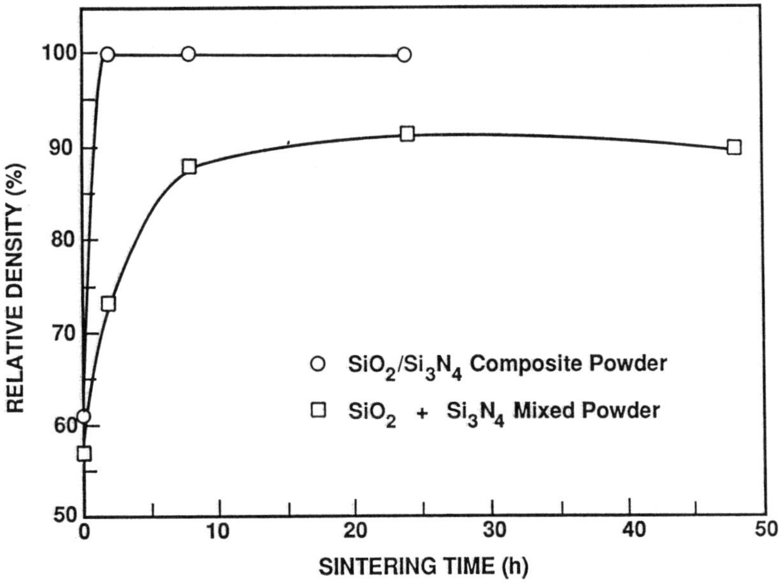

FIG. 3. Plots of relative density vs. sintering time at 1300°C for two types of ~60 vol% SiO_2/~40 vol% Si_3N_4 powder compacts. Compacts were prepared from (i) a mixture of SiO_2 particles and Si_3N_4 particles and (ii) amorphous silica-coated Si_3N_4 particles.

RESULTS AND DISCUSSION

Silica/Silicon Nitride Composites

Densification behavior was directly compared for powder compacts prepared with the silica-coated Si_3N_4 composite particles and the mixture of SiO_2 particles and Si_3N_4 particles.* Figure 3 shows plots of relative density vs. sintering time for ~60 vol% SiO_2/~40 vol% Si_3N_4 samples. The sample prepared with the microcomposite particles reached nearly 100% relative density after heat treatment for 2 h at 1300°C. In contrast, the sample prepared with the powder

---

* An effort was made to maintain similar particle sizes and green compact densities for the two types of samples so that the spatial distribution of the phases (SiO_2 and Si_3N_4) was the key variable influencing densification behavior. The composite particles were slightly larger (median Stokes' diameter of ~0.3 μm compared to a value of ~0.25 μm for the particle mixture), while the green compact prepared with them had slightly higher packing density (~61% vs. ~57%).

mixture reached only ~73% relative density under the same sintering conditions. Even with prolonged sintering times, the maximum relative density achieved was only ~90%. This result suggests that densification was inhibited due to the formation of a rigid network of "non-sinterable" Si_3N_4 particles. The Si_3N_4 concentration (~40 vol%) significantly exceeds the percolation threshold value for a random mixture of particles.[22] Si_3N_4 particle-particle contacts were evident in the polished microstructure of the sintered (1300°C, 2 h) powder mixture sample (Fig. 4A). In contrast, the Si_3N_4 particles were dispersed well in the SiO_2 matrix of the corresponding sample prepared with the microcomposite particles (Fig. 4B). Microstructure observations also showed that the sample was essentially pore-free.

Green compacts prepared with a mixture of ~88 vol% silica-coated Si_3N_4 particles and ~12 vol% PS particles were sintered at 1300°C (0.4-0.5 h). This resulted in shrinkage and elimination of the fine interparticle pores associated with the packing of the microcomposite particles. In contrast, the larger (~2 μm) spheroidal voids created by pyrolysis of the PS particles underwent relatively little shrinkage at this temperature. However, as densification of the SiO_2/Si_3N_4 matrix occurred, the spheroidal voids were closed off from the sample surface. Figure 5A shows a fracture surface of the sintered sample which illustrates that pores are closed and mostly isolated. A higher magnification micrograph of a polished and etched surface (Fig. 5B) shows that SiO_2/Si_3N_4 matrix is almost fully dense and that the Si_3N_4 particles are well-dispersed in the siliceous glass phase.

Table I lists dielectric properties for the ~60 vol% SiO_2/~40 vol% Si_3N_4 samples that were prepared with the microcomposite particles. A fully dense sample (sintered at 1300°C for 2 h) had a dielectric constant of 4.9 and a loss tangent of 2×10^{-4} (at 1 MHz). As expected, the incorporation of ~10% closed pores resulted in lower dielectric constant (4.4 at 1 MHz). The loss tangent was unchanged (2×10^{-4}).

Table I also lists mechanical properties for the SiO_2/Si_3N_4 composites with and without closed porosity. The fully dense sample had an average flexural strength of 183 MPa, average Vickers hardness of 8.7 GPa, and average fracture toughness of 2.3 MPa•m$^{1/2}$. These properties are significantly better than values reported for silicate glasses[10,23,24] and are comparable to values reported for various silicate glass/crystalline ceramic composites and crystallized glass-ceramics (e.g., cordierite) with low dielectric constant.[3-13] The incorporation of ~9% closed porosity in the SiO_2/Si_3N_4 composites resulted in a lower average hardness (6.5 GPa), but the average flexural strength remained the same (183 MPa).* The latter result indicates that the closed, spheroidal pores are not the

* The fracture toughness could not be determined on the porous sample because the cracks generated during indentation were not well-defined.

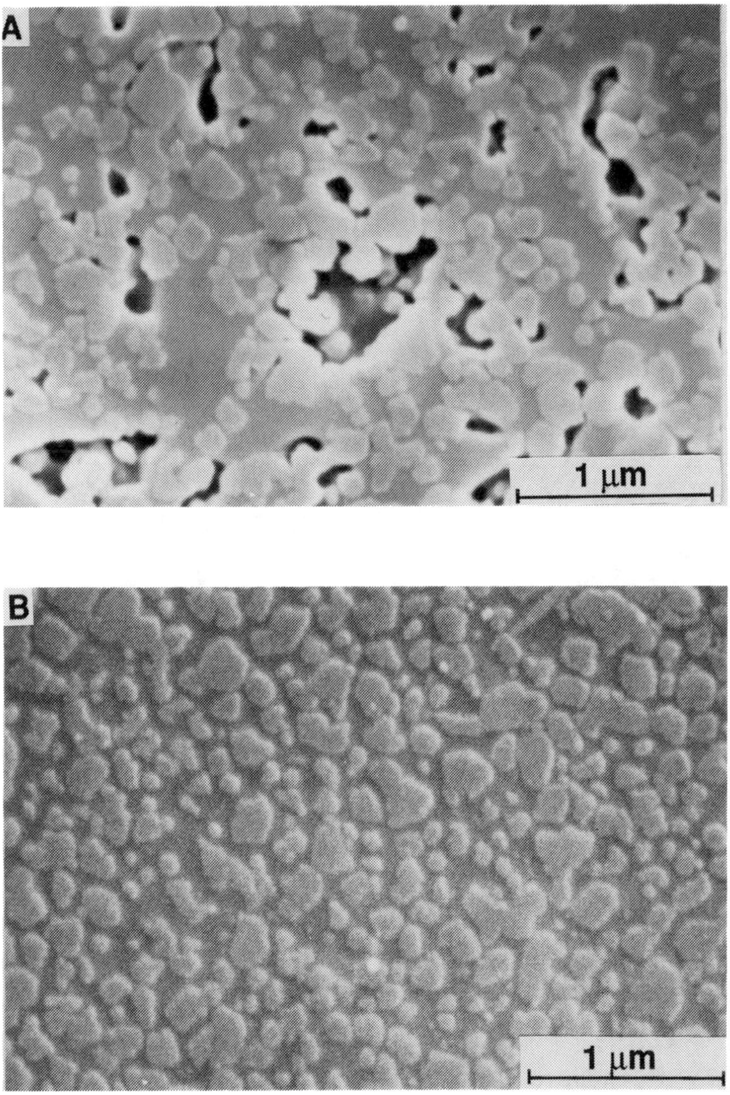

FIG. 4. SEM micrographs of polished and etched SiO_2/Si_3N_4 samples (sintered at 1300°C, 2h) that were prepared from the (A) mixture of two types of particles and (B) microcomposite particles.

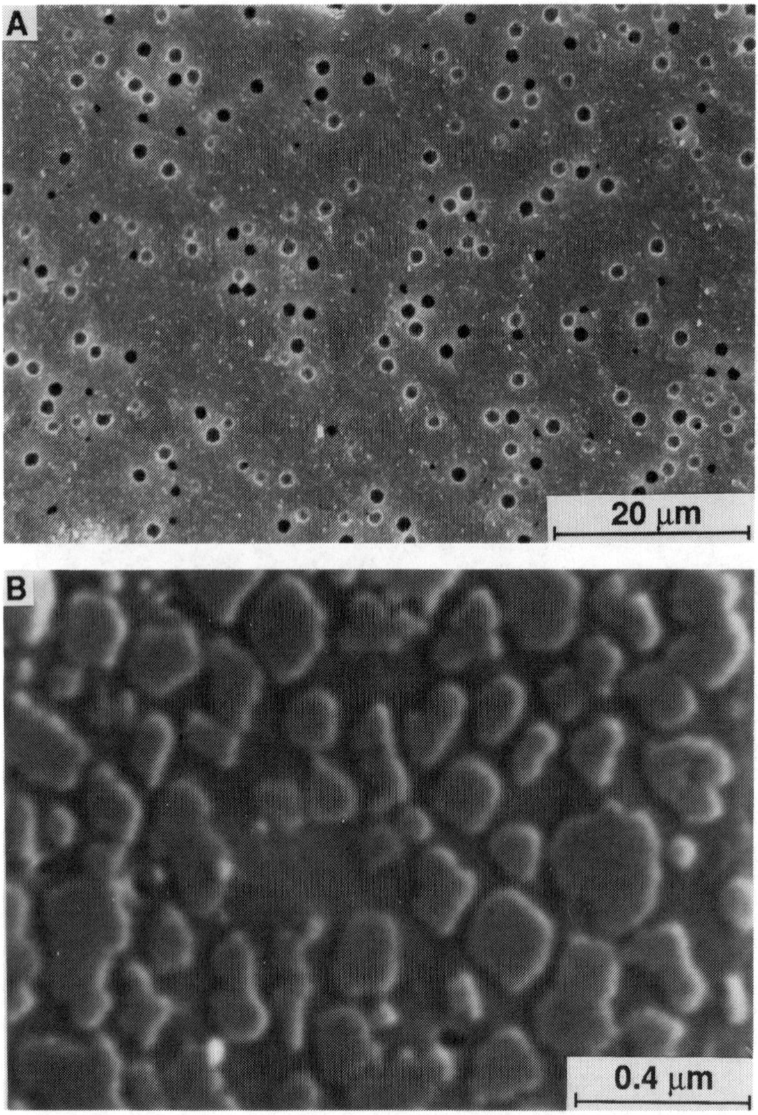

FIG. 5. SEM micrographs of SiO_2/Si_3N_4 sample (sintered at 1300°C, 2 h) prepared from a mixture of microcomposite particles and PS particles. (A) Lower magnification micrograph of a fracture surface shows the closed, spheroidal pores. (B) Higher magnification micrograph of a polished and etched surface shows the dense matrix in which the Si_3N_4 particles are dispersed in the siliceous glass phase.

Table I. Properties of 60 vol% Silica/40 vol% Silicon Nitride Samples

	Zero Porosity	Microporous*
Flexural Strength (MPa)	183 (±33)	183 (±43)
Vickers Hardness (GPa)	8.7 (±0.2)	6.5 (±0.1)
Fracture Toughness (MPa•m$^{1/2}$)	2.3 (±0.2)	
Dielectric Constant (at 1 MHz)	4.9	4.4
Loss Tangent (at 1 MHz)	2 x 10^{-4}	2 x 10^{-4}

* The samples used in the mechanical and dielectric property measurements had ~9% and ~10% closed porosity, respectively.

FIG. 6. SEM micrograph of a fracture surface of a SiO_2/Si_3N_4 sample (sintered at 1300°C, 2 h) showing a large processing defect (powder agglomerate).

strength-limiting flaws in the SiO_2/Si_3N_4 composites. Preliminary observations of fracture surfaces suggests that the strength may be limited by uncontrolled processing defects, e.g., powder agglomerates which were not broken down during processing. Figure 6 shows an example of a sintered powder agglomerate which apparently initiated fracture in one of the pore-free samples.

Borosilicate/Alumina and Borosilicate/Silicon Nitride Composites

Figure 7A shows a plot of open porosity vs. sintering temperature for compacts prepared with coated Al_2O_3 particles containing 0, ~10, and ~17 wt% B_2O_3 additions to the siliceous phase. The sintering temperatures decreased dramatically as the B_2O_3 content increased. The sample with ~17 wt% B_2O_3 in the siliceous coating reached nearly zero open porosity at only 950°C, i.e., ~350°C lower than the sample with no B_2O_3 addition. The enhanced densification is attributed to the much lower viscosity of the borosilicate glass. It was noted in the experimental section that the Al_2O_3/silicate glass ratio in these samples decreased slightly with increasing amount of B_2O_3 addition. In conventional sintering of glass/ceramic particle mixtures, a decrease in the crystalline phase content would be expected to increase the densification rate. In the present study, however, the change in Al_2O_3 content should have minimal effect on the densification kinetics because the process is dominated by viscous flow of the amorphous siliceous coating. This was confirmed by the observation that the densification behavior for the sample prepared with no B_2O_3 (~59 vol% SiO_2/~41 vol% Al_2O_3) was virtually the same as obtained in a previous study[17] in which the silica-coated composite particles had much higher Al_2O_3 contents, i.e., ~52-73 vol%.

Figure 7B shows a plot of open porosity vs. sintering temperature for the compacts prepared with the ~60 vol% SiO_2/~40 vol% Si_3N_4 composite particles and ~65 vol% borosilicate/~35 vol% Si_3N_4 particles. Due to the high B_2O_3 content (~23 wt%) in the silicate coating, the latter sample reaches nearly zero open porosity at only ~750°C, i.e., ~500°C lower than the sample prepared with no B_2O_3. Again, this is attributed to the low viscosity of the silicate glass phase. (The difference in crystalline phase content should have minimal effect on the densification rate because sintering is dominated by viscous flow of the amorphous phase.) Figure 8 shows an SEM micrograph of a polished and etched surface of a sintered (800°C) borosilicate/Si_3N_4 sample. The sample shows no porosity and the fine silicon nitride particles are dispersed in the borosilicate glass matrix. The dielectric constant and loss tangent for this sample were 4.4 and 1 x 10^{-4}, respectively, at 1 MHz.

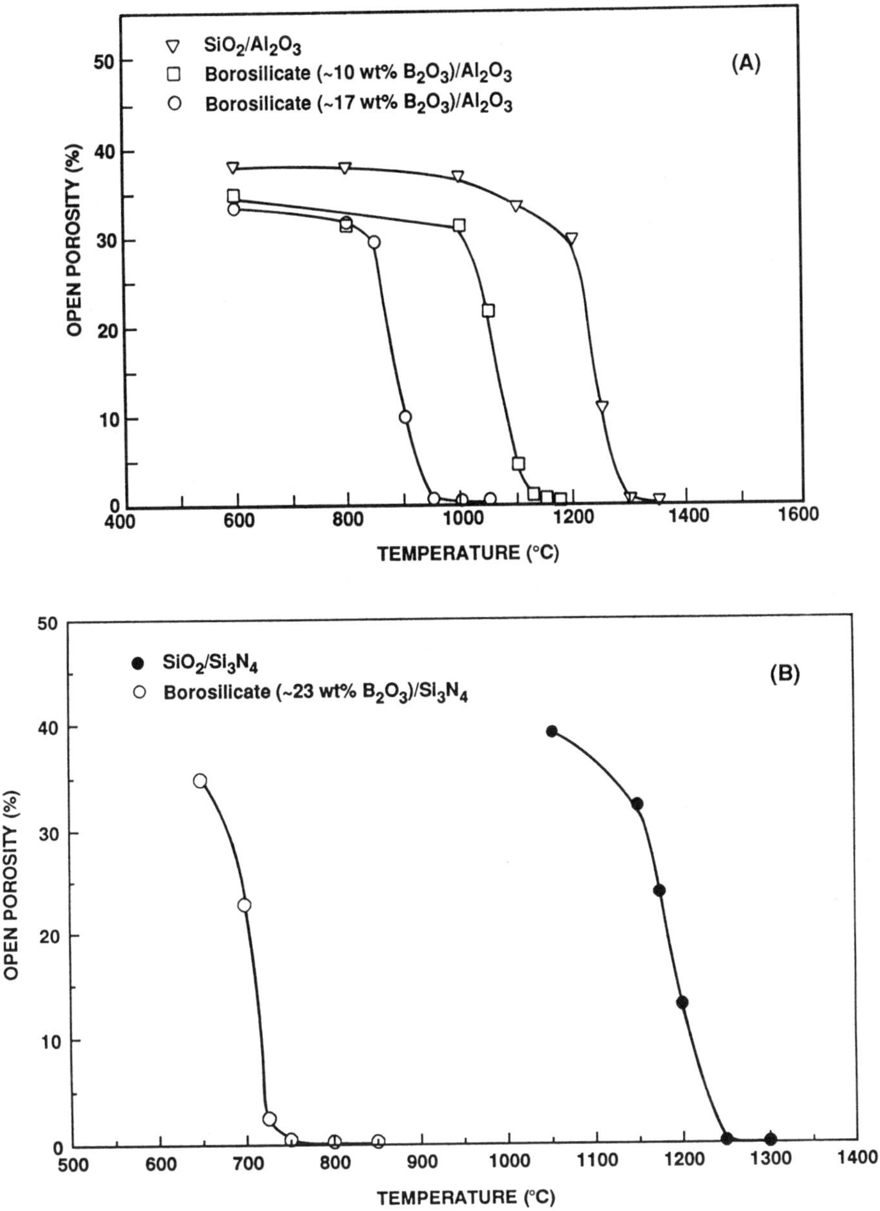

FIG. 7. Plots of open porosity vs. sintering temperature for compacts prepared with (A) silica-coated Al_2O_3 particles and borosilicate-coated Al_2O_3 particles and (B) silica-coated Si_3N_4 particles and borosilicate-coated Si_3N_4 particles.

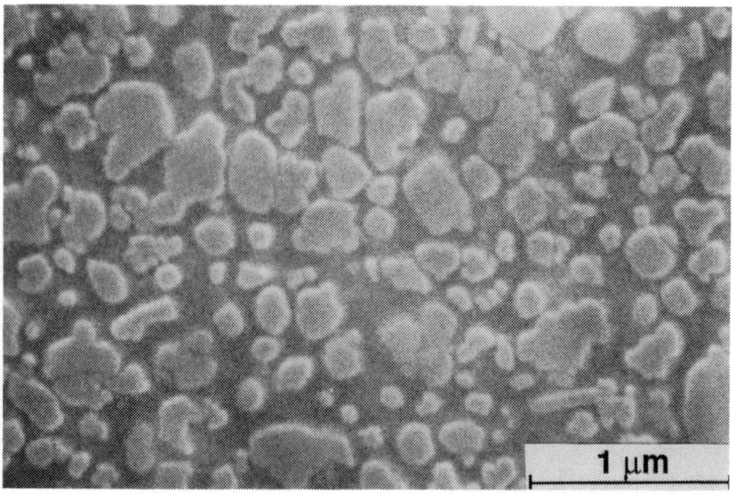

FIG. 8. SEM micrograph of a polished borosilicate/Si_3N_4 sample (sintered at 800°C, 2 h).

CONCLUSION

A novel process was developed in which bulk glass/ceramic composites were fabricated using microcomposite particles consisting of inner cores of a crystalline phase (Al_2O_3 or Si_3N_4) and outer coatings of a glassy phase (SiO_2 or B_2O_3-SiO_2). Powder compacts prepared with these microcomposite particles showed enhanced densification. This was demonstrated by directly comparing the sintering behavior of compacts prepared with silica-coated Si_3N_4 composite particles vs. a mixture of SiO_2 + Si_3N_4 particles; a composite powder compact with ~60 vol% SiO_2/~40 vol% Si_3N_4 reached almost full density within 2 hours at 1300°C, while a mixed powder compact had only ~90% relative density after 48 h at 1300°C. The increased densification rate for the former sample was attributed to two factors: (i) densification occurred by viscous flow of the amorphous coating of the microcomposite particles and (ii) the formation of a connected network of the non-sinterable inclusions was avoided by encapsulating the crystalline particles with the siliceous phase.

Samples with ~60 vol% SiO_2/~40 vol% Si_3N_4 and ~9-10 vol% closed, isolated, spheroidal pores (~2 µm dia.) were prepared by using a mixture of microcomposite particles and polystyrene particles. The incorporation of microporosity resulted in a lower dielectric constant (~4.4 at 1 MHz) compared to a fully dense sample (~4.9 at 1 MHz) with the same SiO_2/Si_3N_4 ratio. Both samples had low loss tangent values ($\sim 2 \times 10^{-4}$). The average flexural strengths were also the same (183 MPa), but the porous sample had lower Vickers hardness (8.7 vs. 6.5 GPa). The mechanical and dielectric properties for these composites are similar to or better than reported values for many other glass + ceramic and glass-ceramic systems of interest for microelectronics packaging applications.[1-13] It may be possible to incorporate a larger amount of closed pores into the samples (and thereby lower the dielectric constant even further) by optimization of the powder characteristics and processing conditions (e.g., the size distributions for the microcomposite particles and the PS particles, sintering conditions, etc.). It may also be possible to obtain higher strength levels in these samples by reducing the concentration of large processing-related flaws (e.g., powder agglomerates).

Glass/ceramic composites were also prepared using borosilicate-coated microcomposite particles. Powder compact sintering temperatures decreased significantly as the concentration of B_2O_3 in the siliceous coating increased. This was attributed to the decreased viscosity of the coating as the B_2O_3 content increased. A sample with $\sim$23 wt% B_2O_3 in the coating and an overall borosilicate glass/Si_3N_4 volume ratio of $\sim$65/35 was sintered to almost full density at only $\sim$800°C. The sample had dielectric constant and loss tangent values of 4.4 and 1×10^{-4}, respectively, at 1 MHz.

In summary, silica-coated and borosilicate-coated microcomposite particles were used to fabricate bulk glass/ceramic composites with good dielectric and mechanical properties. It should be possible to produce other bulk composites with a much broader range of properties (dielectric, mechanical, thermal, etc.) by preparing microcomposite particles with different chemical compositions of the core and coating phases.

ACKNOWLEDGMENT

Experimental contributions from M.M. Amini, C.S. Khadilkar, Y.J. Lin, D. Parker, and A. Ulicny are gratefully acknowledged. This work was supported by the Defense Advanced Research Projects Agency (MDA972-88-J-1006 and N0014-91-J-4075); National Science Foundation, Division of Materials Research, Ceramics and Electronic Materials Program (DMR-8451916); Florida High Technology and Industry Council; and IBM Corp.

REFERENCES

1. A.H. Kumar, P.W. McMillan, and R.R. Tummala, "Glass-Ceramic Structures and Sintered Multilayer Substrates Thereof with Circuit Patterns of Gold, Silver, and Copper," U.S. Pat. No. 4 301 324, 1981.

2. N. Kamehana, K. Niwa, and K. Murakawa, "Packaging Material for High Speed Computers," pp. 388-392 in IEEE 33rd Electronics Components Conference Proceedings, IEEE, Washington, DC, 1983.

3. Y. Shimada, K. Utsumi, M. Suzuki, H. Takamizawa, M. Nitta, and T. Witari, "Low Firing Temperature Multilayer Glass-Ceramic Substrate," IEEE Trans. Components, Hybrids, Manufacturing Technol., 6 [4] 382-388 (1983).

4. S. Nishigaki, S. Yano, J. Fukuta, M. Fukuya, and T. Fuwa, "A New Multilayered, Low-Temperature-Fireable Ceramic Substrate," pp. 225-234 in Proceedings of the 1986 International Symp. on Microelectronics, ISHM, Reston, VA, 1985.

5. K. Kondo, M. Okuyama, and Y. Shibata, "Low Firing Temperature Ceramic Materials for Multilayer Ceramic Substrates," pp. 77-87 in Advances in Ceramics, Vol. 19. Edited by J.B. Blum and W.R. Cannon. American Ceramic Society, Westerville, OH, 1986.

6. K. Kawakami, M. Takabatake, T. Minowa, J. Chiba, and M. Sasaki, "A Low-Temperature Cofired Multilayer Ceramic Substrate," ibid., pp. 95-102.

7. Y. Shimada, Y. Yamashita, Y. Shiozawa, M. Suzuki, and H. Takamizawa, "Low Dielectric Constant Multilayer Glass-Ceramic Substrate with Ag-Pd Wiring for VLSI Package," pp. 398-405 in IEEE 37th Electronics Components Conference Proceedings, IEEE, Washington, DC, 1987.

8. H. Emura, K. Onituka, and H. Maruyama, "Multilayered Ceramic Substrates with Low Dielectric Constants," pp. 375-385 in Advances in Ceramics, Vol. 26. Edited by M.F. Yan, K. Niwa, H.M. O'Bryan, Jr., and W.S. Young. American Ceramic Society, Westerville, OH, 1987.

9. R.R. Tummala, "Ceramics and Glass-Ceramic Packaging in the 1990s," J. Am. Ceram. Soc., 74 [5] 895-908 (1991).

10. K. Kata, Y. Shimada, and H. Takamizawa, "Low Dielectric Constant New Materials for Multilayer Ceramic Substrate," IEEE Trans. Components, Hybrids, Manuf. Technol., 13 [2] 448-453 (1990).

11. J.-Y. Hsu, N-C. Wu, and S-C. Yu, "Characterization of Material for Low-Temperature Sintered Multilayer Ceramic Substrates," J. Am. Ceram. Soc., 72 [10] 1861-1867 (1989).

12. R.N. Master, L.W. Herron, and R.R. Tummala, "Cosintering Process for Glass-Ceramic/Copper Multilayer Ceramic Substrate," IEEE Trans. Components, Hybrids, Manuf. Technol., 14 [4] 780-783 (1991).

13. G. Partridge, C. Elyard, and H.D. Keatman, "Glass Ceramic Materials for Use in Substrate and Packaging Applications," Glass Technol., 30 [6] 215-219 (1989).

14. M.D. Sacks, M.S. Randall, G.W. Scheiffele, R. Raghunathan, and J.H. Simmons, "Processing of Silicate Glass/Silicon Nitride Composites with Controlled Microporosity," pp. 407-420 in Ceramics Transactions, Vol. 19. Edited by M.D. Sacks. American Ceramic Society, Inc., Westerville, OH, 1991.

15. M.N. Rahaman and L. De Jonghe, "Effect of Rigid Inclusions on the Sintering of Glass Powder Compacts," J. Am. Ceram. Soc., 70 [12] C-348 - C-351 (1987).

16. M.D. Sacks, G.W. Scheiffele, N. Bozkurt, and R. Raghunathan, "Fabrication of Ceramics and Composites by Viscous and Transient Viscous Sintering of Composite Particles," pp. 437-455 in Ceramics Transactions, Vol. 22. Edited

by S.-I. Hirano, G.L. Messing, and H. Hausner. American Ceramic Society, Inc., Westerville, OH, 1991.

17. M.D. Sacks, N. Bozkurt, and G.W. Scheiffele, "Fabrication of Mullite and Mullite-Matrix Composites by Transient Viscous Sintering of Composite Powders," J. Am. Ceram. Soc., 74 [10] 2428-2437 (1991).

18. H.-W. Lee "Suspension Processing and Pressureless Sintering of Silicon Carbide Whisker-Reinforced Alumina-Matrix Composites," Ph.D. Dissertation, University of Florida, 1990.

19. M.D. Sacks, G.W. Scheiffele, N. Bozkurt, R. Raghunathan, and Y.J. Lin, unpublished work.

20. Y.Y. Lu, M.S. El-Aasser, and J.W. Vanderhoff, "Dispersion Polymerization of Styrene in Ethanol: Monomer Partitioning Behavior and Locus of Polymerization," J. Polymer Sci., Part B: Polymer Physics, 26, 1187-1203 (1988).

21. Anstis, G.R., Chantikul, P., Lawn, B.R., and Marshall, D.B., "A Critical Evaluation of Indentation Techniques for Measuring Fracture Toughness: I, Direct Crack Measurements", J. Am. Ceram. Soc., 64 [9] 533-38 (1981).

22. R. Zallen. The Physics of Amorphous Solids. Wiley, New York, 1983.

23. L.A. Ketron, "Fiber Optics: The Ultimate Communications Media," Am. Ceram. Soc. Bull., 66 [11] 1571-1578 (1987).

24. L.E. Sanchez and D. Ngo, "Electrical Characterization of Matched Impedance Cofired, High Speed, Microelectronic Packages Made From Low Dielectric Constant, Boron-Doped, Sol-Gel Silica," Int. J. Hybrid Microelectronics, 12 [2] 95-100 (1989).

DENSIFICATION INHIBITOR OF BINARY BOROSILICATE GLASS COMPOSITE

Jau-Ho Jean and Tapan K. Gupta
Alcoa Electronic Packaging, Inc.
Alcoa Center, PA 15069

ABSTRACT

With added alumina, the cristobalite formation in the binary borosilicate glass composite, containing low-softening borosilicate glass (BSG) and high-softening high silica glass (HSG), is known to be completely prevented. In this paper, the effect of alumina on densification kinetics and mechanism of this binary borosilicate glass composite is investigated. A slower and smaller densification with a larger activation energy is observed when a small amount of alumina, 2-10 vol%, is present as a dopant.

I. INTRODUCTION

In the first paper of this series [1], a crystalline phase of cristobalite is found to continuously precipitate out of the initial amorphous binary glass mixture, containing low-softening borosilicate glass (BSG) and high-softening high-silica glass (HSG), at temperatures ranging from 800 to 1200°C. Based upon activation analysis, it is concluded that the transport of alkali ions from the BSG to the HSG controls the crystal growth kinetics and acts as the rate-controlling step. In the second paper of this series [2], we found that by adding a small amount of alumina, 1-10 vol%, into the binary glass mixture, the cristobalite formation is completely inhibited at the same temperature range. The above result has been attributed to a strong coupling reaction between alkali ions from BSG and aluminum ion from alumina, thus causing a diversion in the transport of alkali ions from HSG to alumina. In this paper we investigate the effect of the above coupling reaction on densification kinetics and mechanism of the binary borosilicate glass composite. Effects of alumina content and sintering temperature on densification are also studied.

II. RESULTS

The materials and experimental methods used in this study have been described previously [1] and are not repeated here. The alumina powder used in this investigation has a median size of 0.6 μm. Since the composition of 40 vol% BSG-60 vol% HSG has been used for previous devitrification and densification studies [1], we continue to use the same composition to conduct various experiments that are described in this paper. To reveal the effect of alumina on densification of the binary borosilicate glass composite, only the HSG is replaced with alumina for comparison because the low-softening BSG acts as densification aid during firing.

The densification behavior is described as densification factor versus time, and the densification factor, which represents the porosity removed during densification, is defined by the following equation [3]

$$DF = (D_s - D_g)/(D_{th} - D_g) \quad (1)$$

where D_s is the instantaneous sintered density, D_g the green density, and D_{th} the theoretical density calculated by the mixing rule. Although the green density of the compacts after binder burnout should be used in Eq. (1), there is difficulty in obtaining the reliable densities because of their fragility, and hence we used the density of the as-pressed compacts.

The densification results for three volume fractions of alumina, 2, 5 and 10 vol%, fired at 875°C are shown in Fig. 1, along with that of alumina-free for comparison. Clearly, a slower and smaller densification is observed with alumina present, even as little as 2 vol%, than that of alumina-free. For example, a densification factor of 0.5 is obtained in 7-10 min for the alumina-free samples and in 240 min for those of alumina-doped. It is also noted in Fig. 1 that the alumina-doped samples exhibit two stages of densification; a rapid increase initially, followed by an even slower rise in densification. Moreover, the densification shows little dependence on alumina content from 2 to 10 vol%. The same phenomenon is also observed at other temperatures from 850 to 925°C. Thus, while alumina addition is beneficial in preventing the formation of cristobalite [2] in a mixture of BSG and HSG, it is highly detrimental to the achievement of densification.

Temperature dependence on densification of the alumina-added system is shown in Fig. 2. For clarity of the data, only the results with 2 vol% alumina are shown as an example. As expected, a faster and larger densification is observed at a higher sintering temperature. The same phenomenon is also observed for other alumina contents from 2 to 10 vol%.

III. DISCUSSION

The results shown in Fig. 1 clearly indicate that the densification of binary boroslicate glass composite is degraded in the presence of alumina. It is further noted that even a small amount of alumina, 2 vol%, can dramatically impede the binary borosilicate glass composite from densification. However, the densification shows no dependence on the amount of alumina present from 2 to 10 vol% at temperatures investigated. These results are attributed to a coupling reaction between Al^{+3} from alumina and alkali ions from BSG, as illustrated in Fig. 3. Fig. 3a illustrates a typical SEM micrograph for the sample with 2 vol% alumina fired at 925°C for 240 min, along with elemental mapping of K^+ (Fig. 3b), Al^{+3} (Fig. 3c), and Si^{+4} (Fig. 3d). It is clear from these micrographs that Al^{+3} from alumina and K^+ from BSG are strongly coupled and exhibit the same geometric configuration in the microstructure. A similar phenomenon was also observed in a kinetic study of interfacial reaction between BSG and sapphire substrate [4]. Since the above coupling reaction between BSG and alumina promotes a continuous migration of the alkali ions from BSG to the interface of alumina particles during sintering, it correspondingly depletes the concentration of alkali ions in BSG. Accordingly, the resultant loss of alkali ions from BSG will raise the viscosity of the BSG [5], causing a significant reduction in densification kinetics because the densification is mostly controlled by viscous flow of the low-softening BSG [1] at the sintering temperatures investigated. The little dependence between alumina content and densification in Fig. 1 can be interpreted by recalling the fact [4] that the Al^{+3} and alkali ion-rich reaction layer around the alumina particle in Fig. 3 is formed in a time frame much shorter than that required for viscous-flow controlled densification kinetics. For example, the required time to complete the coupling reaction between alumina and BSG at temperatures from 850 to 925°C is found to increase from 30-140 sec for 10 vol% to 100-400 sec for 2 vol% alumina [2]. However, the time needed to complete the densification of alumina-free samples, determined from densification results reported previously [1], ranges from 10,800 sec at 850°C to 1,000 sec at 925°C. As noted, the calculated reaction times, even for the sample with 2 vol% alumina, are much less than those of viscous-flow controlled densification kinetics, indicating that the alkali ion-rich reaction layer around the alumina particle is formed far too rapidly, compared to the viscous-flow controlled densification kinetics.

Support for the above remarks can also be provided by the activation analysis, in which the controlling kinetics of densification can be determined. Since sintering is a thermally activated process in nature, the activation energy of the rate-controlling step during densification can be determined from an Arrhenius plot. The plots are constructed, based upon Kingery's analysis of liquid-phase sintering [6], as the logarithm of the specific rate ($T\partial DF/\partial t$) at a constant densification factor versus 1/T. The activation energies are then determined from the slopes by a least-squares fit method. Typical results at DF = 0.1-0.35 are shown in Fig. 4 for the system with 2 vol% alumina. As noted, straight lines are observed and the slope of these lines increases with increasing densification factor, suggesting that the apparent activation energy of densification increases with increasing densification factor. This is further evidenced by the results of activation energy as a function of densification factor in curves b-d of Fig. 5 for the system with 2, 5, and 10 vol% alumina, respectively. The activation energy as a function of densification factor for the alumina-free sample [1] was also determined by the same analysis, and is illustrated in curve a of Fig. 5. For comparison, moreover, the activation energies for viscous flow of BSG [7] and HSG [7] are also included in Fig. 5.

To determine the rate-limiting step of densification, let's first review the salient features of densification mechanism of the alumina-free sample in curve a of Fig. 5 as described elsewhere [1], and then compare it with that of the alumina-doped system. No significant change in activation energy ranging from 240 to 250 kJ/mol over the densification factors from 0.1 (relative sintered density = 70%) to 0.7 (relative sintered density = 90%) is observed. These values are very close to the activation energy for the viscous flow of BSG (220-240 kJ/mol), suggesting that the densification kinetics is controlled by viscous flow of pure BSG. For DF > 0.7, however, the activation energy increases gradually with increasing DF from 270 kJ/mol at DF = 0.7 to 380 kJ/mol at DF = 0.85. Since these activation energies are much lower than those potential dissolution- or diffusion-controlled kinetics of HSG in BSG [1], it is concluded that the predominant densification mechanism in these regions is likely to be viscous flow of a changing composition of (reacted) BSG, resulting from the continuous dissolution of HSG in BSG.

As stated earlier, the compositional change in BSG caused by the coupling reaction between Al^{+3} from alumina and alkali ions from BSG forms a reaction layer adjacent to alumina particle. Since the composition of the reaction layer is known to be rich in Al^{+3} and alkali ions (Fig. 3b and 3c), the alkali ions content in BSG is continuously decreased during sintering. Accordingly, the resultant loss of alkali from BSG causes a rise in viscosity of BSG [5], thus slowing down the densification kinetics and increasing the activation energy of densification. Alternatively, it can be argued that the densification is controlled by viscous flow of the amorphous reaction layer around alumina particles.

For the alumina-doped samples, the activation energy is much more sensitive to DF than that of alumina-free samples, and increases dramatically with increasing DF, from 180-220 at DF = 0.1-0.15 (relative sintered densities, 70-72%) to 380-400 kJ/mol at DF = 0.3-0.35 (relative sintered densities, 77-79%). The activation energy, however, remains relatively unchanged with increasing alumina content from 2 to 10 vol%, suggesting that the alternative mechanism as stated above is unlikely to be the rate-controlling step. It follows then that the lower activation energy observed at DF < 0.15 is attributed to viscous flow of pure BSG, and the increased activation energy with DF at DF > 0.15, viscous flow of a changing composition of BSG which is formed by the coupling reaction taking place between alumina and BSG as described above.

IV. CONCLUSIONS

Effect of alumina on densification kinetics and mechanism of binary glass composite, containing low-softening borosilicate (BSG) and high-softening high silica (HSG) glasses, has been studied. Both densification and instantaneous densification rate are significantly reduced, but the activation energy of densification at a given densification factor is dramatically increased with the addition of a small amount of alumina, 2-10 vol%, in the binary glass mixture of BSG and HSG. However, no significant difference in densification with increasing alumina contents from 2 to 10 vol% is observed. These results are attributed to a strong coupling between alkali ions in BSG and Al^{+3} in alumina, thus forming an Al^{+3}- and alkali ion-rich reaction layer around alumina particles. Since the above reaction continuously depletes the concentration of alkali ions in BSG during sintering, it causes a rise in viscosity of BSG and thus slows down the densification kinetics and increases the activation energy of densification.

ACKNOWLEDGMENTS

The technical contributions of C. Sell and K. J. Tipinski are greatly appreciated.

REFERENCES

[1] J. H. Jean and T. K. Gupta, J. Mater. Res., 7, 3103 (1992).

[2] J. H. Jean and T. K. Gupta, to J. Mater. Res., (1993, in press).

[3] H. S. Cannon and F. V. Lenel, Proc. Plansee Semin., ed. F. Benesovsky, (Metallwerk Plansee, Reutte, 1953), p. 106.

[4] J. H. Jean and T. K. Gupta, J. Mater. Res., 7, 2514 (1992).

[5] R. H. Doremus, Glass Science (John Wiley & Sons, New York, NY, 1973), Chap. 6.

[6] W. D. Kingery, J. Appl. Phys., 30, 301 (1959).

[7] W. Espe, Materials of High Vacuum Technology, Vol. 2, (Pergamon Press, Oxford, 1968), Chap. 10.

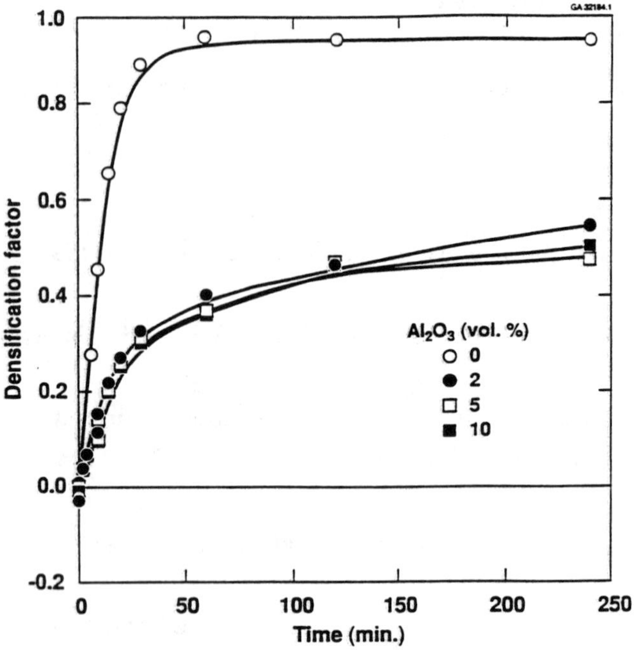

Figure 1 Densification factor as a function of time for the samples with various alumina contents from 2 to 10 vol%, fired at 875°C.

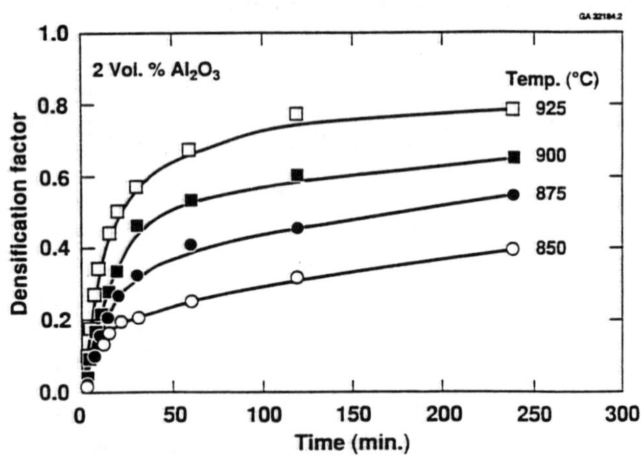

Figure 2 Densification factor as a function of time for the sample with 2 vol% alumina, fired at various temperatures from 850 to 925°C.

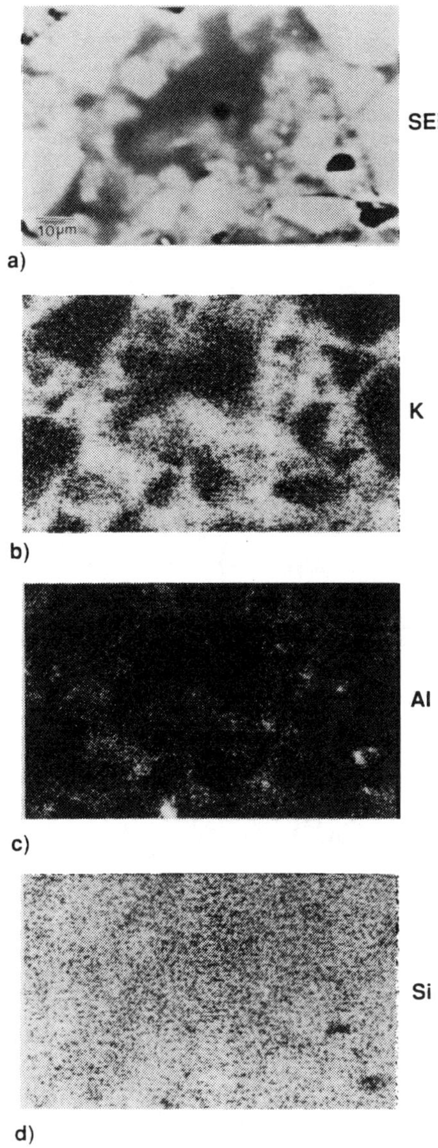

Figure 3 (a) Microstructure for the sample with 2 vol% alumina, fired at 925°C for 240 min, and elemental mapping of (b) K, (c) Al, and (d) Si.

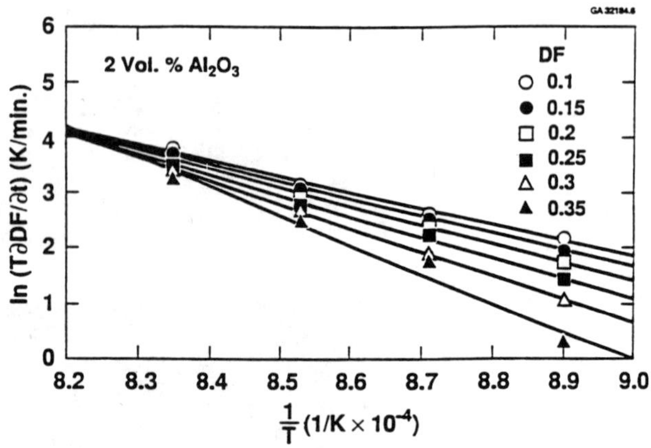

Figure 4 Logarithm of specific densification rate (ln (T∂DF/∂t)) versus 1/T at a given densification factor for the sample with 2 vol% alumina.

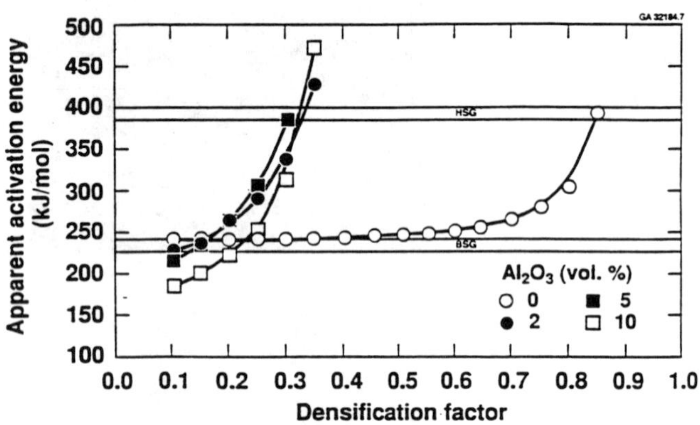

Figure 5 Activation energy as a function of DF for the samples with (a) 0, (b) 2, (c) 5, and (d) 10 vol% alumina, and activation energies for viscous flow of BSG and HSG.

NITROGEN-CONTAINING CORDIERITE GLASS-CERAMICS FOR CERAMIC PACKAGES

H. Unuma, J. Nagao, H. Ohara*, K. Asaka* and T. Yamamoto*
Government Industrial Development Laboratory, Hokkaido
2-17 Tsukisamu-Higashi, Toyohira-ku, Sapporo 062, JAPAN
*Department of Mechanical Engineering, Hokkaido Insitute of Technology
Maeda, Teine-ku, Sapporo 006, JAPAN
keywords: cordierite, package, oxynitride glass.

INTRODUCTION

Since the ceramics comprising cordierite ($2MgO \cdot 2Al_2O_3 \cdot 5SiO_2$) as the major phase have low dielectric constant, low dielectric loss and low thermal expansion coefficient, they are promising materials for ceramic packages. Many attempts have been made to fabricate cordierite-based ceramic packages. Sintering and subsequent crystallization of cordierite-based glass powders makes it possible to fabricate fully dense sintered body at relatively low temperatures, which may enable us to embed base metal conductors during the process of sintering of the package.

It is known that cordierite may form substitutional solid solutions by substituting Mn or Fe for Mg, Ga for Al, Ge for Si, or F for O[1,2]. The properties of the resultant ceramics such as the thermal expansion coefficient, dielectric constant and Young's modulus may be varied by the substitution. The substitution can be made by crystallizing the glasses containing substituting ions, for example.

In the present study, it was attempted to substitute a part of oxygen atoms by nitrogen in cordierite by crystallizing a nitrogen-containing cordierite glass, and to study the properties of the resultant ceramics from a viewpoint of the application of cordierite-based ceramics to ceramics package. In this paper, some preliminary results are described.

EXPERIMENTAL

A glass of the composition $2MgO \cdot 1.75Al_2O_3 \cdot 5SiO_2 \cdot 0.5AlN$ (denoted as N-Cor.) was prepared according to the following manner. The oxide part was melted beforehand to prepare an oxide glass, and the mixture consisting of the oxide glass powder and AlN powder was melted at 1600℃ for 4 hours in a molybdemun crucible in a flow of dry nitrogen gas. The resultant glass was free from crystalline phases as examined by powder X-ray diffraction (XRD).

Differential thermal analysis (DTA) was made for a block of the glass ($3x3x1mm^3$) at a heating rate of 10℃/min up to 1200℃. Dry argon gas was passed in the DTA equipment to avoid the oxidation.

The glass was ground to powder and ball-milled for 24 hours in water. The powder was shaped into pellets and plates with a CIP equipment under a pressure of 200MPa, and was sintered under a flow of helium at temperatures from 900 to 1300℃ for 3 hours. The heating rate was 5℃/min.

The water absorption of the sintered specimens was determined from the weight change of the specimens before and after immersing in hot water for 5 hours. The precipitated crystalline phases were identified by XRD. The average linear thermal expansion coefficient from room temperature to 500℃ was measured with a push rod dilatometer at a heating rate of 5℃/min. The 3-point bending strength was measured for bars $3x4x30mm^3$ in size.

For the measurements of the real density, dielectric constant, dielectric loss and Young's modulus, a fully dense sintered body is required. For these purposes, a hot press sintering of the glass powder was also made at 900 to 1200℃ for 0.5 to 1 hour under a pressure of 30MPa. The dielectric constant and the dielectric loss were measured at room temperature at a frequency of 1MHz.

The nitrogen content of the glass and glass-ceramics was determined by the Kjeldahl method[3], and it was ascertained that nitrogen was incorporated in the glass and glass-ceramics 90% of the intended. Nitrogen was not lost during the milling and sintering.

RESULTS AND DISCUSSION

Fig. 1 shows the DTA profile of the glass. A DTA curve for a nitrogen-free oxide glass powder of cordierite composition (denoted as Cor.) is also shown in the figure for comparison. The glass transition temperature and the onset of crystallization becomes higher as a result of the nitrogen incorporation.

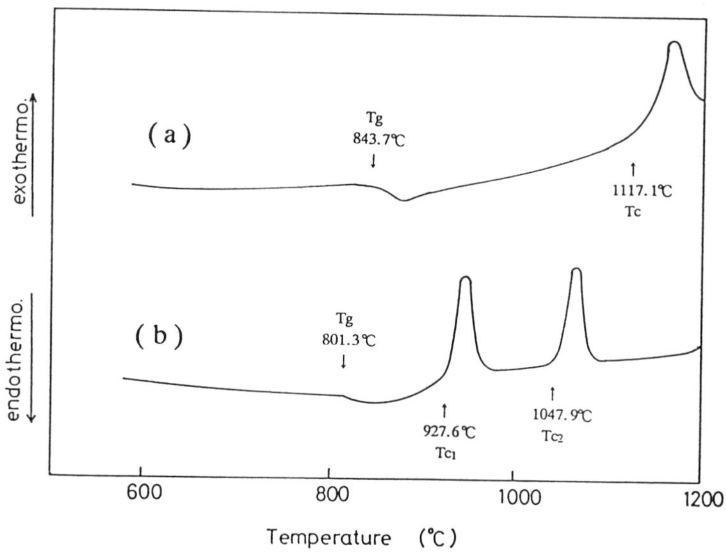

Fig. 1 DTA profiles of (a) N-containing cordierite glass and (b) N-free cordierite glass.

Fig. 2 shows the change of the warter absorption with the sintering temperature. The sintering is promoted by the incorporation of nitrogen in the glass. In general, precipitation of crystals in the glass powder hinders the sintering. In the nitrogen-containing glass, the temperature for the precipitation of crystals is higher as shown in Fig. 1, which makes the sintering easier taking advantage of the viscous flow of glass.

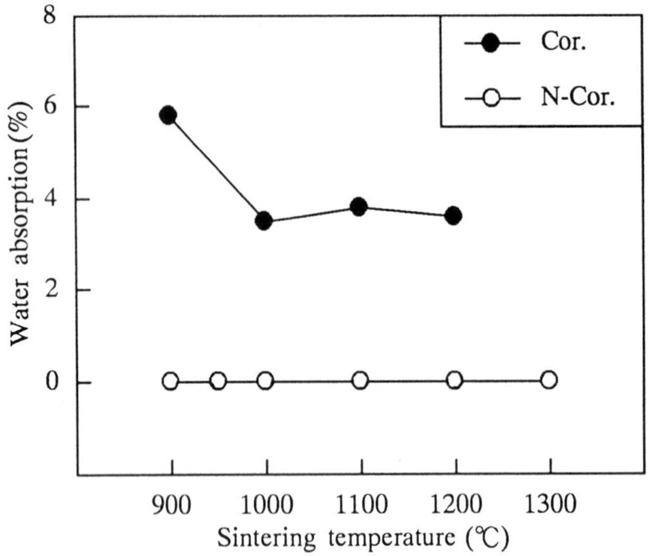

Fig.2 Change of the water absorption with sintering temperature.

Fig. 3 shows the XRD profiles of the nitrogen-containing and nitrogen-free cordierite glass-ceramics sintered at various temperatures. For nitrogen-free glass, μ-cordierite is precipitated at 900℃ and the transformation to α-cordierite phase is completed at 1000℃. In contrast, only a small amount of μ-phase is presipitated from the nitrogen-containing glass-ceramics sintered at 900℃ and the transformation to α-phase is completed at 1200℃. In the ceramics crystallized at 1200℃, a certain amount of β-forsterite ($2MgO \cdot SiO_2$) is precipitated. There is a possibility that nitrogen atoms may be incorporated in the crystal structure of μ- and α-cordierite phase. That is, if Si-N bonds are formed in the crystal structure of μ-phase, more thermal energy would be required to rearrange atoms from μ-phase to α-phase because the Si-N bond has a higher bond dissociation energy than the Si-O bond[4]. It is still uncertain, however, whether nitrogen is incorporated in the α-phase after the phase transformation.

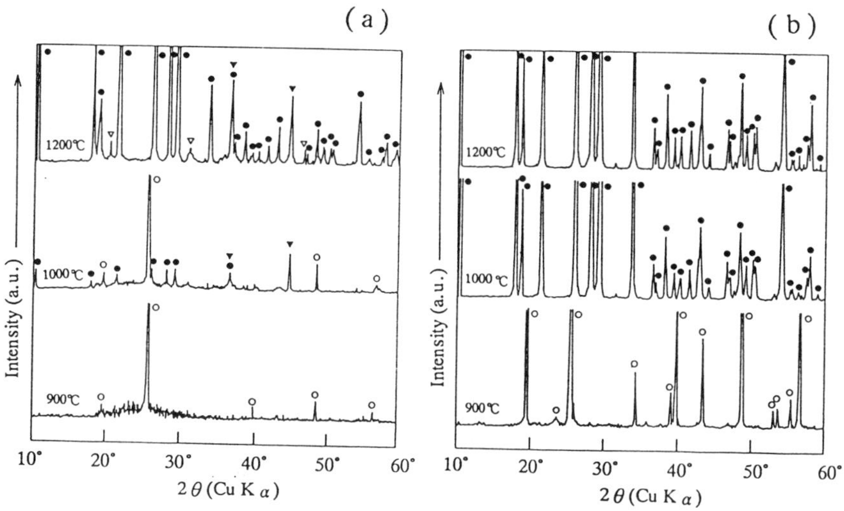

Fig. 3 XRD profiles of (a) N-containing and (b) N-free glass-ceramics. (●) α-cordierite, (○) μ-cordierite, (▼) β forsterite and (▽)unidentified phase.

Fig. 4 shows the change of the thermal expansion coefficient of the glass-ceramics with the sintering temperature. In the nitrogen-free cordierite glass-ceramics, the phase transformation from μ- to α-phase occurs at around 950-1000℃ and the glass-ceramics sintered at temperatures higher than 1000℃ have a low thermal expansion coefficient of about 0.5×10^{-6}/K. On the other hand, in the nitrogen-containing glass-ceramics crystallized at temperatures lower than 1100℃ show somewhat higher thermal expansion coefficient of about 3.5×10^{-6}/K. When the nitrogen-containing glass-ceramics were sintered at temperatures higher than 1200℃, the crystalline phase, μ-phase, transformed to the α-phase, and the thermal expansion coefficient decreases to about 1.5×10^{-6}/K, which is still higher than that of nitrogen-free glass-ceramics. The impurity phases of β-forsterite and an unidentified phase in the nitrogen-containing glass-ceramics may cause an increase in the thermal expansion coefficient.

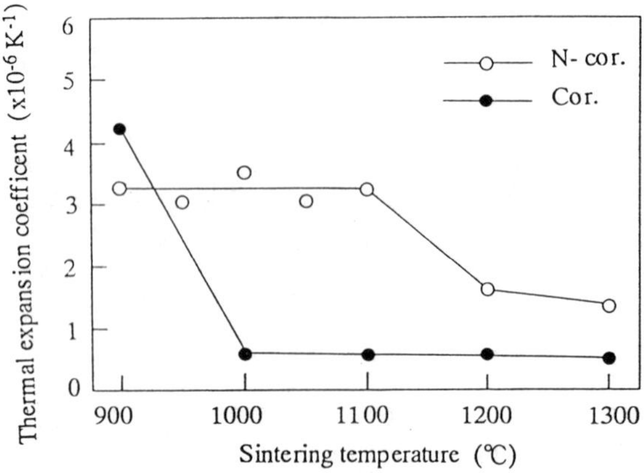

Fig. 4 Change of the thermal expansion coefficient with the sintering temperature

Table 1 shows the density and Young's modulus of the hot-pressed nitrogen-containing glass-ceramics.

Table 1 The density and Young's modulus of hot-pressed nitrogen-containing glass-ceramics.

Hot-press temperature (℃)	Density (g/cm^3)	Young's modulus(GPa)	cryst. phase
950	2.68	111	glass
1000	2.75	119	μ >glass
1200	2.73	135	α >forst.

The glass-ceramics hot-pressed at 1200℃, consisting of α-phase and a certain amount of β-forsterite, has a higher density than that of pure α-cordierite (2.512g/cm^3). The Young's modulus becomes higher as the hot-press temperature is raised.

Table 2 shows the dielectric constant and dielectric loss of the hot-pressed nitrogen-conatining ceramics.

Table 2 The dielectric constant, ε, and dielectric loss, $\tan\delta$ of hot-pressed nitrogen-containing glass-ceramics.

Hot-press temperature (℃)	ε	$\tan\delta$ (%)
950	6.680	0.439
1000	6.547	0.489
1200	5.633	0.197

The glass-ceramics hot-pressed at temperatures lower than 1000℃, consisting of μ-phase, have higher dielectric constant and dielectric loss than that hot-pressed at 1200℃. The dielectric properties of the glass-ceramics are improved after the hot-pressing at 1200℃ accompanied by the phase transformation from μ- to α-phase.

Table 3 shows the bending strength of nitrogen-containing glass-ceramics sintered under ambient pressure and in a hot-press.

Table 3 The bending strength of nitrogen-containing glass-ceramics.

Sintering temperature (℃)	Bending strength (MPa) ambient pressure	hot-press
950	-	182.3
1000	112.8	187.0
1100	92.2	-
1200	115.5	46.2

The bending strength exhibit an anomalous dependence on the sintering temperature and sintering pressure. A nitrogen-free glass-ceramic of near-cordierite composition has a higher bending strength of about 170-200MPa[5] although it may contain a glassy phase in the grain boundary and its Young's modulus is only 110GPa. The present nitrogen-containing glass-ceramics sintered at 1200℃ have poorer bending strength in spite of the absence of a glassy phase. The reason for the poor bending strength is still uncertain. Also it is difficult to discuss the reason for the variation of the bending strength with the sintering pressure. A plausible explanation may be given by assuming that there remains a intergranular stress between α-cordierite and β-forsterite in the glass-ceramics sintered at 1200℃, and the residual stress would be enhanced by the hot-press.

High Young's modulus of the present nitrogen-containing glass-ceramics may imply that the intrinsic mechanical strength of the nitrogen-containing glass-ceramics should be higher. Although we have not yet succeeded in the fabrication of mechanically strong glass-ceramics, there is a possibility that the mechanical strength may be improved by modifying the cationic composition and the nitrogen content of the mother glass and by optimizing the sintering process so that β-forsterite is not precipitated.

CONCLUSION

The properties of the cordierite-based glass-ceramics derived from nitrogen-conatining cordietire glass have been investigated. The nitrogen-conatining glass-powder could be sintered more easily than nitrogen-free glass powder. The phase transformation of precipitated μ-phase to α-phase is suppressed when nitrogen was incorporated in glass. It has been suggested that nitrogen may be incorporated in the precipitated μ-cordierite phase. The glass-ceramics comprising α-phase as the major phase were obtained by sintering the glass powder at temperatures higher than 1200℃. The glass-ceramics sintered at 1200℃ has a dielectric constant of 5.633, a dielectric loss of 0.197 and Young's modulus of 135GPa. Although the bending strength is not satisfactory for practical use, it may be improved by optimizing the composition of the mother glass and the sintering process.

Acknowledgement

The present authors wish to thank Mr. Masayuki Ishihara of Matsushita Electric Works Ltd. for his collaboration in preparation and measurements of the properties of hot-pressed specimens.

REFERENCES

1. H. Ikawa, T. Otagiri, O. Imai, K. Urabe and S. Udagawa, "Thermal Expansion of High Cordierite and Its Solid Solutions," J. Ceram. Soc. Jpn., **94**, 344-350(1986).
2. G. M. Singer and M. Tomozawa, "Cordierite-based Oxyfluoride Glasses and Its Glass-Ceramics," Phys. Chem. Glass., **30**, 86-95, 95-102 and 102-109(1989).
3. K. Kamiya, M. Ohta and T. Yoko, "Nitrogen-Containing SiO_2 Glass Fibers Prepared by Ammonolysis of Gels Made from Silicon Alkoxides," J. Non-Cryst. Solids, **83**, 208-222(1986).
4. H. Unuma, K. Miura, T. Furusaki and K. Kodaira, "Preparation and Properties of Glass-Ceramics Derived from Nitrogen-Containing β-Spodumene Glasses," J. Am. Ceram. Soc., **74**, 1291-1295(1991).
5. M. Nawa, K. Makio, M. Ishihara, A. Kamiya and K. Nakano, "Mechanical Properties of Whisker-Reinforced Cordierite Glass Ceramic Composites," J. Ceram. Soc. Jpn., **100**, 784-790(1992).

STRUCTURE OF NA-SI-O-N OXYNITRIDE GLASSES

H. Unuma, H. Maekawa*, H. Kiyono*, K. Kawamura* and T. Yokokawa*
Government Industrial Development Laboratory, Hokkaido.
2-17 Tsukisamu-Higashi, Toyohira-ku, Sapporo 062, JAPAN
*Department of Chemistry, Faculty of Science, Hokkaido University.
N-10, W-8, Kita-ku, Sapporo 060, JAPAN
Key words: structure, oxynitride glass, molecular dynamics, ^{29}Si NMR.

ABSTRACT: The structure of Na-Si-O-N oxynitride glasses has been studied by the molecular dynamics (MD) calculations and ^{29}Si MAS NMR. The MD calculations have shown that nitrogen atoms bonded to one, two and three silicon atoms may coexist in the structure, and that the mean of the number of silicon around nitrogen ranges from 2.4 to 2.1, decreasing with increasing Na_2O content from 15 to 30 mol%. The analysis of the NMR spectra has lead to a similar result that about 2.5 silicon atoms are bonded to a nitrogen on the average. It has been assumed that the nitrogen atoms bonded to two or less silicon atoms may be formed when nitrogen atoms are substituted for non-bridging oxygen atoms.

INTRODUCTION

Oxynitride glasses contain nitrogen as well as oxygen as anions in glass structure. The hardness, dielectric constant, thermal expansion coefficient and fracture toughness increase on incorporating nitrogen in glass[1-3]. Unuma et al.[4] have prepared Li-Al-Si-O-N glass-ceramics from nitrogen-containing β-spodumene glasses, and the glass-ceramics have higher hardness, higher resistance to phase transformation and lower thermal expansion coefficient than nitrogen-free glass-ceramics. Oxynitride glasses and glass-ceramics, therefore, are promising materials for industrial use, such as high-modulus glass-fiber or matrix of ceramic composites. Also, it has been attempted to prepare cordierite-based glass-ceramics having a high Young's modulus from an oxynitride glass[5]. In order to understand better the properties of oxynitride glasses and derived glass-ceramics, it is essential to investigate the structure of oxynitride glasses.

The change of the properties of glasses and glass-ceramics on incorporating nitrogen has been explained by many authors[2, 6, 7] by using a structural model proposed by Mulfinger[8], (1), in which nitrogen atoms are bonded to three network-forming(NWF) cations.

$$\text{-Si-}\underset{\underset{|}{\text{Si}}}{\overset{}{\text{N}}}\text{-Si-} \tag{1}$$

On the other hand, some researchers[9, 10] claimed that there may be nitrogen atoms bonded to two or less NWF cations (2, 3).

$$\text{-Si-N}^{-}\text{-Si-} \tag{2}$$

$$\text{-Si-N}^{2-} \tag{3}$$

The clear picture of the structure of oxynitride glasses is still uncertain. In the present study, the structure of Na-Si-O-N oxynitride glasses has been studied by the molecular dynamics calculations and ^{29}Si MAS NMR.

EXPERIMENT

Molecular Dynamics Calculations

The compositions studied were $30Na_2O \cdot (70\text{-}3x)SiO_2 \cdot xSi_3N_4$ (x=0, 2, 5 and 10) and $yNa_2O \cdot (85\text{-}y)SiO_2 \cdot 5Si_3N_4$. The interatomic pair potential used was the Born-Mayer-Huggins form (4) [11].

$$\Phi_{ij} = f_0(b_i+b_j)\exp\{(a_i+a_j-r)/(b_i+b_j)\} + Z_iZ_je^2/4\pi\varepsilon_0 r \tag{4}$$

where r is the distance between atoms, ε_0 is the dielectric constant of vacuum and $f_0 = 6.9472 \times 10^{-11}$N. a_i and b_i are the potentoal parameters which are related to the size and softness of i th atom, respectively. The values of the parameters are shown in Table 1. The computations were initiated from a random distribution of the atoms in the basic cells containing about 1300 atoms, to which three dimensional periodic boundary condition were applied. A constant pressure of 1 atm and constant temperatures of 4000, 3000, 2500, 2000, 1500 and 1000K were achieved, and the system was kept at each temperature for 20 to 100 ps. The structural data were accumulated at 1000K. Fig 1 shows the snapshot of the structure. The structure was analyzed in terms of the number of silicon atoms around nitrogen, the bond angle distribution around nitrogen and the pair correlation function. The pair correlation function, $g_{ij}(r)$, was defined as follows;

$$g_{ij}(r) = n_{ij}(r_m) / 4\pi(N_iN_j/V)r_m^2\Delta r \tag{5}$$

where $n_{ij}(r_m)$ is the number of the pair of i and j atoms found in the distance between $r_m+(\Delta r/2)$ and $r_m-(\Delta r\text{-}2)$, Δr is 0.001 nm, r_m ranges from 0.0005 nm to 0.6 nm at an interval of 0.001 nm, N_i is the number of i atoms in the basic cell and V is the volume of the basic cell.

Table 1 The parameter set used in the present calculations.

	Z	a(Å)	b(Å)
Si	4	1.012	0.080
Na	1	1.260	0.080
O	-2	1.626	0.085
N	-3	1.713	0.080

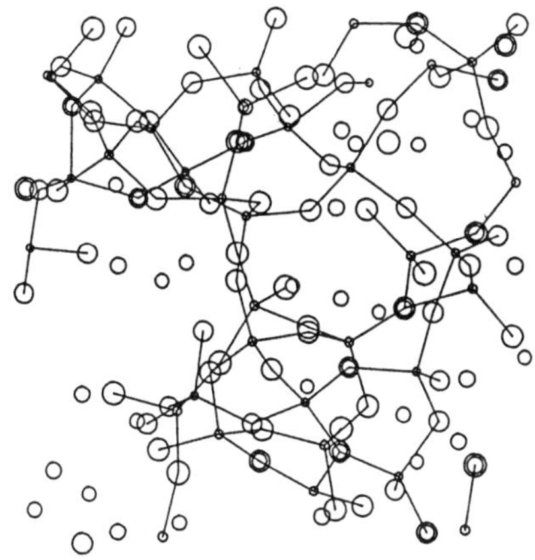

Fig. 1 A snapshot of the structure of glass.

^{29}Si MAS NMR

Oxynitride glasses of the compositions $20Na_2O \cdot (80\text{-}3x)SiO_2 \cdot xSi_3N_4$ (x=0, 2 and 4) were prepared. 0.05 mol% of Gd_2O_3 was added to all the glasses to accerelate the spin relaxation of Si nuclei. The ^{29}Si NMR spectra were recorded on a Bruker MSL-200 spectrometer operating at 4.7T. The powdered glass specimens stuffed into alumina or zirconia sample holders were spun at the magic angle (54.74 degree) with respect to the external magnetic field at a rate of 3500Hz. The spectrum was accumulated 1000 times with a repetition interval of 2.5 sec.

RESULTS

Molecular Dynamics Calculations

Table 2 shows the percentage of nitrogen atoms bonded to n silicon atoms (represented as $N_{(n)}$) together with the mean value, M_{Si}. The nitrogen atoms bonded to three silicon atoms, $N_{(3)}$, are formed in the glass structure. However, the major species is those bonded to only two silicon atoms, $N_{(2)}$. M_{Si} ranges from 2.4 to 2.1, decreasing with increasing Na_2O content.

Table 2 Percentage of the nitrogen atoms bonded to n silicon atoms, $N_{(n)}$, and the mean of the number of silicon around nitrogen, M_{Si}.

glass	$N_{(1)}$	$N_{(2)}$	$N_{(3)}$	M_{Si}
	$30Na_2O\cdot(70-3x)SiO_2\cdot xSi_3N_4$ glasses			
x=2	2.8	77.8	19.4	2.17
x=5	13.5	61.8	24.7	2.11
x=10	6.2	64.7	29.1	2.23
	$yNa_2O\cdot(85-y)SiO_2\cdot 5Si_3N_4$ glasses			
y=30	13.5	61.8	24.7	2.11
y=20	0.0	74.0	26.0	2.26
y=15	0.0	58.7	40.3	2.41

Fig. 2 shows the bond angle distribution of <Si-N-Si in $30Na_2O\cdot 40SiO_2\cdot 10Si_3N_4$ glass. The dotted and undotted area in the figure correspond to <Si-$N_{(2)}$-Si and <Si-$N_{(3)}$-Si, respectively. It is known that the bond angle around oxygen in oxide glasses distributes from 120 to 180 degree, around 144 degree. The bond angle around nitrogen is smaller than that around oxygen.

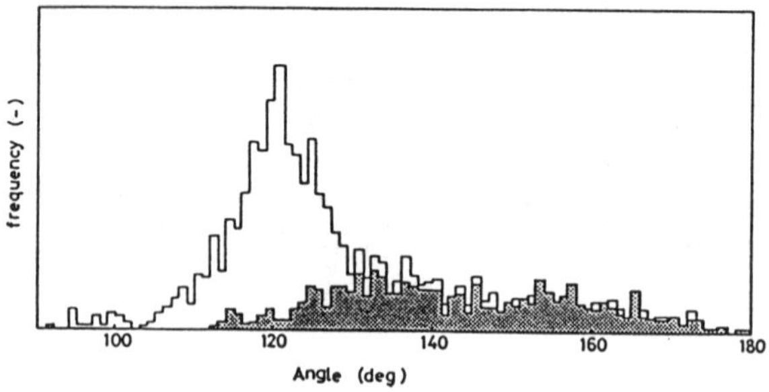

Fig. 2 The bond angle distribution of < Si-N-Si in $30Na_2O\cdot 40SiO_2\cdot 10Si_3N_4$ glass.

Fig. 3 shows the pair correlation functions of $30Na_2O\cdot 70SiO_2$ glass and $30Na_2O\cdot 40SiO_2\cdot 10Si_3N_4$ glass. The most distinctive feature is that Si to Si distance is shortened in the nitrogen-containing glass.

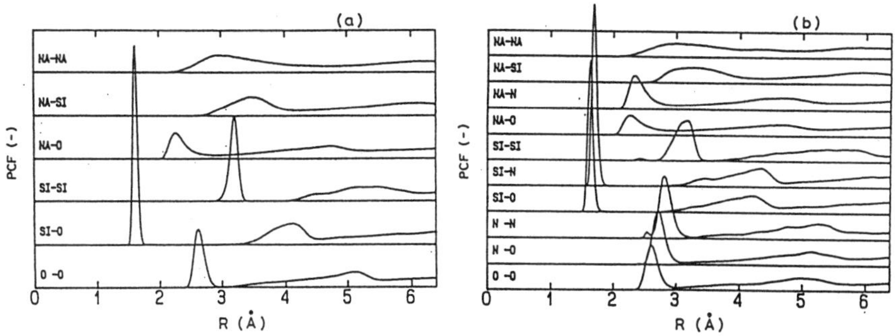

Fig. 3 The pair correlation functions of (a)$30Na_2O.70SiO_2$ and (b)$30Na_2O.40SiO_2.10Si_3N_4$ glasses.

^{29}Si MAS NMR

Fig. 4 shows the NMR spectra of $20Na_2O\bullet(80\text{-}3x)SiO_2\bullet xSi_3N_4$ glasses. Each spectrum can be separated to a number of Gaussian curves as shown in the figure, and each Gaussian curve can be assigned to a particular silicate tetrahedral unit (Qn and Qn'). The notation Qn represents a silicate tetrahedral unit containing n bridging oxygens. The notation Qn' is used for nitrogen-containing glasses to distinguish the curves from those of nitrogen-free glass. The relative abundance of Qn(') species, W[Qn(')], is calculated from the relative intensity area of the Gaussian curves.

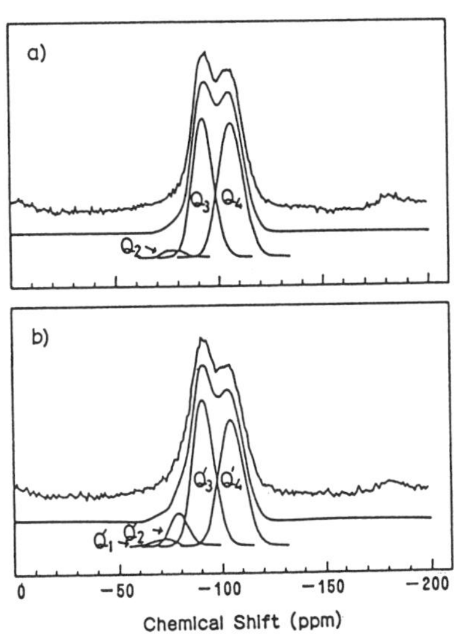

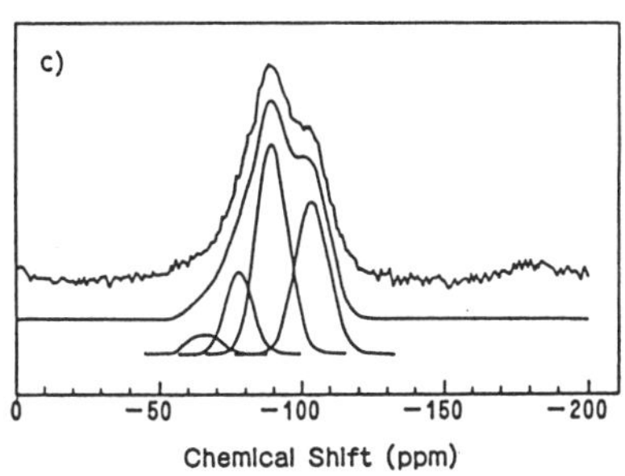

Fig. 4 ^{29}Si NMR spectra of $20Na_2O\bullet(80\text{-}3x)SiO_2\bullet xSi_3N_4$ glasses. (a) x=0, (b) x=2 and (c) x=4.

Table 3 shows the relative abundance of Qn(') species. It is recognized from Fig. 4 and Table 3 that the intensities of less shielded peaks (less negative chemical shift) increase with increasing nitrogen content, that is, W[Q_1'] and W[Q_2'] increases and W[Q_4'] decreases.

Table 3 Observed Q_n distributions in oxynitride glasses

Sample	Q_4	Q_3	Q_2	Q_1
x=0	50.0	48.0	3.0	0.0
x=2	44.2	45.9	8.7	1.2
x=4	38.8	44.7	12.7	3.9

The number of silicon atoms around nitrogen may be estimated by assuming the extent of the shielding effect of nitrogen on silicon. In this relation, it is considered that the resonance of silicon nuclei shifts toward less shielded direction by about 15ppm when an adjecent oxygen is replaced by nitrogen[6]. On that basis, we calculated W[Qn'] of nitrogen-containing glasses using M_{Si} as a parameter. The detailed procedure af the calculation is described elsewhere[12].

Fig. 5 shows the calculation of W[Qn'] as a function of nitrogen content. The measured W[Qn'], the solid circles, agree well with the calculation made on the assumption that M_{Si} is 2.5. This means that about a half of nitrogen atoms are bonded to three silicon atoms and the rest are bonded to two silicon atoms.

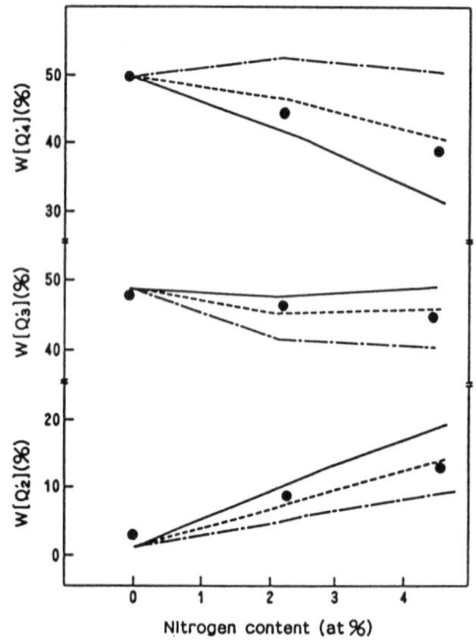

Fig. 5 Calculation of the change of W[Q_n'] in oxynitride glasses as a function of the nitrogen content, in which the parameter M_{Si} is set at 3(——), 2.5(----) and 2(—·—). Solid circles represent the measured values.

DISCUSSION

There are some studies on the structure around nitrogen in Na-Si-O-N oxynitride glasses[6, 7, 9, 11, 12]. Hater et al.[6] estimated that all the nitrogen atoms would be bonded to three silicon atoms on the basis of ^{29}Si NMR. On the other hand, Clabes et al.[9] suggested the existence of the nitrogen atoms of structural group (2) on the basis of N_{1s} XPS. Jin et al.[10] found that about 66% of nitrogen are bonded to only two silicon atoms and MSi is 2.3 on the basis of the neutron diffraction analysis.

In the present study, both the molecular dynamics calculations and the NMR have given a consistent result that not all the nitrogen atoms are bonded to three silicon atoms but about a half or more of nitrogen atoms are bonded to two or less silicon atoms.

The nitrogen atoms bonded to two or less silicon atoms, $N_{(2)}$ and $N_{(1)}$, may be formed when nitrogen atoms substitute for non-bridging oxygen atoms as shown below;

$$Si\text{-}O^- + Si\text{-}\underset{\substack{| \\ Si}}{N}\text{-}Si \qquad\qquad Si\text{-}O\text{-}Si + Si\text{-}N^-\text{-}Si \qquad (6)$$

$$Si\text{-}O^- + Si\text{-}N^-\text{-}Si \qquad\qquad Si\text{-}O\text{-}Si + Si\text{-}N^{2-} \qquad (7)$$

These equations imply that the number of silicon around nitrogen decreases with increasing number of non-bridging oxygen atoms. This is consistent with the present molecular dynamics calculations, which have shown that M_{Si} decreases with increasing Na_2O content because the number of non-bridging oxygen atoms increases with increasing Na_2O content.

The present idea of the formation of $N_{(2)}$ and $N_{(1)}$ is analogous to the formation of non-bridging oxygen atoms on oddition of alkali oxides to silicate glasses, and $N_{(2)}$ and $N_{(1)}$ atoms would accompany one and two charge compensating Na^+ ions, respectively.

The bond length of Si-N and the bond angle distribution of $<Si\text{-}N_{(3)}\text{-}Si$ calculated in the present study, Figs. 1 and 2, are consistent with the molecular orbital calculations made by Murakami and Sakka[13]. Taking into account that the Si-$N_{(2)}$-Si bonds would be more flexible for bending than Si-$N_{(3)}$-Si bonds, the calculated bond angle distribution of $<Si\text{-}N_{(2)}Si$ would be reasonable.

The distance of Si-Si in nitrogen-containing glasses is shorter than that in nitrogen-free glass as shown in Fig. 3. That is related to the small bond angle around $N_{(3)}$. The density of glass increases when nitrogen is incorporated[1]. The shortened Si to Si distance correspond to the increase in the density.

CONCLUSION

The local structure in Na-Si-O-N oxynitride glasses has been studied. The molecular dynamics calculations have shown that the nitrogen atoms bonded to one, two and three silicon atoms may coexist in the glass structure, and the mean of the number of silicon atoms ranges from 2.1 to 2.4 depending on the Na_2O content. The analysis of ^{29}Si MAS NMR spectra has lead to a similar result to that of the molecular dynamics calculations that about 2.5 silicon atoms are bonded to nitrogen on the average. It was assumed that the nitrogen atoms may substitute for non-bridging oxygen as well as bridging oxygen, resulting in the decrease in the number of silicon atoms around nitrogen.

REFERENCES

1. S. Sakka, "Oxynitride Glasses," Ann. Rev. Mater. Sci., **16**, 29-46(1986).
2. S. Hampshire, R. A. Drew and K. H. Jack, "Oxynitride Glasses," Phys. Chem. Glass., **26**, 182-186(1985).
3. R. E. Loehman, "Preparation and Properties of Oxynitride Glasses," J. Non-Cryst. Solids, **56**, 123-134(1983).
4. H. Unuma, K. Miura, T. Furusaki and K. Kodaira, "Preparation and Properties of Glass-Ceramics Derived from Nitrogen-Containing β-Spodumene Glasses," J. Am. Ceram. Soc., **74**, 1291-1295(1991).
5. H. Unuma, J. Nagao, H. Ohara, K. Asaka and T. Yamamoto, paper presented to this Conference.
6. W. Hater, W. M-Warmuth and G. H. Frischat, ^{29}Si MAS NMR Studies of Alkali-Silicate Oxynitride Glasses," Glastech. Ber., **62**, 328-334(1989).
7. C. Schrimpf and G. H. Frischat, "Property-Composition Relations of N_2-Containing Na_2O-CaO-SiO_2 Glasses," J. Non-Cryst. Solids, **56**, 153-160(1983).
8. H. O. Mulfinger, "Physical and Chemical Solubility of Nitrogen in Glass Melts," J. Am. Ceram. Soc., **49**, 462-467(1966).
9. J. G. Clabes, R. E. Fern and G. H. Frischat, "Characterization of Nitrogen-Containing Glasses by X-ray Photoelectron Spectroscopy," J. Vac. Sci. Technol. A, **4**, 1580-1584(1986).
10. J. Jin, T. Yoko, F. Miyaji, S. Sakka, T. Fukunaga and M. Misawa, "Neutron Diffraction Study on the Structure of Na-Si-O-N Oxynitride Glasses," J. Am. Ceram. Soc., in press.
11. K. Kawamura, "A Molecular Dynamics Simulation of $Na_2O \cdot 2SiO_2$-$K_2O \cdot 2SiO_2$ Melts - Effect of Basic Cell Size," Mol. Simu., **6**, 245-255(1991).
12. H. Unuma, H. Maekawa, H. Kiyono, K. Kawamura. T. Maekawa and T. Yokokawa, "^{29}Si MAS NMR of Na-Si-O-N Oxynitride Glasses," J. Ceram. Soc. Jpn., **100**, 1292-96(1992).
13. M. Murakami and S. Sakka, "*Ab Initio* Molecular Orbital Calculation of the Interatomic Potential and Force Constants in Silicon Oxynitride Glass," J. Non-Cryst. Solids, **101**, 271-279(1988).

TEM AND X-RAY DIFFRACTION STUDIES OF CRYSTALLIZED CORDIERITE POWDER

Vivekanand Ramakrishnan and Edward Goo, Department of Materials Science and Engineering, University of Southern California, Los Angeles, CA 90089-0241.

ABSTRACT

In this study, cordierite glass powders were compacted and then sintered in the temperature range of 825 to 1250 °C, and studies were made using x-ray diffractometry and electron microscopy. It was found that α-$Mg_2Al_4Si_5O_{18}$ - a phase called indialite, which has a hexagonal crystal structure was the predominant phase through all of the sintering range, with protoenstatite appearing at high sintering temperatures, and that the lattice parameter of indialite changed with sintering temperature.

INTRODUCTION

Ceramics and glass-ceramics are used in electronic packaging (1) to provide power and signal distribution via metallic interconnects, to protect the circuit mechanically and chemically, and in certain applications, to provide heat dissipation. The properties the ceramic or glass-ceramic material have to satisfy in high performance applications are a low dielectric constant, accommodation of a high wiring density, thermal expansion match with silicon, good dimensional control and reasonable strength (2). Cordierite, with a chemical formula of $2MgO.2Al_2O_3.5SiO_2$, offers the best combination of all these properties and maybe sintered at a low enough temperature to allow metallization with copper.

In the case of cordierite, glass particles crystallize during sintering to get a completely crystalline solid. The study of crystallization of cordierite with the addition of catalysts has been extensively investigated [3-9]. In this study, an effort has been made to characterize crystallographically the nucleation and growth of crystals without the addition of catalysts. The increased surface area due to the use of powder leads to higher chance of nucleation. A green compact was made with the powder, and then sintered at different times and temperatures. The phases that were formed were studied using Transmission Electron Microscopy (TEM) and X-ray diffractometry.

EXPERIMENTAL PROCEDURE

A commercial glass ceramic powder, supplied by Dr. E.A. Giess from IBM was used in this study. The average powder size was about 3 to 3.5μm. The powder was compacted under a pressure of 30,000 psi in a hydraulic press. The samples were then inserted in a horizontal tube furnace (Minibrute MB2) which was preheated to the sintering temperature. Five minutes were added to the sintering time to allow the sample to reach the sintering temperature. The samples were then sintered in air and the sintering temperature range was from 825 to 1250 °C, and the time was varied from 15 minutes to four hours. The samples were then pulled out and allowed to cool in air . TEM (EM 420, Philips Instruments, Mahwah NJ) samples were prepared by conventional means. The samples were coated with carbon to prevent them from charging. X-ray studies were done on the Rigaku x-ray powder diffractometer(Rigaku Co., Tokyo, Japan). No special sample preparation procedure was used to overcome preferred orientation.

The samples are labeled to by a combination of heat treatment temperature and time. "825-5" refers to a sample sintered at 825 °C for 5 hours.

RESULTS AND DISCUSSION

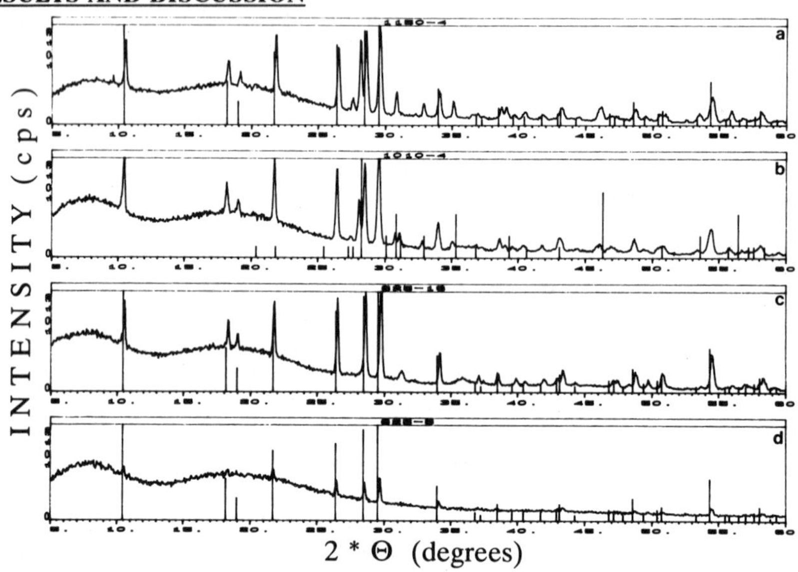

Figure 1. X-ray diffraction data from (a) 1150-4, (b) 1010-4, (c) 825-16 and (d) 825-5. Expected positions of peaks from the JCPDS files for indialite are marked in (a), (c) and (d) and for protoenstatite in (b).

X-ray results obtained from samples 1150-4, 1010-4, 825-16 and 825-5 are shown in figures 1a to 1d. The high background intensity at low angles arises from the Scotch tape that was used to mount the sample to the holder. X-ray data from 1150-4 and 1010-4 show the presence of two crystalline phases, protoenstatite and indialite. Protoenstatite is an orthorhombic phase of $MgSiO_3$, and its space group is Pbcn. The structure of protoenstatite can be thought of as a unit cell containing 8 molecules [10] having lattice parameters of a=9.23, b=8.64 and c=5.32 Å. These values for the lattice parameters were determined from x-ray data using regression analysis. The other phase, indialite, was present in all samples that were crystalline. It is referred to as a "disordered" hexagonal ∝-phase of cordierite, and its space group is P/6mcc. Its crystal structure has been discussed elsewhere [11]. It appears from the x-ray data that the lattice parameters of indialite when protoenstatite is present (a=9.75 Å, c=9.29 Å) is slightly different from the lattice parameters of indialite when protoenstatite is not present (a=9.68 Å, c=9.35 Å). It can also be noticed that the intensities of the indialite peaks from the sample 825-5 are very low as shown in figure 1d, and this is because there is a considerable amount of glassy phase in the sample.

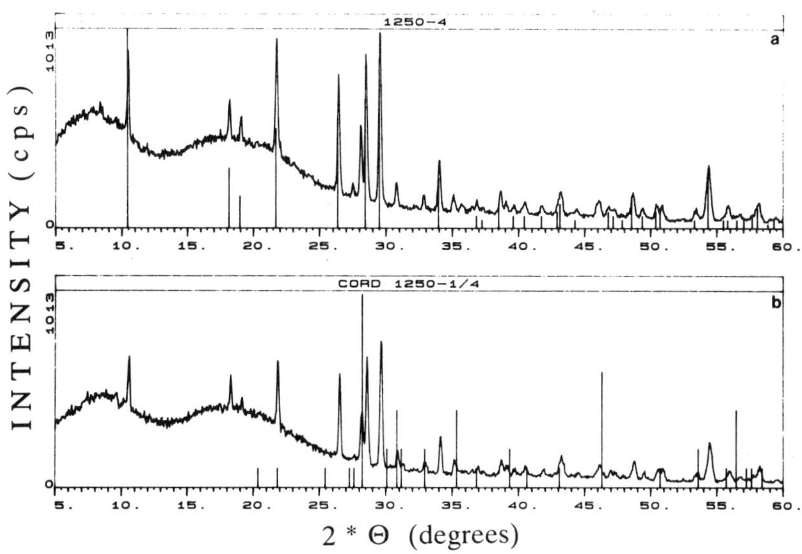

Figure 2. X-ray diffraction data from (a) 1250-4 and (b) 1250-1/4. Expected positions of peaks from the JCPDS files for indialite are marked in (a), and for protoenstatite in (b).

Figures 2a and 2b show x-ray diffraction data from 1250-4 and 1250-¼ (¼ = 15 minutes). The intensity of the peaks are quite high for the sample 1250-¼, and this indicates a high degree of crystallinity, which was confirmed by TEM studies. When sintered at 825 °C, the sample takes up to 8 hours to become crystalline. The size of the grains were determined from TEM data. On average, smaller grains are seen in 825-16 than in 1250-4.

From the Al_2O_3-MgO-SiO_2 equilibrium phase diagram [3], protoenstatite is formed as an equilibrium phase for composition in the cordierite phase region. However, protoenstatite was not observed at lower temperatures even after sintering for long times (up to 65 hours at 825 °C). It is probable that at lower temperatures (below about 950 °C), the kinetics are too slow for protoenstatite to precipitate out, and that there may be a critical temperature below which protoenstatite will not form, irrespective of the sintering time. Grains of protoenstatite were observed using the TEM, and it appears that they form between indialite grains.

TEM studies indicate that the crystals almost always grow from the surface inward, or from the region of intersection of two particles. The morphology of the grains vary depending on the sintering temperature. As mentioned earlier, the average grain size are higher at higher temperatures (1.5-2μm from 1250-4 as compared to 0.7-1μ from 825-16). Also, at high temperatures (above about 1060

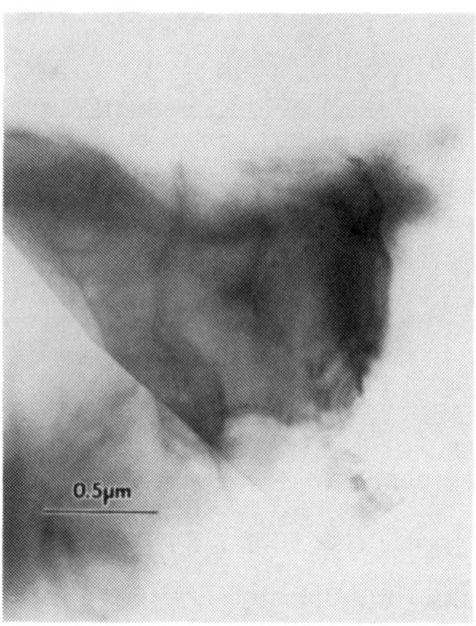

Figure 3. Facet shaped grain from 1250-¼.

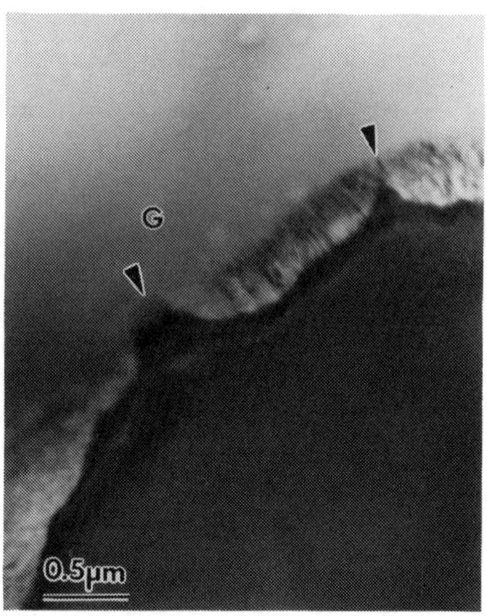

Figure 4. Crystalline region from 825-6. The glassy region is marked G.

°C), facet shaped grains were noticed, and at lower temperatures, these were not seen. A facet shaped grain from 1250-¼ is shown in figure 3. Figure 4 shows the crystalline region marked by arrows from the sample 825-6: the absence of facets may be easily noticed. Previous work [4] done on crystallization in cordierite bulk glasses with B_2O_3 and P_2O_5 indicates that different growth mechanisms maybe operative : layer growth and continuous growth at higher temperatures and continuous growth at lower temperatures. It was also noticed that the samples seem to change on prolonged exposure to the electron beam. The diffraction pattern changes from spots due to crystalline indialite to diffuse rings and spots and the actual grain image became spotted. This effect is shown in figures 5 and 6. Figure 5 shows the diffraction patterns obtained from the same grain, 5a at the beginning, and 5b after tilting it to another zone axis. The corresponding images are seen in figures 6a and 6b. The presence of rings in figure 5b (the pattern has been overexposed to show the rings clearly-they were not observed at all when pattern 5a was photographed) can be noticed. More work needs to be done, but it appears that the crystalline material seems to become amorphous under prolonged exposure to the electron beam.

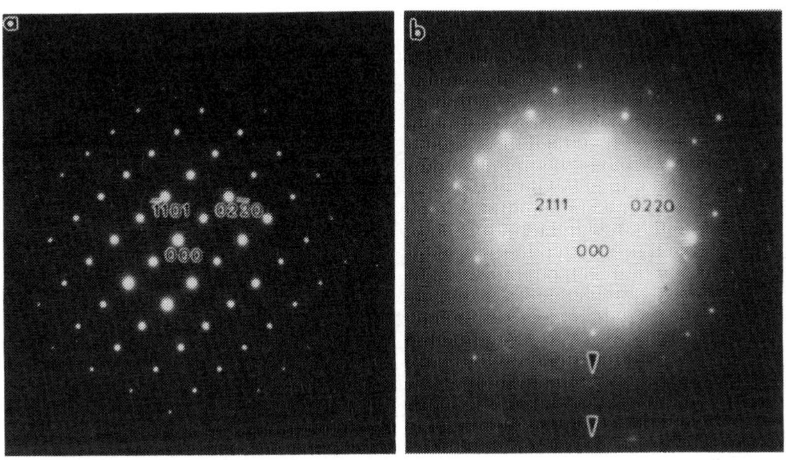

Figure 5. Diffraction pattern along the a) [10$\bar{1}$0] and b) [10$\bar{1}$2] zone axis from the same grain are shown. Figure 5b) was obtained after tilting the sample from the orientation given in 5a). The arrows point to the presence of rings.

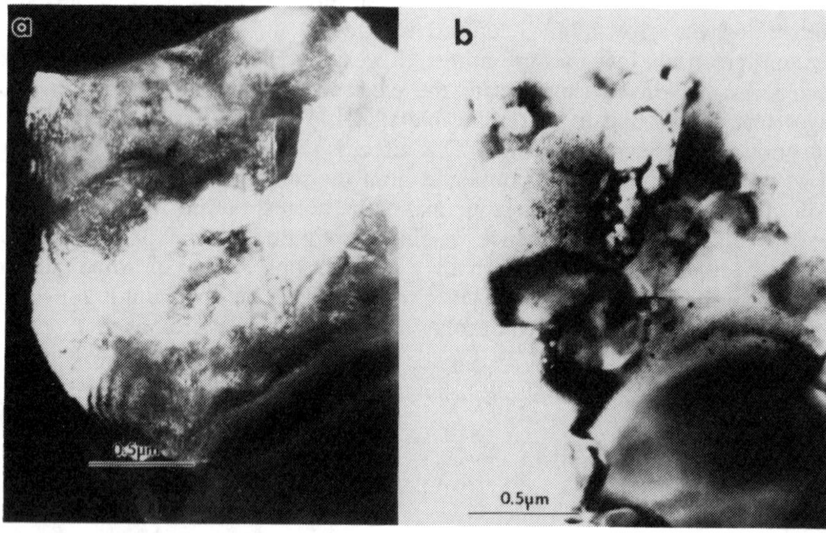

Figure 6. Dark field images from (02$\bar{2}$0) spot for the orientations described in Figure 5 a) and b). The image has become "spotted" in figure 6b) and the actual area imaged in the dark field has become smaller.

CONCLUSION

1. The crystallization behavior of compacted glass powders in the cordierite phase region has been studied at different sintering times and temperatures. The crystalline phases that appear are indialite, and for samples sintered above about 950 °C, protoenstatite is noticed

2. The growth rate is higher at higher temperatures of sintering.

3. The mechanism of growth seems to depend on sintering temperatures. At higher temperatures (above 1100 °C), facet shaped crystals, indicative of layer growth mechanism are seen. At lower temperatures, these are not noticed, but rather cell-like crystals, indicative of continuous growth mechanism are formed.

4. The samples seem susceptible to damage by the electron beam irradiation. Initial results suggest that crystalline areas may turn amorphous.

RFERENCES

1. R. R. Tummala, "Ceramics in Microelectronic Packaging"; pp. 3-16 in Advances of Ceramics Vol.26, *Ceramic Substrates and Packages for Electronic Applications*, Edited by M. F. Yan, K. Niwa, H. M. O'Bryan Jr., W. S. Young, American Ceramics Society, Columbus, Ohio,1989

2. R. R. Tummala, "Ceramic and Glass-Ceramic Packaging in the 1990s", *J. Am. Ceram. Soc.*, **74** [5] 895-908 (1991).

3. W. Wogel and W. Holand, "Nucleation and Crystallization Kinetics of a MgO-Al_2O_3-SiO_2 Base Glass with Various Dopants," pp. 125-145 in Advances in Ceramics, Vol. 4, *Nucleation and Crystallization in Glasses*, Edited by J. H. Simmons, D. R. Uhlmann, and G. H. Beall, American Ceramic Society, Columbus OH, 1981.

4. K. Watanabe, E. A. Giess and M. W. Shafer, "The Crystallization Mechanism of High-Cordierite Glass," *J. Mater. Sci.*, **20** [2] 508-515 (1985)

5. A. G. Gregory and T. J. Veasey, "Review: The Crystallization of Cordierite Glass,"*J. Mater. Sci.*, **6** [10] 1312-21 (1971).

6. M. McCoy, W. E. Lee, and A. H. Heuer, "Crystallization of Mg0-Al_2O_3-SiO_2-ZrO_2 Glasses," *J. Am. Ceram. Soc.*, **69** [3] 292-296 (1986).

7. W. Zdaniewski, "Crystallization and Structure of a MgO-Al_2O_3-SiO_2-TiO_2 Glass-Ceramic," *J. Mater. Sci.*, **8** [2] 192-202 (1973).

8. W. Zdaniewski, "DTA and X-Ray Analysis Study of Nucleation and Crystallization of MgO-Al_2O_3-SiO_2 Glasses Containing ZrO_2, TiO_2 and CeO_2," *J. Am Ceram. Soc.*, **58** [5-6] 163-169 (1975).

9. R. Chaim and A. H. Heuer, "Crystallization in a Barium-Containing Magnesium Aluminosicate Glass-Ceramic," *J. Amer. Ceram. Soc.*, **75** [6] 1512-21 (1992).

10. R. W. G. Wyckoff, "Miscellaneous Inorganic Compounds, Silicates and Intermetallic Compounds," *Crystal Structures*, Vol. 4, Interscience Publishers Inc., New York 1967.

11. E. P. Meagher and G. V. Gibbs, "The Polymorphism of Cordierite :II. The Crystal Structure of Indialite," *Can. Mineral.*, **15,** 43-49 (1977).

IMPROVED AlN POWDERS FOR ELECTRONIC APPLICATIONS

Thomas J. Mroz, Jr. and Eric A. Groat
Advanced Refractory Technologies, Inc., Buffalo, N.Y. 14207

ABSTRACT:

Aluminum nitride (AlN) is a leading candidate material for high thermal conductivity electronic substrates. Improved utilization of the material requires, in part, a reduction in processing costs. By improving the sinterability of AlN powders, the processing requirements and their costs can be minimized.

Sinterability of AlN powders can be enhanced by increasing the surface area of the powder. However, by increasing the powder surface area, the oxygen content is also increased. High oxygen contents typically lead to poor thermal conductivity in sintered parts.

The role of surface area as it relates to AlN sinterability is examined. The source of oxygen contamination, as well as methods of reducing the oxygen content prior to densification are described. The effect of these treatments on thermal conductivity are discussed.

INTRODUCTION:

Aluminum nitride has been receiving increased attention for thermal management applications in recent years. Many of the issues related to repeatably achieving useful properties have been resolved. Further expansion of aluminum nitride into this applications area will require reduction of costs associated with the final part. While cost reduction in the area of powder synthesis has been discussed for several years, cost reduction in processing is a relatively new topic.

The processing conditions required for producing acceptable AlN parts from existing commercial powders are well understood. However, in order to further reduce the costs associated with processing AlN, the cost components of the firing cycle must be reduced. Firing costs include the cost of the energy and gases used during the cycle, costs associated with the furnace construction, and the cost of replacement furnace parts. In each case, these costs can be shown to decrease with lower sintering temperature requirements.

In general, the reduction of the required sintering temperature of a powder mixture can be obtained by increasing the surface area of the powder. Higher surface areas enhance the drive for sinterability by increasing the free energy of the particle surfaces, and thus increasing the free energy of the entire system .[1]

This effect has been demonstrated previously for AlN powders.[2-3] In one study[2], increased sinterability was demonstrated for fine powders produced by classification as reproduced in Figure 1. The effect of increasing surface area was further separated from the effect of particle size distribution in this study by producing several different distributions with constant surface areas. This was facilitated by mixing appropriate portions of three different size distributions produced by classification. Table 1 shows the formulation for these experimental powders, as well as the distribution of each final blend as characterized by the difference between the D_{10} and D_{90} size values.

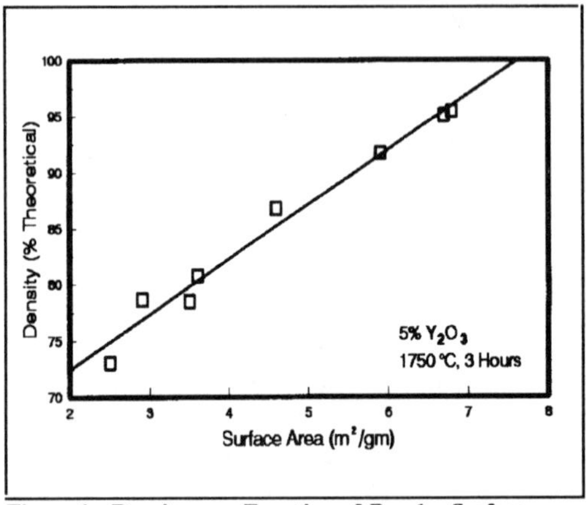

Figure 1. Density os a Function of Powder Surface Area for Classified AlN Powders

Table 1 AlN Powder Blend formulations for Sinterability Experiment.

	Sample A	Sample B	Sample C	Sample D	Sample E
% Coarse	---	16 %	32%	49%	Ball
% Fine	20%	30%	41%	---	Milled
% X-fine	80%	54%	27%	51%	Sample
Surface Area	3.92	3.97	3.88	4.07	3.86
D_{50}	1.3	1.3	1.8	2.0	1.4
D_{90}	2.3	2.4	3.5	3.8	3.2

The change in size distribution was found to have no effect on sinterability. Thus, any method of increasing surface area in powders such as increased milling, classification of fines, or reducing particle size during synthesis is expected to result in powders with increased sinterability.

In AlN powders however, the increase of surface area has also been shown to result in an increase in oxygen content.[2] As before, the various sample powders were prepared by classification. The raw powder feed to the classifier was obtained from the same source with surface area differences arising from various classifier settings. Thus, the effect of surface area on powder oxygen content was observed directly.

Figure 2 shows the oxygen content of AlN powders over a range of surface areas. The curve can be further broken down to describe several sources of oxygen in AlN powders. By extrapolating the best fit line back to a surface area value of zero, the lattice oxygen content can be determined. This value is comparable to values determined previously for AlN powders.[4] It is this oxygen that must be reduced in order to achieve high thermal conductivity.

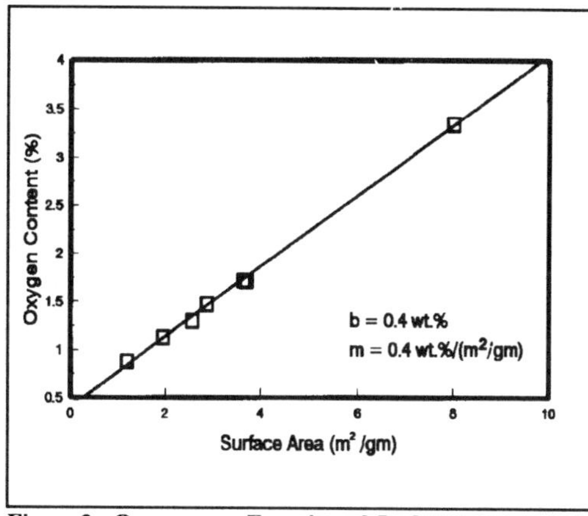

Figure 2. Oxygen as a Function of Surface Area for AlNel AlN Powder

The slope of the line describes the surface oxygen increase as it relates to increased surface area. The linear relationship found for oxygen as a function of surface area suggests that the oxygen layer thickness on each particle is independent of particle size or surface area. This is consistant with earlier statements proposing that the oxide layer on an AlN particle surface builds to a stable, self-protecting thickness when exposed to air at room temperature.[5-6]

Therefore, in order to utilize the potential cost savings of using finer AlN powders, it is necessary to determine processing methods that will reduce the detrimental effects of increased oxygen content. In this study, the effects of various sintering additives on the sintered phase composition and thermal conductivity will be examined.

EXPERIMENTAL:

A fine grade of AlN powder* was used exclusively in this study. The powder is prepared by a milling/classification method. The surface are of this powder is 4.0 m^2/gm, and the associated oxygen content is 2.0 percent. Table 2 gives the general characteristics of the powder.

---

* AlNel F, Advanced Refractory Technologies, Inc., Buffalo, NY, 14207

Table 2 AlNel F Powder Characteristics	
AlNel F	Characteristics:
Particle Size (D_{50})	1.3 um
Surface Area (m^2/gm)	4.0 m^2/gm
Oxygen Content (%)	2.0 %
Carbon Content (%)	0.1 %
Total Metals (ppm)	< 300 ppm

Powder mixtures were made by dry ball milling a combination of AlN, sintering additive, and binder together for several hours. The binder used was a blend of stearic and oleic acids. After an initial blending period of 2 hours, the entire sample within the mill jar was placed into a drier heated to 80 °C for one hour. The heat melts and partially vaporizes the binder, which then further segregates into the powder mixture. Prior to discharging the mill, the powder is again milled for 1 hour to homogenize the blend.

Several sintering additives were mixed with the powder at various concentrations. The additives; Y_2O_3, YF_3, and carbon were selected for their varied ability to incorporate and/or remove oxygen from the AlN body during sintering. The yttrium compounds were added at concentrations of 5% and 8%. Fine carbon powder at concentrations of 0.5% and 1% was added to some samples also containing Y_2O_3 at a standard concentration of 5%.

Powder samples were pressed at 10 kpsi in a uniaxial die with nominal dimensions 13 mm dia. by 3 mm high. Pressed powder samples were placed into a BN enclosure and loaded into a graphite resistance heated furnace. The binders were removed by slowly ramping the furnace temperature to 500 °C under nitrogen. After the binder burnout, the samples were sintered at 1850 °C for three hours.

The residual carbon resulting from binder burnout was characterized by direct determination via the fusion technique** on standard AlN - 5% Y_2O_3 samples exposed only to the burnout conditions, and compared to the native carbon content of the powders, and the carbon contents of as-pressed pellets.

Thermal diffusivity was measured on selected samples via a laser flash technique***. One sample at each condition was measured for thermal conductivity. Each sample

** Carbon analyzer, model 761-100, LECO, St. Joseph, MI. 49085

*** Flashline 5000, Anter Laboratories, Pittsburgh, P.A. 15235

was tested 5 times and the diffusion data was calculated from the averaged flash results. Thermal conductivity was calculated by multiplying the diffusivity of the sample with the measured sample density, and the heat capacity of the sample. The heat capacity for all samples was assumed to be 72 J/gm· K.

Various sintered samples were crushed to a relatively fine powder using a Spex mill. The resulting powder was characterized for bulk phases using X-ray diffraction (XRD)[****]. Concentrations of yttrium aluminate phases were determined quantitatively by comparing the most dominant peak of each yttrium aluminate species present against the sum total intensity of these peaks. This method assumed that the yttrium aluminate phases identified by XRD were the only ones present at any concentration, and that no free aluminum oxide was present other than the residual within the grains.

RESULTS/DISCUSSION:

The effect of oxygen content on thermal conductivity has been discussed elsewhere.[7-12] The general consensus suggests that it is the oxygen remaining within the AlN lattice which most significantly reduces the thermal conductivity of the bulk sample.

While the reaction between Y_2O_3 and the native Al_2O_3 surface layer is known to be the deoxidation reaction when sintering AlN, the effect of the composition of the $Y_2O_3 \cdot Al_2O_3$ on the deoxidation kinetics of the bulk has not been completely described. In several papers,[10-12] it is suggested that the increase in Al_2O_3 content in the $Y_2O_3 \cdot Al_2O_3$ phase results in a decreasing diffusion gradient from the bulk AlN grain to the grain boundary. At a point where the $Y_2O_3 \cdot Al_2O_3$ phase reaches the specific stoichiometry of $3Y_2O_3 \cdot 5Al_2O_3$, the grain boundary phase is saturated with oxygen, and the diffusion gradient through the grain is eliminated. All additional deoxidation must occur through a much slower process of reduction at the surface of the macroscopic body, which only occurs under specific furnace atmosphere conditions. Elimination of the lattice oxygen in the bulk grain is imperative for thermal conductivity development, as it is the dominant source of phonon scattering in the material.[7]

It can be speculated, conversely, that the exact composition of the $Y_2O_3 \cdot Al_2O_3$ phase within an AlN body *after* sintering might be correlated with the oxygen content within the AlN grains, and in turn, the thermal conductivity of the bulk sample. The phase compositions within an AlN body sintered with Y_2O_3 have been described in the form of a pseudoterinary phase diagram between AlN, Al_2O_3 and Y_2O_3. The specific stoichiometries of $2Y_2O_3 \cdot Al_2O_3$, $Y_2O_3 \cdot Al_2O_3$, and $3Y_2O_3 \cdot 5Al_2O_3$ (named in reference to their crystal structures as yttrium aluminum monoclinic (YAM), yttrium aluminum perovskite (YAP) and yttrium aluminum garnet (YAG) respectively, or by their yttria:alumina ratios as 2:1, 1:1, and 3:5 respectively) are

[****] Washington Mills, Niagara Falls, N.Y. 14302

shown to form at high temperatures under the appropriate conditions. The room temperature compositions are limited to YAM and YAG, as the YAP structure is unstable. Of course, under low oxygen conditions, Y_2O_3 may also be found in the sintered body.

In order to facilitate the removal of lattice oxygen, maximization of the oxygen concentration gradient across the grain is essential. By controlling the chemistry of the boundary phase to result in a low oxygen concentration after reaction with the native Al_2O_3 surface, this concentration gradient can be maximized.[10,12] Compositions that result in grain boundary phases of YAM, or Y_2O_3 and YAM should promote maximum thermal conductivity, as their relatively high reactivity with oxygen promotes the necessary diffusion gradient within the AlN grains. This conclusion has been suggested in several previous papers, and is seemingly the basis for a series of patents describing an AlN-based composition field .[14-17]

Control of the sintered composition can be facilitated several ways.[8-10] In some cases, control of the oxygen content of the AlN powder is used. However, in the case of high surface area powders, this means of control is not available.

In this work, the composition of the sintered body has been modified by selection of the additive chemistry. By adding additional Y_2O_3 beyond the 5% considered standard by the industry, the amount of oxygen that can be chemically combined is increased. YF_3 is also used, as it first converts to Y_2O_3 by reducing a portion of the surface alumina to AlN, then further converts to a yttrium aluminate upon reacting with additional alumina. It therefore has a higher oxygen affinity than Y_2O_3 and results in yttrium aluminates with lower alumina concentrations.

Carbon is used to carbothermically reduce some of the oxygen from the surface of the AlN grain prior to densification. In this case, the amount of oxygen requiring combination by the yttrium compound added as a sintering aid is reduced. Thus, an yttrium aluminate with lower alumina concentration can be produced at constant yttria concentrations.

The thermal conductivity of the experimental samples can be found in Table 3. In general, samples prepared with YF_3 were found to exhibit higher thermal conductivities than those sintered with the same weight of Y_2O_3. Samples sintered with additions of carbon had higher thermal conductivities than those sintered with only yttrium additives.

Figure 3 shows the thermal conductivity of all samples as a function of the ratio between

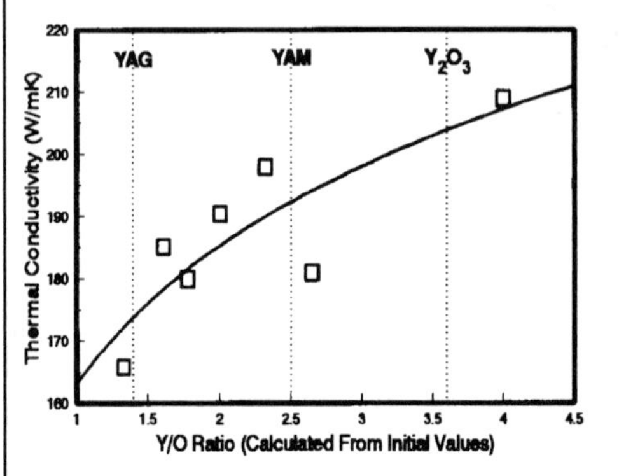

Figure 3. Thermal Conductivity as a Function of the Calculated Y/O Ratio

yttrium to oxygen as calculated from the initial powder composition. Expected transitions in the phase composition of the second phase are shown as they relates to the calculated Y/O ratio. As expected, the thermal conductivities of the experimental samples generally increased as the calculated Y/O ratio increased. While the thermal conductivity increases with increasing yttrium concentration, the relationship is non-linear with the rate of change decreasing with increasing yttrium concentration.

Table 3 Estimate Y/O Ratio and Thermal Conductivity Data Related to Sample Composition.

Additive	Estimated Y/O Ratio	Thermal Conductivity
5% Y_2O_3	1.33	166
8% Y_2O_3	1.78	180
5% YF_3	1.61	185
8% YF_3	2.65	181
5% Y_2O_3 + 0.5% C	2.0	190
5% Y_2O_3 + 1.0% C	4.0	209

XRD characterization of the sintered samples showed that all of the sintering aid in every case was in either the YAM or YAG form, or a combination of both. No free Y_2O_3 was found in any sample, however its presence has been observed in unrelated samples prepared under more extreme reducing conditions.

Figure 4 shows the thermal conductivity of all samples as a function of the amount of YAM phase detected in the sample. As shown before, the thermal conductivity of the samples increases with increasing Y/O ratio, signified here by increasing YAM phase content. Now, however, the relationship can be seen to be linear, with the exception of one unexplained data point. Thermal conductivities as high as 200 W/mK were achieved despite the high oxygen content of the starting powder using carbon as an internal reducing phase prior to densification.

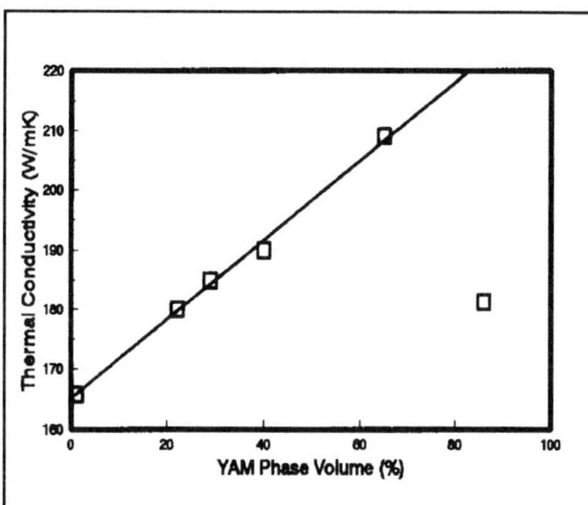

Figure 4. Thermal Conductivity as a Function of the Volume of YAM Phase

Figure 5. shows the actual Y/O ratio related to the calculated ratio. The actual ratio was calculated from the XRD measurements of the YAM and YAG phases. Though the relationship is linear (with the exception of one point), the actual concentration is lower than the calculated concentration. This effect is the subject of future experimentation.

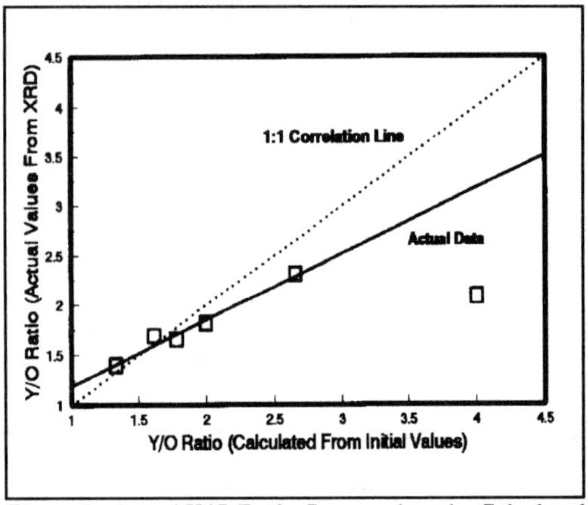

Figure 5. Actual Y/O Ratio Compared to the Calculated Y/O Ratio

CONCLUSION:

The sinterabilty of AlN powders can be increased by increasing the powder surface area. However, this results in higher powder oxygen contents. Several methods of reducing this additional oxygen with increased amounts of sintering aids, or with more efficient reducing aids have been demonstrated. A relationship between the composition of the second phase in sintered parts and the thermal conductivities of those parts was established. Sinterable powders with 2% oxygen were processed to produce parts with thermal conductivities as high as 200 W/mK.

REFERENCES:

1.) W. Kingery, H. Bowen, D. Uhlman, Introduction to Ceramics, John Wiley & Sons, Inc., New York, 1976

2.) T. Mroz, Jr., E. Groat, W. Hauth III, "New Electronic Grade Aluminum Nitride Powders and Resulting Parts Properties," Presented at the 94th Annual Meeting of the American Ceramic Society, Minneapolis, Minn, April 15, 1992

3.) N. Hashimoto, H. Yoden, "Sintering Behavior of Fine Aluminum Nitride Powder Synthesized From Aluminum Polynuclear Complexes," *J. Am. Ceram. Soc.,* **75** 8 (1992)

4.) A. Thomas, G. Muller, "Alteration of Oxygen Content During Powder Conditioning and Sintering of AlN Ceramics, *cfi/Ber. DKG,* **67** [4] 146-49 (1990)

5.) G. Slack, T. McNelly, "Growth of High Purity AlN Crystals," *J. Cryst. Growth,* **34** 263-279 (1976)

6.) P. Bowen, J. Highfield, A. Mocellin, T. Ring, "Degradation of Aluminum Nitride Powder in an Aqueous Environment," *J. Am. Ceram. Soc.*, **73** [3] 724-28 (1990)

7.) G. Slack, "Nonmetallic Crystals with High Thermal Conductivity," *J. Phys. Chem. Solids,* **34** 321-335 (1973)

8.) R. Lee, "Development of High Thermal Conductivity Aluminum Nitride Ceramic," *J. Am. Ceram. Soc.,* **74** [9] 2242-49 (1991)

9.) K. Watari, M. Kawamoto, K. Ishizaki, "Sintering Chemical Reactions to Increase Thermal Conductivity of Aluminum Nitride," *J. Mat. Sci.,* **26** 4727-4732 (1991)

10.) H. Buhr, G. Muller, H. Wiggers, F. Aldinger, P. Foley, A. Roosen, "Phase Composition, Oxygen Content, and Thermal Conductivity of $AlN(Y_2O_3)$ Ceramics," *J. Am. Ceram. Soc.,* **74** [4] 718-23 (1991)

11.) W. Lee, S. Chiang, D. Readey, R. Donn, P. Shaffer, "Relation Between Thermal COnductivity, Sintering Mechanism, and Microstructure of ALN with Yttrium Aluminate Grain Boundary Phases, *J. Mat. Sci; Mater. in Electr.,* **3** 93-101 (1992)

12.) A. Virkar, T. Jackson, R. Cutler, "Thermodynamic and Kinetic Effects of Oxygen Removal on the Thermal Conductivity of Aluminum Nitride," *J. Am. Cer. Soc.*, **72**, [11], 2031-42, 1989.

13.) W. Sun, Z. Huang, T. Tien, T. Yen, "Phase Relationships in the System Y-Al-O-N," *Mat. Lett.,* **11** [3-4] 67-69 (1991)

14.) Huseby, I., High Thermal Conductivity Aluminum Nitride Ceramic Body., U.S. Patent 4,533,645, Aug. 6, 1985.

15.) Huseby, I., High Thermal Conductivity Aluminum Nitride Ceramic Body., U.S. Patent 4,547,471, October 15, 1985.

16.) Huseby, I., Pressureless Sintering Process to Produce High Thermal Conductivity Ceramic Body of Aluminum Nitride., U.S. Patents 4,578,232, 4,578,233, 4,578,234, 4,578,364, 4,578,365, March 25, 1986.

17.) Huseby, I., Process of Pressureless Sintering to Produce Dense, High Thermal Conductivity Aluminum Nitride Ceramic Body. U.S. Patent 4,478,785, October 23, 1984.

SINTERING AND GRAIN BOUNDARY PHENOMENA OF Y_2O_3 DOPED AlN CERAMICS

Nobuyasu Mizutani, Kazuo Shinozaki and Takeshi Yagi
Department of Inorganic Materials, Faculty of Engineering,
Tokyo Institute of Technology,
2-12-1, O-okayama, Meguro-ku, Tokyo 152, Japan

ABSTRACT

In the sintering process of AlN ceramics with 5-10wt% Y_2O_3 as a sintering aid, the formation of grain boundary phase and the change of AlN grain were investigated by XRD and TEM. The compounds of grain boundary phase cooled from high temperature were well crystallized and did not contain either amorphous regions or Al_2O_3 and Y_2O_3 as a single oxide.

The sintered AlN ceramics with Y_2O_3 were further heated in a carbon gas reducing atmosphere at 1900°C. By this processing, the grain boundary phase migrated from the bulk to the surface of the specimen and finally was completely taken off from the specimen. The elimination of the grain boundary phase resulted in the shrinkage of the specimen and the grain growth of AlN.

By heating a hot pressed laminate consisting of a) AlN ceramics without Y_2O_3 and b) Y_2O_3 doped AlN ceramics in a carbon gas reducing atmosphere at 1900°C, the diffusion of liquid grain boundary phase into the hot pressed AlN ceramics took place and brought about the formation of some Y_2O_3-Al_2O_3 compounds by the reaction of the grain boundary phase and Al_2O_3 on the surface of AlN grain.

Finally, the increases of the thermal conductivity of AlN ceramics by the elimination of grain boundary phase and by passage of the grain boundary phase into AlN ceramics were discussed on the base of the interaction of grain boundary and grain.

1. INTRODUCTION

Aluminum nitride (AlN) is one of the promising materials with a very high thermal conductivity. The theoretical thermal conductivity of single crystal AlN is 319W/mK[1,2], which is 10 times higher than that of alumina. However, owing to low sinterability and cation and/or oxygen impurities, the thermal conductivity of AlN ceramics was low. It was found that some additives such as Y_2O_3, CaO and $Ca(NO_3)_2$[3,4], which densified AlN to nearly theoretical density, were very effective for the increase of thermal conductivity of AlN ceramics by scavenging lattice oxygen in AlN. It is very important for the improvement of properties and processing to understand the grain boundary.

In this paper, our recent studies concerning the grain boundary of AlN are presented. Especially, the characterizations of grain boundary phases using TEM[5,6], the migration of liquid grain boundary phase by the reduction with carbon gas at high temperature, the diffusion of grain boundary into hot pressed AlN body without additive from the pressureless sintered body with Y_2O_3, and the effect of the amount of the grain boundary present on the thermal conductivity are discussed.

2. PHASE CHEMISTRY OF GRAIN BOUNDARY REGION

Various percent Y_2O_3 powder was added to AlN powder and mixed with Al_2O_3 balls in a ball mill for 4h using butanol as a solvent. Binder was added after the solvent dried. The mixed powder was formed in a metal mold of 20 mm diameter with uniaxial pressure of 300kg/cm^2. Then the powder thus formed was dewaxed at 500°C for 3h in N_2 before it was densified at 1800°C for 1h in N_2.

Figure 1 shows the compounds present in the grain boundary versus the amount of the added Y_2O_3. By increasing Y_2O_3 content, the compound in the grain boundary changes from $Y_3Al_5O_{12}$ to $Y_4Al_2O_9$. $Y_3Al_5O_{12}$ is the compound with the largest Al_2O_3 content in the system of Y_2O_3-Al_2O_3. The kind of compounds present in the grain boundary depends on the ratio of the oxygen content of AlN starting powders as impurity and the amount of the added Y_2O_3.

The bright-field micrograph and energy dispersive X-ray spectra of specimen sintered at 1800°C, 2h are shown in Figure 2. From this figure, the microstructure of this sample composed of grains and grain boundaries. The EDX spectra of grain consist of only AlKα and suggested that the grain is AlN and does not contain Y, while the grain boundary contains both Al and Y from EDX spectra. Grains are round and their average size is 2-3μ m. The electron diffraction patterns of grain boundary phase of AlN ceramics with 10%Y_2O_3(10Y) are shown in Figure 3. The pictures (a), (b) and (c) in Figure 3 can be indexed as [100] for $Y_4Al_2O_9$, [129] for $YAlO_3$ and [157] for $Y_3Al_5O_{12}$. Figure 4 is the lattice image picture of $Y_4Al_2O_9$ taken using the diffraction pattern of Figure 3(a).

These rounded grain and the formation of grain boundary phase suggest that the sintering proceeds by the mechanism of the liquid phase sintering. That is, the compounds in Y_2O_3-Al_2O_3 system are formed by the reaction of Y_2O_3, Al_2O_3 in the starting AlN powder and oxygen present in the AlN particles as impurity. These grain boundary compounds form liquid

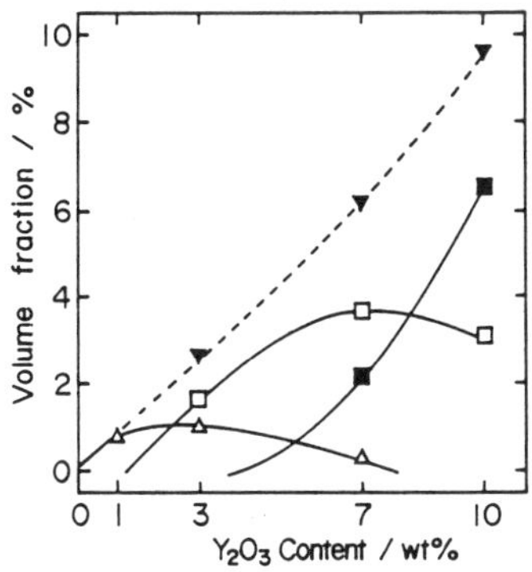

Figure 1 Relationship between amount of Y_2O_3 and volume fraction of various grain boundary compounds. △:$Y_3Al_5O_{12}$, □:$YAlO_3$, ■:$Y_4Al_2O_9$, ▼:sum of grain boundary compounds

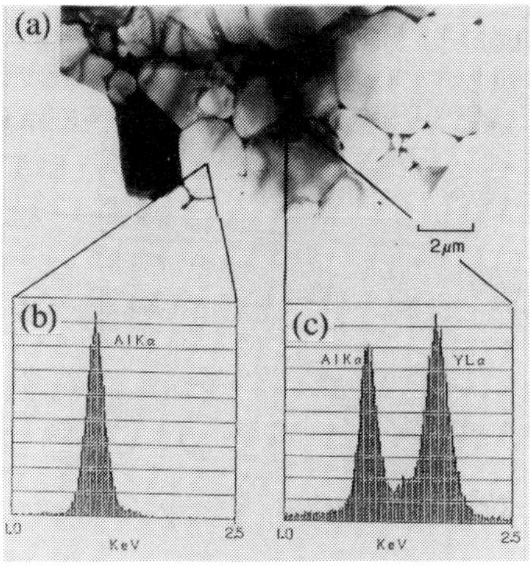

Figure 2 (a) Bright field image and EDX spectra of (b) inter-grain phase and (c) grain boundary phase of samples added with 7wt% of Y_2O_3 and sintered at 1800°C for 2 hours.

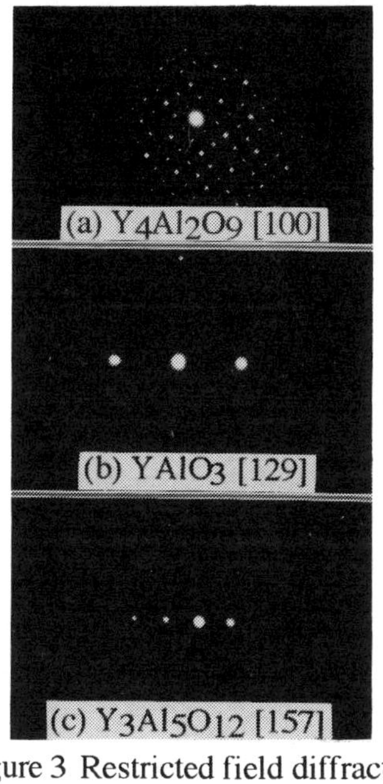

Figure 3 Restricted field diffraction pattern of grain boundary phase of the sample.

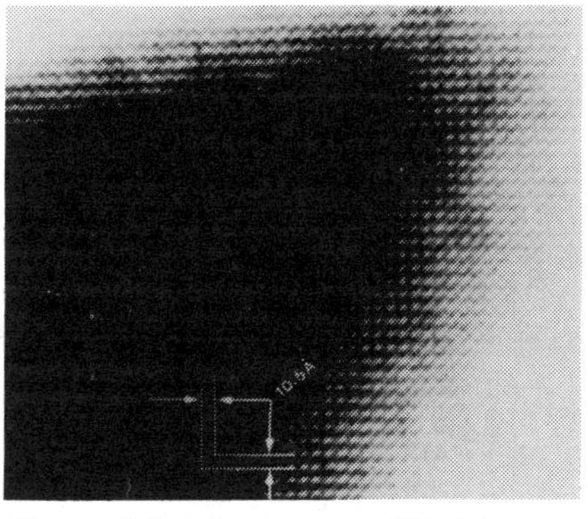

Figure 4 Lattice image of $Y_4Al_2O_9$ grain boundary phase.

phase at sintering temperature around 1600-1700°C[7] and on cooling crystallized into each compound again. The liquid phase promotes the rearrangement of particles into densification and also to bring about the grain growth by dissolution and precipitation.

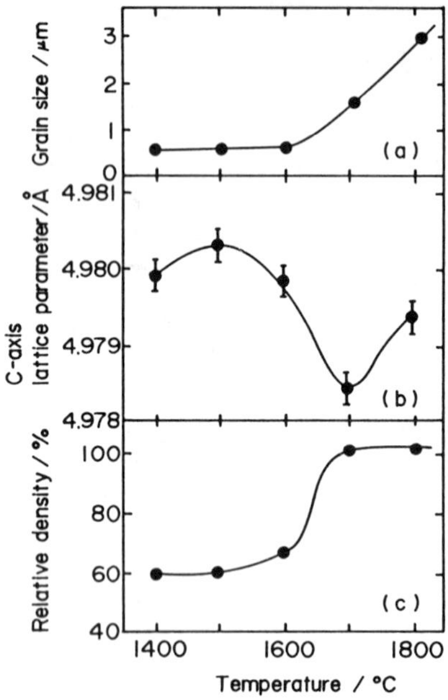

Figure 5 (a) Grain size, (b) c-axis lattice parameter and (c) relative density of AlN ceramics added with 7wt% of Y_2O_3 and sintered at various temperature.

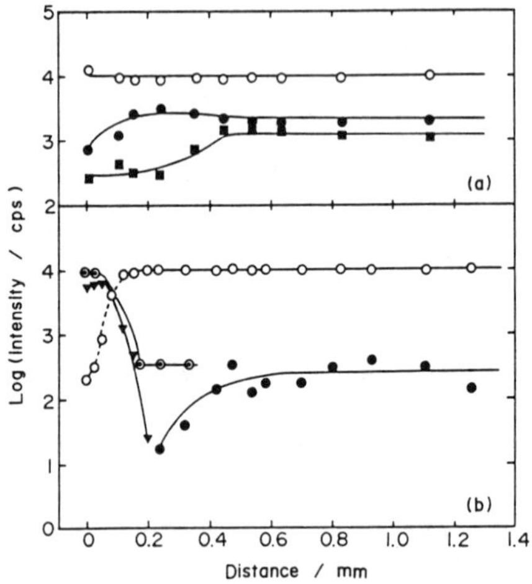

Figure 6 Change of X-ray diffraction intensity of phases in AlN ceramics, added 10wt% Y_2O_3, (a) sintered at 1800°C for 1h in N_2 and (b) sintered at 1800°C for 1h in N_2 and annealed at 1900°C for 6h in reduced atmosphere with the distance from the surface of body.
○:AlN (100), ▼:YN (111)
■:$Y_4Al_2O_9$ (221),
●:Y_2O_3 (222), ⊙:AlN (002)

It is important to investigate the behavior of the AlN grains on sintering. Figure 5 shows the relation of c-axis lattice parameter of AlN, density and grain size versus sintering temperature for 2h. The c-axis is almost constant up to 1500°C and becomes short from 1550°C but lengthens again at 1700°C. While the shrinkage begins slowly from 1450°C and proceeds rapidly around 1600°C and finished about 1700°C[8,9]. The grain size also increased from 1600°C. The formation of the solid solution with AlN and oxygen results in the decrease of the lattice parameter[1,10]. While the liquid phase begins to form at 1600-1700°C and picks up oxygen from the AlN grain, that is "Trapping effect". It is understood from above situation that the change of c-axis is due to competitive reaction both the reaction of AlN and oxygen, and this "Trapping effect". In higher temperature, the grain boundary phase seems to be melt. When liquid phase is cooled, the phase separation occurs. The above situation in this experiment is basically agreed with the phase diagram of Y_2O_3-Al_2O_3[11) but is not always agreed with phase diagram such as the liquidus temperature.

3. MIGRATION OF GRAIN BOUNDARY PHASE TO THE SURFACE OF THE SPECIMEN IN CARBON GAS REDUCING ATMOSPHERE[12)

Figure 6(a) shows the results of X-ray diffraction with distance from the surface of AlN ceramics. Although the ratio of $Y_4Al_2O_9$ to Y_2O_3 in the grain boundary from the surface to 300μ m is not uniform, the inner region deeper than 300μ m is uniform. Both sides of the surface of the AlN ceramics used for annealing in the reduced atmosphere were abraded by polishing by about 0.5mm so as to obtain a uniform phase formation of whole product.

The sintered and polished compact was placed in a carbon container for annealing in the reduced atmosphere at 1900°C for 1, 3, 6 and 24h. In this paper, the sintered compact densified before annealing in the reduced atmosphere is described as the starting specimen.

Figure 6(b) shows the changes in the strength of X-ray diffraction of phases due to the thickness of the AlN ceramics with 10Y when annealed at 1900°C for 6h in the reduced atmosphere. $Y_4Al_2O_9$ of the grain boundary phase that existed after densification was no longer detected at any position of the AlN ceramics. The amount of Y_2O_3 also decreased at any position of the AlN ceramics, and not detectable in the neighborhood of the surface. However, YN was detected in a position most close to the surface, where the strength of (100) peak of AlN was decreased. While the intensity of the (002) peak of AlN was stronger than that in the powder condition, in a position of about 200μ m from the surface. In other words, AlN in this region was oriented in the c-axis, forming a surface layer different from the inner layer.

Figure 7(a)(b) shows the annealing time and the volume fraction of the grain boundary phase of the AlN ceramics. It is clear from this figure that the volume fraction occupied by the grain boundary phase decreased in proportion to the duration for annealing. $Y_4Al_2O_9$ decreased, and Y_2O_3 alone remained when processed for 3 h or more. Horiguchi et al.[13) reported that when a specimen with additional 5wt% Y_2O_3(5Y) was sintered in a reducing atmosphere, the boundary phase was divided into two phases of $Y_4Al_2O_9$ and Y_2O_3 when sintered for less than 6 h, while Y_2O_3 alone remained 12 h later, but even Y_2O_3 no longer existed after 96 h.. In other words, a decrease of grain boundary phases due to the decomposition of $Y_4Al_2O_9$ was predominant at an early stage of annealing up to 3 h, but since the increased amount of Y_2O_3 caused by the decomposition was not removed, the decreased rate of grain boundary phases was smaller.

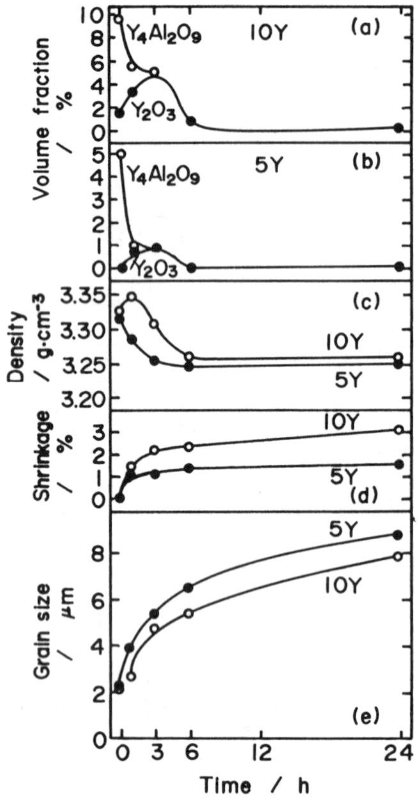

Figure 7 (a)(b) Change of content of grain boundary phases, (c) density, (d) linear shrinkage and (e) grain size of specimen annealed at 1900°C in reduced atmosphere with time.

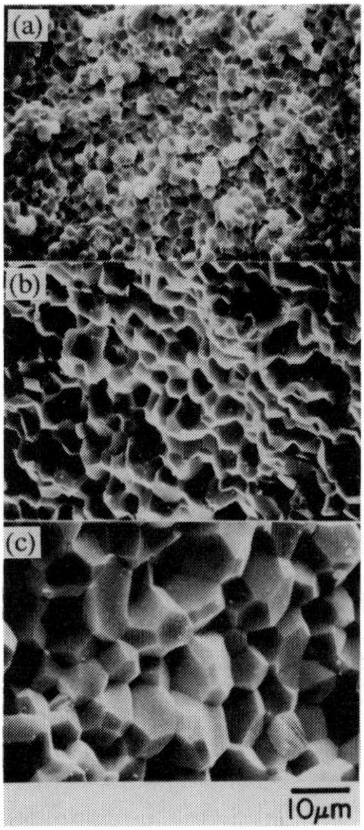

Figure 8 SEM photographs of fracture surfaces of AlN ceramics, containing 5wt% Y_2O_3, (a) before annealing and annealed at 1900°C for (a) 1h and (c) 24h in reduced atmosphere.

After this mixture was dewaxed and densified at 1800°C for 1 h in the N_2 atmosphere, it was cut into half and the section was abraded before the distance between W wires on both sides was measured under an optical microscope. The grain boundary phases with a higher density than that of $Y_4Al_2O_9$ and Y_2O_3 existed in the densified AlN ceramics. Thus, The AlN ceramics before annealing showed 3.26g/cm^3, a density higher than that AlN alone. As shown in Figure 7(c), the density becomes lower in proportion to the duration for annealing, since the grain boundary phases were discharged from the AlN ceramics. The relative density of specimens other than that of starting specimens mixed with 10Y was approximately 100%, when it was based on the density of AlN ceramics obtained from the results of the quantitative determination of grain boundary phases.

Figure 7(d) shows the shrinkage factor of AlN ceramics against the duration of annealing. The AlN ceramics shrink in proportion to the discharge of grain boundary phases, when annealed for a long period. The shrinkage was larger at an early stage of annealing, but it became extremely smaller thereafter, although it remained larger up to 1h, particularly in a case of 5Y.

Figure 7(e) shows the changes of the grain size in the central region of AlN ceramics, which were measured from SEM pictures against the duration for annealing. As for the starting specimens, the grain size of both 10Y and 5Y were almost as large as 2μ m. Both sizes became larger when annealed, but the size of the latter was larger than that of former. Although 5.5vol% of the grain boundary phases existed in 10Y, only 0.9vol% existed in 5Y when processed for 1 h. Thus, the growth of grains was conceivably inhibited, because of higher sprigging effects in the grain boundary for 10Y.

Figure 8 shows the SEM picture of fracture surface of the specimen with 5% Y_2O_3 annealed at 1900°C in a reduced atmosphere for various annealing time.

4. DIFFUSION OF GRAIN BOUNDARY PHASE TO AlN SINTERED BODY WITHOUT Y_2O_3[14)]

The joined samples used in our experiments were prepared by the following procedure. A sintered body of AlN containing Y_2O_3 was prepared by mixing AlN powder (oxygen content 0.09%) with 10% by weight of Y_2O_3 powder, adding binder, and then forming by uniaxial pressing under a pressure of 200kg/cm^2, using a metal mold of 30x30mm size. This body was dewaxed in a N_2 atmosphere at 1800°C for 1 our (this sample is called the sintered body). A hot-pressed AlN body was prepared by mixing. AlN powder without Y_2O_3 and binder, then forming by uniaxial pressing under a pressure of 200kg/cm^2, using a metal mold 30x30mm. After dewaxing, this was densified by hot-pressing in a N_2 atmosphere under a pressure of 400kg/cm^2 at 1800°C for 1h (this hot-pressed body is called the HP body). The surfaces of both the sintered body and HP body were smoothed by polishing with a grinder using a 600# stone.

W wires (20μ m diameter) were placed as markers along the interface between a sintered body and a HP body, as shown in Figure 9, and the sintered body and the HP body were joined by a hot-pressing technique. This joining was carried out at 1800°C under a pressure of 200kg/cm^2 for 10 min. The surface of the joined sample was ground off in the thickness direction by several tens to a hundred and several tens of μ m at a time, and the composition of the phases determined from the X-ray diffraction peaks obtained for each polished surface.

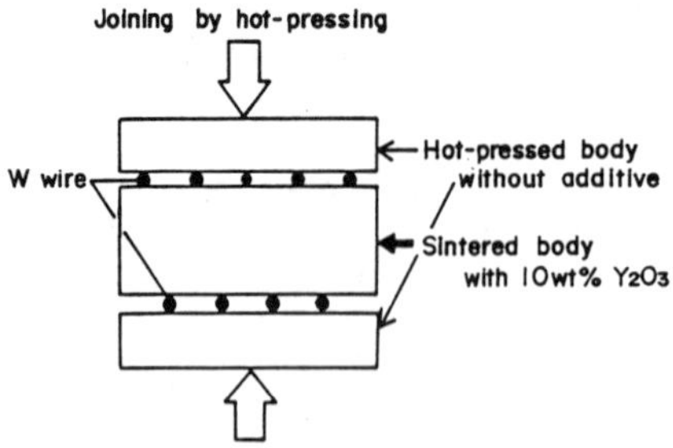

Figure 9 Schematic diagram of joining hot-pressed and sintered body

Figure 10(a) shows the distribution of grain boundary phases in the starting sample in the thickness direction. $Y_4Al_2O_9$ and Y_2O_3, which are the components of the grain boundary phase in the sintered body, were not detected in the HP body. Both $Y_4Al_2O_9$ and Y_2O_3, which constitute the grain boundary phases, were distributed homogeneously in the sintered body before joining, but after joining the Y_2O_3 content decreased and $Y_4Al_2O_9$ content increased in a region about 0.1mm from the joining interface on the sintered body side, in comparison to the components in the inside of sintered body. Since water was used in the process of polishing the surfaces of both the sintered body and the HP body prior to joining, it is possible that aluminum hydroxide was formed in a layer on the polished surface of the sample. It is assumed that this aluminum hydroxide was changed to aluminum oxide at the high temperature during the joining treatment, and this aluminum oxide reacted with Y_2O_3 to form $Y_4Al_2O_9$.

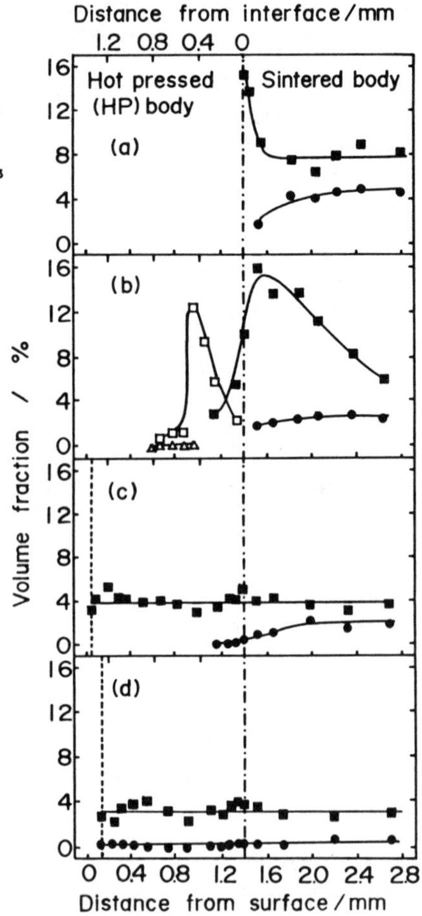

Figure 10 Diffusion profile of joined HP and sintered body annealed at 1900°C for (a) 0min, (b) 1min, (c) 0.5h and (d) 3h.
■:$Y_4Al_2O_9$, ●:Y_2O_3
□:$YAlO_3$, △:$Y_3Al_5O_{12}$

Figure 10(b)-(d) shows the distribution of the grain boundary phases by components in the thickness direction when the joined sample was annealed at 1900°C in a reducing atmosphere for 1min(b), 0.5h(c) and 3h(d). In the case of the sample at 1900°C for 1min, the components of the grain boundary phase were observed to cross the joining interface and to migrate into the HP body as far as about 0.7mm from the joining interface. However, they did not reach the outer surface of the sample. From the distributions in the grain boundary phases, $Y_4Al_2O_9$ increased from the inside of the sintered body toward the joining interface, reached a maximum value, and entered the HP body across the joining interface. In the HP body, the $Y_4Al_2O_9$ content decreased with increased distance from the joining interface, so that it was not detected at a distance of about 0.3mm from the joining interface.

$YAlO_3$ began to be detected in place of $Y_4Al_2O_9$, increased with an increased distance from the joining interface until it reached maximum at a distance of about 0.4mm from the joining interface, and then decreased but was still present at a distance of about 0.7mm from the joining interface. $Y_3Al_5O_{12}$ was present in the region of about 0.4 to 0.8 mm from the joining interface. Y_2O_3 slightly decreased in the sintered sample in the vicinity of the joining interface, and was not detected in the HP body across the joining interface.

Next, in the case of the joined sample annealed at 1900°C for 0.5h, $Y_4Al_2O_9$ reached the outer surface of the joined sample, and Y_2O_3 migrated to a position about 0.2mm beyond the joining interface. In addition, a surface layer was observed which considered of AlN, YN and Y_2O_3.

In the case of the sample annealed for 3h, both $Y_4Al_2O_9$ and Y_2O_3 were evenly distributed through the whole sample. Since both $Y_4Al_2O_9$ and Y_2O_3 had already reached the sample surface after 1h of annealing, the $Y_4Al_2O_9$ concentration decreased due to the decomposition by nitridation, and the Y_2O_3 concentration decreased due to nitridation during the longer annealing time between 1h and 3h, resulting in accelerating the migration of the phases in the inside of the sample. It is observed that AlN grains in the sintered body and in the HP body are continuous at the joining interface, and the grain boundary phases have not migrated selectively along the joining interface.

It is shown, therefore, that the grain boundary phases in the sintered body have migrated to the sample surface by way of the grain boundaries in the HP body. It is further clear those grain boundary phases in the sintered body can migrate through the grain boundaries where the AlN grains are in contact without the involvement of the grain boundary phases.

5. GRAIN BOUNDARY AND THERMAL CONDUCTIVITY

AlN is a high electrical resistance material and has a high thermal conductivity by phonon. In AlN ceramics the average thermal conductivity of compounds present in a grain boundary is below a few W/mK[15] as compared with theoretical value of 340W/mK of AlN crystal[1]. On considering the thermal conductivity of AlN ceramics, it is important to characterize the presence of grain boundary and the oxygen impurity in AlN grain. Figure 11(a)(b) shows change of the c-axis lattice parameter with the reduction time at 1900°C. In both 5 and 10% Y_2O_3, c-axis increased rapidly after 3h-reduction and is almost constant after 6h-reduction.

Figure 11(c) shows the change of the thermal conductivity with reduction time at 1900°C. With the reduction time, the thermal conductivity quickly increases to became constant around 6h. This behavior of thermal conductivity is similar to the change of c-axis. The increase of thermal conductivity up to 3h-reduction and 6h-reduction results in the decrease of grain boundary phase and of oxygen content in grains respectively. In Figure 12, the thermal conductivity of the reduced AlN ceramics increases with the reduction temperature for 24h. The thermal conductivity of 5Y sample is higher than that of 10% Y_2O_3 within this experimental temperature range. This is due to the residue of grain boundary after elimination of grain boundary within the constant reduction time.

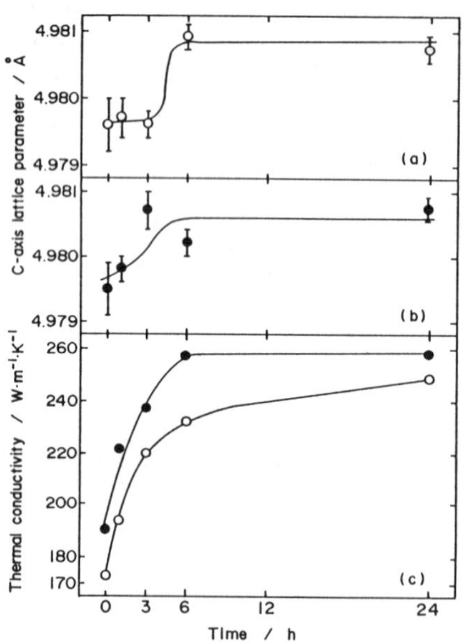

Figure 11 (a)(b) C-axis lattice parameter and (c) thermal conductivity of AlN ceramics containing (○) 10wt% and (●)5wt% Y_2O_3 and annealed at 1900°C in reduced atmosphere for various time.

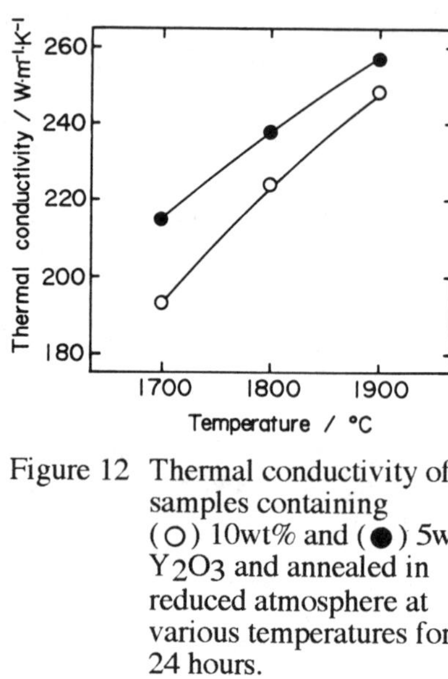

Figure 12 Thermal conductivity of samples containing (○) 10wt% and (●) 5wt% Y_2O_3 and annealed in reduced atmosphere at various temperatures for 24 hours.

Figure 13(a)(b) shows the change of c-axis lattice parameter of the hot pressed AlN body without Y_2O_3 with reduction time in carbon atmosphere at 1900°C. In this case, this HP sample was joined with the AlN body sintered with Y_2O_3 (10wt%). The liquid grain boundary phase passed through the HP sample to be eliminated from the surface of the other side of HP sample as mentioned above. It is suggested to decrease the oxygen

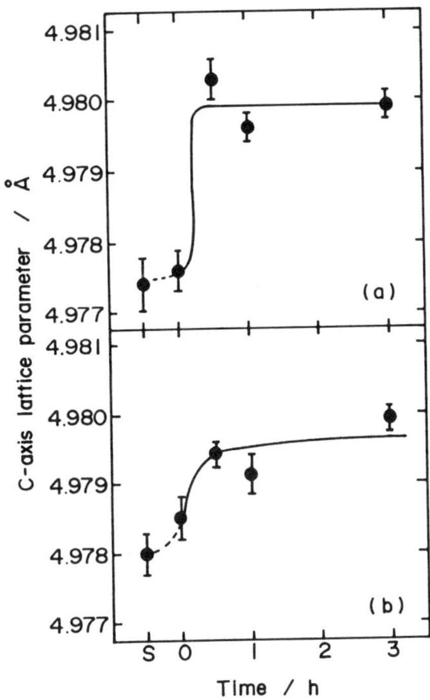

Figure 13 C-axis lattice parameter of the starting AlN sintered body (S) and those of the hot-pressed body derived from that sintered body and annealed at 1900°C in reduced atmosphere for various length of time, samples containing (a) 10 and (b) 5 wt% Y_2O_3 respectively.

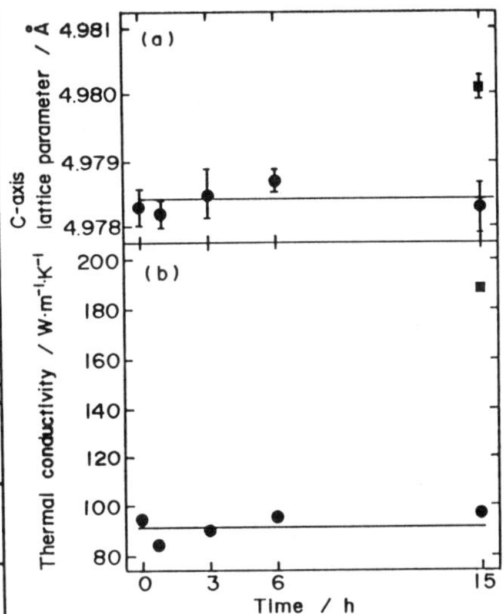

Figure 14 (a) C-axis lattice parameter and (b) thermal conductivity of hot-pressed body annealed at 1900°C in reduced atmosphere for various length of time (●), and those of the hot-pressed portion of the joined samples (10wt% Y_2O_3) annealed at 1900°C in reduced atmosphere for 15 hours (■).

content in AlN grain that the c-axis increased quickly only for 30 min. While in Figure 14(a), the c-axis of the HP sample without Y_2O_3 which was not joined with the Y_2O_3 added sample, seems to be unchanged against a long time reduction. In figure 14(a), the only one point of c-axis (■) corresponding to 15h-reduction is from the HP sample joined with the Y_2O_3 (10wt%) added sample. In this sample, the c-axis increased to the almost same value as maximum value in Figure 13(a). Figure 14(b) shows the change of thermal conductivity of samples in Figure 14(a) with reduction time. The thermal conductivity of unjoined HP sample did not increase but that of joined HP sample increased up to 190W/mK for 15h reduction.

From the experimental results mentioned above, it can be concluded that the liquid grain boundary phases pick up the oxygen from the AlN grain, decrease the oxygen content and finally increases the thermal conductivity.

CONCLUSIONS

(1) The added Y_2O_3 reacts with oxygen present as amorphous Al_2O_3 on the surface of AlN grain to form some compounds in the system of Y_2O_3-Al_2O_3. These compounds are melt to form the liquid phase at 1600-1700°C. These liquid phases play a role of the rearrangement of AlN particles and also pulling off oxygen from AlN grain and finally on cooling are crystallized into each grain boundary compound. The kind of compounds formed in the grain boundary depends on the ratio of Y_2O_3 and impurity oxygen. Typical compounds are $YAlO_3$, $Y_4Al_2O_9$ and $Y_3Al_5O_{12}$.
(2) The AlN sintered specimens with Y_2O_3 were annealed in a carbon gas reduced atmosphere. The grain boundary phases immigrate from the bulk to the surface. $Y_4Al_2O_9$ was decomposed into Y_2O_3 and Al_2O_3, which were nitrided on the surface to YN and AlN respectively. AlN of the surface layer was oriented in a c-axis. With the elimination of grain boundary phase at 1900°C, the volume shrinkage occurred and the grain grew.
(3) A HP body of AlN without any additives and a sintered body of AlN containing 10wt% Y_2O_3 were joined and annealed at 1900°C in a carbon gas reduced atmosphere. The grain boundary phases in the sintered body migrated to the HP body across the joining interface due to the concentration gradient of the grain boundary phases between the HP body and sintered body. After the grain boundary phases reached the outer surface of HP body, the grain boundary phases were decomposed and subsequently nitrided to form different phases in a surface layer.
(4) The decrease in grain boundary phase phase must be accompanied by removal of more oxygen from AlN lattice. According to Slack[1)], phonon mean free path is 0.01μm and hence grain boundary by itself should not affect thermal conductivity.

(5) When the liquid grain boundary phase passed through the HP body of AlN without Y_2O_3, whose thermal conductivity was very low, the thermal conductivity of the HP body was improved into 180W/mK.

ACKNOWLEDGMENTS

Authors thank for Prof. Y. Sawada of Tokyo Institute of Polytechique as a coworker. and Toshiba Co. for the measurement of the thermal conductivity. This study was partially supported by the Research Foundation for Materials Science and the Iwatani Memorial Foundation.

REFERENCES

1) G. A. Slack, J. Phys. Chem. Solids, 34 321 (1973), "Nonmetallic Crystals with High Thermal Conductivity"
2) G. A. Slack, R. A. Tanzilli, R. O. Pohl and J. W. Vandersande, J. Phys. Chem. Solids, 48 641 (1987), "The Intrinsic Thermal Conductivity of AlN"
3) K. Komeya and H. Inoue, J. Mat. Sci., 4 1045 (1969), "Sintering of Aluminum Nitride : Particle Size Dependence of Sintering Kinetics"
4) K. Komeya, H. Inoue and A. Tsuge, J. Ceram. Soc. Japan, 89 330 (1981), "Effect of Various Additives on Sintering of Aluminum Nitride"
5) T. Yagi, K. Shinozaki, N. Ishizawa, N. Mizutani, M. Kato and A. Tsuge, J. Am. Ceram. Soc., 71 C334 (1988), "Effect of Silicon Dioxide on the Thermal Diffusivity of Aluminum Nitride Ceramics"
6) T. Yagi, K. Shinozaki, N. Mizutani and M. Kato, J. Mat. Sci., 24 1332 (1989), "Thermal Diffusivity of SiO_2 and Y_2O_3 Added AlN Ceramics"
7) M. Kasori, F. Ueno and A. Tsuge, Proc. Jpn. Ceram. Soc. the 27th Ceram. Basic Sci. Meeting, pp171 (1989), "Sintering behavior and thermal conductivity of AlN"
8) E. Streichter, T. Chartier, P. Boch, M. Denanot and J. Ralier, J. Europ. Ceram. Soc., 6 23 (1990), "Densification and thermal conductivity of low-sintering-temperature aluminum nitride materials"
9) T. Troczynski and P. S. Nicholson, J. Am. Ceram., Soc., 72 1488 (1989), "Effect of Additives on the Pressureless Sintering of Aluminum Nitride between 1500° and 1800°C"
10) J. Harris, R. Youngman and R. Teller, J. Mater. Res., 5 1763 (1990), "On the nature of the oxygen-related defect in aluminum nitride"
11) N. A. Torov, I. A. Bondavi, F. Y. Galokov, K. S. Nikogosyan and N. V. Ninigradova, "Phase Diagram for Ceramics", Fig. 2344, Am. Ceram. Soc. Inc., (1964)
12) T. Yagi, K. Shinozaki, N. Mizutani, M. Kato and Y. Sawada, J. Ceram. Soc. Japan, 97 1372 (1989), "Migration of Grain Boundary Phases in AlN Ceramics by Heating in a Reducing Atmosphere"

13) A. Horiguchi, F. Ueno, M. Kasori, K. Shinozaki and A. Tsuge, 25th Basic Ceramic Meeting, Proceeding Abstract, p155 (1987), "Effect of Sintering Atmosphere on the Thermal Conductivity and Microstructure of AlN Ceramics"
14) T. Yagi, K. Shinozaki, Y. Sawada and N. Mizutani, J. Ceram. Soc. Japan, 98 198 (1990), "Migration of Grain Boundary Phase of AlN Ceramics in Joined Sample of Sintered and Hot Pressed Body"
15) T. Sakai, Yogyo-Kyoukai-Shi, 86 125 (1978) (J. Ceram. Soc. Jpn.), "Hot Pressing of the AlN-Al2O3 System"

PHOSPHATE GLASS-POLYMER EMULSIONS

George H. Beall and Candace J. Quinn, Corning Incorporated, Sullivan Park FR-51, Corning, NY 14831

INTRODUCTION

A new family of hybrid organic-inorganic materials has recently been developed through melt mixing and injection molding of high-temperature thermoplastics and low-temperature phosphate glasses.(1) Up to 85 wt% or nearly 70 vol% of the inorganic glass can be successfully compounded with the polymer because the glass softens at the processing temperature. This contrasts with a practical processing limit for injection molding with typical inorganic materials used as fillers at approximately 20-40 vol%.(2) The key to development of these fine mixtures called Cortem™ glass/polymer blends was the discovery of a family of low-temperature zinc alkali phosphate glasses with good chemical durability and sufficiently low softening temperatures to be blended as viscous liquids with many of the higher temperature thermal plastic polymers including polyarylether ketones, aromatic liquid crystal polyesters (LCPs), polyaryl sulfones, perfluor-alkoxy resins, and polyetherimides. Because of the high loadings and fine microstructures involved, useful homogeneous properties such as high elastic modulus, good thermal and dimensional stability, and increased hardness and flame resistance over standard polymer blends have been achieved.

NEW LOW-TEMPERATURE GLASS COMPOSITIONS

Glass polymer blends are based on durable low-temperature phosphate glasses in the basic P_2O_5-ZnO-R_2O system, where R is mixed alkali. These glasses have transition temperatures in the range of 275° to 375°C, and fall within the preferred composition range in mol%, 30-38 P_2O_5, 30-55 ZnO, 0-4 Al_2O_3, and 10-30 R_2O, where R_2O consists of 3-12 Li_2O, 4-18 Na_2O, and 0-12 K_2O. Significant substitution of SnO or PbO (up to 10%) for ZnO is allowed. Two typical compositions are listed in Table 1. Thermal expansion coefficients are typically in the 130-160 x 10^{-7}/°C range, Young's Modulus is about 40 GPa, and densities generally range from 2.9 to 3.5 g/cc.

EXPERIMENTAL

The glasses are initially melted in silica or refractory crucibles at temperatures of 900°C to 1200°C. Typical batch materials that have been used include ammonium phosphate, phosphoric acid, alkali phosphates and carbonates, zinc oxide, zinc orthophosphate, stannous oxide and aluminum metaphosphate. The viscosity-temperature curves are steep with viscosities typically ranging from less than 10 poises at 1000°C to around 10,000 poises at 500°C. The softening point is typically near or below 400°C.

Table 1. Some Mixed Alkali Zinc Pyrophosphate Glass Compositions

mole % batched	Compositions #1	#2
P_2O_5	33	33
Al_2O_3	2	2
Li_2O	6	7
Na_2O	7	8
K_2O	7	5
ZnO	43	40
SnO	-	5
SiO_2	2	-
suitable thermoplastic	PEK, PEEK	LCP-Xydar®

LCP-Xydar® = Liquid crystal polymer-Xydar® is trademark of Amoco[3]
PEK = polyether ketone[4]
PEEK = polyether ether ketone[5]

In order to provide feedstock for glass-polymer blends, the molten glass is quenched in water or between chilled rollers to produce a coarse granular material for the compounding operation. The compounding process illustrated in Figure 1 consists of melt mixing glass granules with polymer pellets or powders in a heated twin screw extruder at temperatures of 380°C to 425°C and is analogous to forming polymer/polymer blends or alloys. Blends prepared in this way can be subsequently extruded as sheet or injection molded into parts. The data reported in

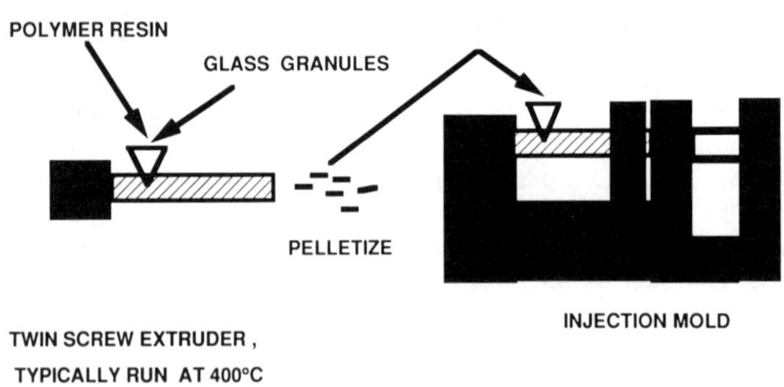

Figure 1. Schematic of process for producing glass-polymer blends (after Ref. 1).

this paper are from injection molded ASTM samples containing up to 80 wt% (62 vol%) glass. Scanning electron microscopy (SEM) was used to examine the microstructures of fracture surfaces and polished sections of the blends.

GLASS STRUCTURE AND DURABILITY

Raman spectra studies[6] on alkali zinc pyrophosphate glasses (Figs. 2, 3) reveal a broad phosphate anion distribution in binary zinc pyrophosphate glass with bonds characteristic of long chains (>1150 cm^{-1}) and simple orthophosphate groups (950 cm^{-1}) which is narrowed towards $(P_2O_7)^{-4}$ dimers (1030 cm^{-1}) as sodium or lithium is substituted for zinc. Surprisingly, aqueous durability of the binary pyrophosphate glass is not compromised by alkali-for-zinc substitution even though the glass transition is decreased substantially. Mixed alkali further improves durability (Fig. 4).

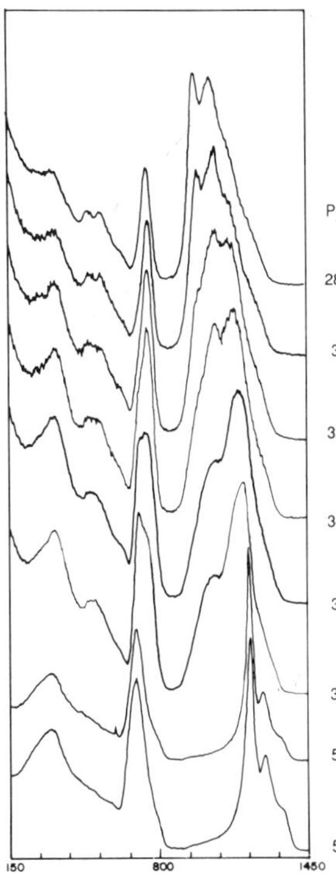

Figure 2. Raman spectra of binary ZnO-P_2O_5 glasses (after Ref. 3).

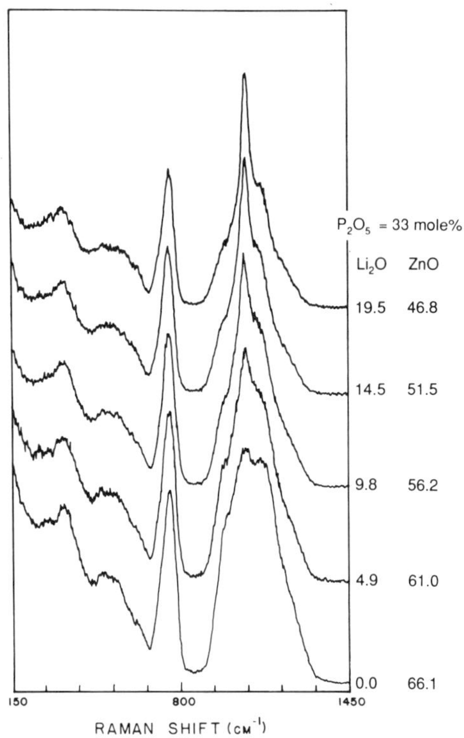

Figure 3. Raman spectra showing narrowing of phosphate anion speciation with Li_2O for ZnO substitution in pyrophosphate glass (after Ref. 3).

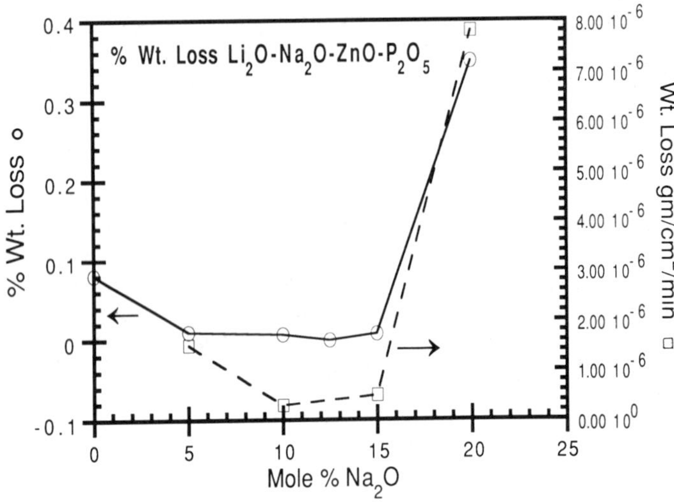

Figure 4. Mixed alkali effect on aqueous durability at P_2O_5 33, ZnO 47, R_2O 20 [100°C] (after Ref. 3).

It appears that shorter average phosphate chain length favors increased durability, probably because the middle phosphorous-oxygen-metal units in trimers and longer chains are most easily hydrolyzed. In this regard, the lower charge on oxygen favors increased durability. This correlates with electronegativity of the cation in the order Zn>Li>Na>K.

Measured dissolution rates at 100°C in boiling water can be below 5 x 10^{-7} g/cm^2 min, comparable to those of soda lime silicate glass and matched by other phosphate glasses only with high alumina and Tgs (glass transition temperatures) above 650°C. Rare earth additions are effective in improving the durability in alkali solutions.

THE CHOICE OF GLASS COMPOSITION FOR A SPECIFIC POLYMER

The choice of glass composition to be blended with a specific thermal plastic can depend on several factors. If the glass is to be compounded at high levels (50% or more by volume), its viscosity at the processing temperature should be significantly higher than that of the polymer, typically by a factor of 10^3 in order to avoid phase inversion. This ensures that the polymer remains the continuous matrix phase imparting desirable strength and toughness to the composite. Secondly, the glass composition should be chosen so that it bonds well to the polymer at the desired forming temperature (low interfacial tension), and thirdly, care must be taken that ingredients in the glass do not enter the polymer network in such as way as to cause chemical degradation or cross-linking. Table 2 shows two glass compositions designed to be compounded with polyether ketone and the liquid crystal polymer Xydar®, respectively. In both cases, the viscosity of the glass at the working temperature is about 10^7 poise.

Table 2. Mechanical Properties of Some Cortem™ Glass/Polymer Blends

Glass Blend Vol % Glass	Modulus x 10^6 psi	Tensile Strength	% Strain	Flexural Strength	Notched Izod Impact
PEK* 70G (#1)	4.3	11,000	0.3	17,900	0.52
PEK 62G (#1)	3.5	13,600	0.6	23,000	0.92
PEK 48G (#1)	1.6	12,500	1	17,000	0.50
XYDAR® 62G (#2)	3.3	7,100	0.3	11,200	0.60
XYDAR® 48G (#2)	2.9	11,800	0.6	16,000	
PES* 48G	2.3	6,700	0.4	13,100	0.65

PES = polyethersulfone *obtained from ICI

DEVELOPMENT OF MICROSTRUCTURE

The fluid nature of the glass at the polymer processing temperature allows the use of much higher glass levels than with conventional filled systems at reasonable processing viscosities. Figure 5 shows viscosity-shear rate curves of some representative glass/polymer blends and the neat resins at 405°C, a temperature within the range of processing for these materials.

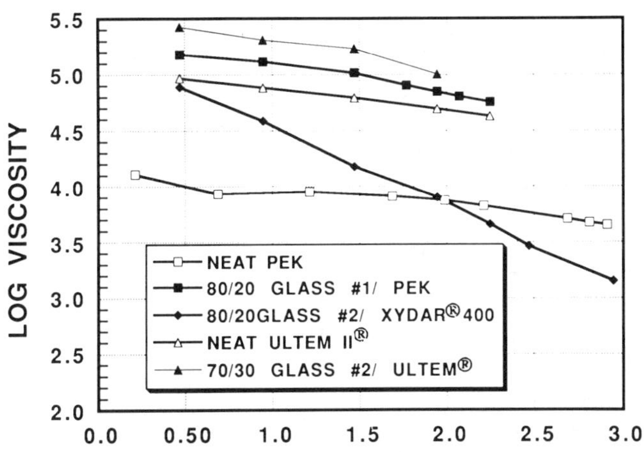

ULTEM®, trademark of GE is polyetherimide.

Figure 5. Viscosities of glass-polymer blends and neat polymers at 405°C.

In compounding, the polymer, which has a lower heat capacity and typically a lower viscosity, melts first and forms the matrix phase. The glass then softens and begins to deform as indicated in Figure 6. With continued shear at elevated temperatures the glass is further elongated until the strands become unstable and break up into "strings of pearls". Thus, the microstructure evolves towards a fine dispersion of glass droplets in a polymer matrix as illustrated in Figure 7, which

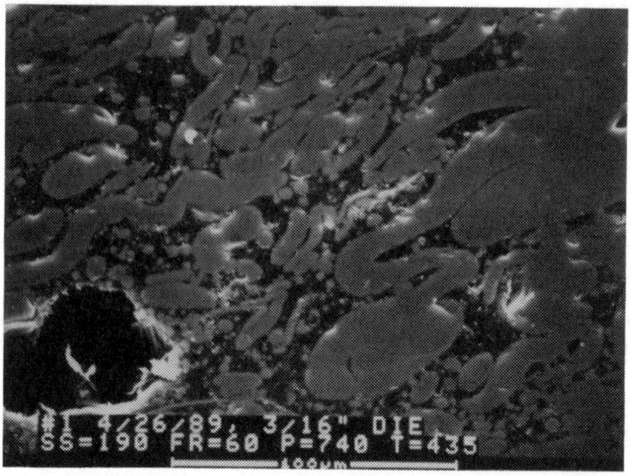

Figure 6. Development of microstructure in glass/polymer blends. Scanning electron micrograph (SEM) of polished section showing dark polymer matrix and light glass particles as they begin to soften and elongate early in the process (after Ref. 1).

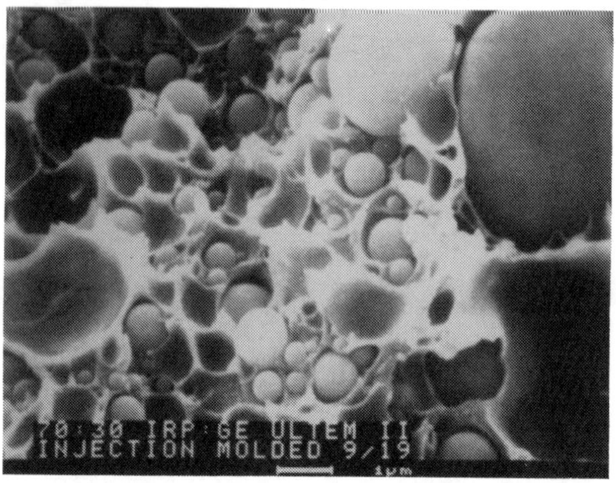

Figure 7. SEM of fractured surface of 70 wt% glass/30 wt% polyetherimide blend (after Ref. 1).

depicts a fracture surface of a blend of polyetherimide and 70 wt% glass. Many of the glass droplets are submicron in size, in fact transmission microscopy reveals particles down to several hundred microns in size. Note, however, that the adhesion of the polymer to this particular glass is poor. Figure 8 shows a polished section of a polyethersulfone/glass blend revealing a more intricate microstructure with several generations of polymer and glass droplets contained within each other. The interfacial tension is apparently low in this system and adhesion between phases is good.

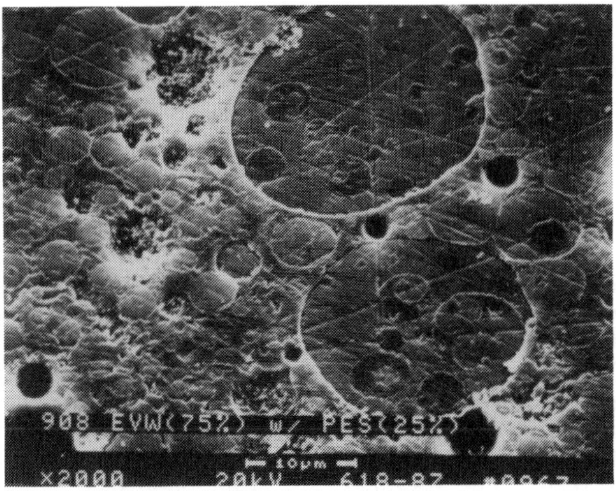

Figure 8. SEM of polished section of 75 wt % glass/25 wt % polyethersulfone blend (after Ref. 1).

Co-continuous or interpenetrating microstructures have also been observed. Figure 9 represents a polished section of a 25wt% polyether ether ketone/75wt% glass blend having a continuous polymer matrix and a highly interconnected glass phase.

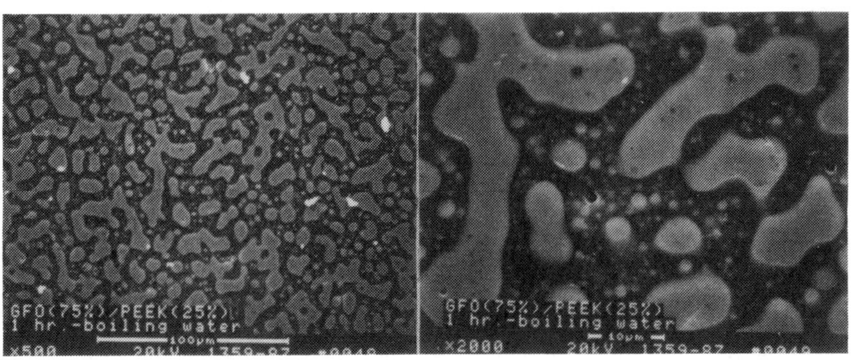

Figure 9. SEM at different magnifications of polished section of 75 wt % glass/25 wt % polyether ether ketone with continuous polymer matrix (dark) and highly connected glass droplets (light).

This bimodal microstructure may be the result of coalescence of an initial dispersion of fine particles. Another possibility is that, above 400°C, there was partial miscibility between the glass and the polymer with more glass dissolved in the polymer than vice versa. Upon cooling, the dissolved phases may have precipitated out to give the fine particles observed.

Figure 10. SEM of fractured section of same blend as shown in Figure 9. Smooth areas are glass, rough areas are polymer.

A microphoto of a fracture surface of this material is shown in Fig. 10. The transgranular value of the fracture contrasts with that of Fig. 7, and shows a well-bonded interface. Note the rough nature of the filetted polymer fracture surface.

PROPERTIES OF GLASS POLYMER BLENDS

Thermal Properties: The thermal expansions of the blends described above are intermediate between those of the polymer and the glass and a good match for many metals. For example, the glass/polyether ketone system shows a linear relationship between the coefficient of thermal expansion and the volume % glass, ranging from 14 x 10^{-6}/°C for the glass to 108 x 10^{-6}/°C for the polymer to 300°C. The effect of glass content on heat deflection temperature is also linear for this system, ranging from 185°C to above 300°C over the interval of 0% to 80 wt% glass.

Perhaps the most significant improvement in thermal performance of the Cortem™ glass/polymer blends over the base polymers is in flame resistance. One measure of flame resistance is limiting oxygen index which reports the level of flowing oxygen at which the sample becomes self extinguishing. Figure 11 gives comparative data for glass/polymer blends and conventionally filled resins. Three mechanisms contribute to this dramatic improvement in flame resistance. First, phosphate compounds are known to be chemically active as flame retardants[7].

Second, at the high glass loadings possible with the low melting glasses, the amount of polymer available is further reduced. And, most importantly, the low temperature glasses flow in a flame to form an excellent oxygen barrier, preventing further combustion of the sample.

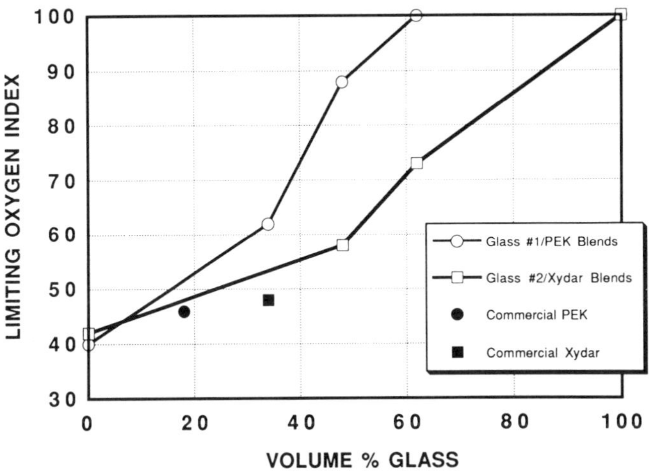

Figure 11. Limiting oxygen index of some polymer-glass blends and conventionally filled polymers (after Ref. 1).

Mechanical Properties: (see Table 2) Very high glass loadings lead to increases in modulus and values in excess of 4 million psi have been obtained on injection molded bars. As the modulus increases, the toughness, as measured by notched Izod impact, decreases. Nonetheless, even the most rigid blends have strains at rupture more than an order of magnitude greater than that of glass. Since the glass in these melt mixed blends is typically present as fine dispersed particles, it does not act as a reinforcing agent and therefore does not improve the tensile strength. The microstructure and the adhesion at the glass polymer interface both influence the mechanical properties which are not fully optimized in the data presented here.

Electrical Properties: The electrical resistivity of the alkali zinc pyrophosphate glass (composition #1) is surprisingly high, about 10^{10} ohm•cms at 200°C and over 10^{13} ohm•cms at 100°C. The resistivity-temperature curve is shown in Figure 12 and is compared to some standard Corning silicate glasses of differing alkali content: soda lime glass, sodium borosilicate Code 7740 Pyrex®, and 96% SiO_2 Vycor® Code 7900. These glasses contain 17, 4, and less than 1 weight percent Na_2O, respectively. The phosphate glass contains 20 mole % combined alkali, or an equivalent weight percent of 13 Na_2O if the lithia and potassium were replaced by soda.

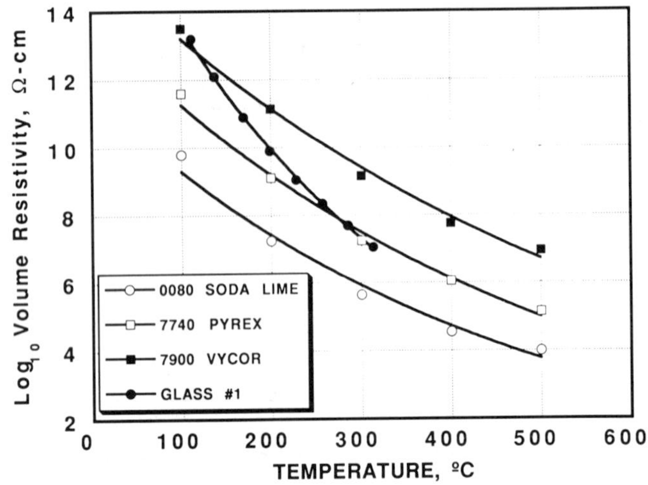

Figure 12. Resistivity vs. temperature curve of low temperature alkali zinc pyrophosphate glass compared to some standard silicate glasses.

The slope of the phosphate resistivity-temperature curve is much steeper than those of the silicate glasses reflecting the steep viscosity-temperature relationship in this range, only 15°C per order of magnitude viscosity change. Nevertheless, in the range of interest of temperature from 0 to 300°C, the resistivity is high, between those of Vycor® and Pyrex® glasses, glasses with less than 1% and 4% Na_2O. This surprising result indicates that the mobility of alkali ions is low in the pyrophosphate glass structure. This may be explained by a short chain structural model where alkali ions are specifically assigned to individual phosphorous-oxygen-alkali groups along chains as follows:

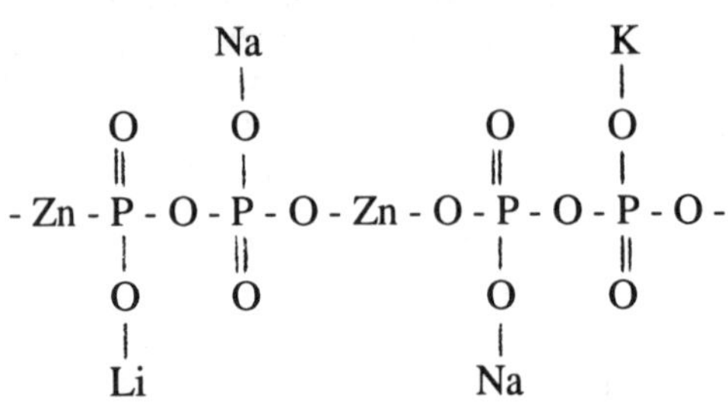

The dielectric properties of alkali zinc pyrophosphate glass #1 are given in Table 3 as a function of temperature and frequency. The dielectric constant (K) at room temperature is near 7.5 at the measured frequencies and increases to over 8 at 170°C. The dissipation factor (D.F.) at room temperature ranges from 0.0025 at low frequencies to less than 0.001 at 100 KHz. It approaches 0.01 at elevated temperature (170°C) at low frequencies.

Table 3. Dielectric Properties, Glass #1

Temp. °C	100 Hz K	100 Hz D.F.	1 KHz K	1 KHz D.F.	10 KHz K	10 KHz D.F.	100 KHz K	100 KHz D.F.
22	7.600	.0025	7.578	.0011	7.568	.0012	7.548	.0008
69	7.695	.0034	7.668	.0017	7.653	.0015	7.631	.0011
89	7.778	.0051	7.739	.0025	7.717	.0020	7.690	.0014
112	7.931	.0102	7.856	.0048	7.817	.0031	7.780	.0020
138	8.151	.0242	7.999	.0097	7.927	.0052	7.875	.0030
170	8.802	.0963	8.328	.0301	8.141	.0123	8.041	.0059

The dielectric properties of some experimental glass blends at 25°C are shown in Table 4. The dielectric constant of the composite can be observed to have been decreased as expected by the polymer in the case of polyether ketone, but the dissipation factor has doubled from that of the glass.

Table 4. Electrical Properties of Some Experimental Cortem™ Glass/Polymer Blends

Material Polymer/ Vol% Glass	Dielectric Constant 100 Hz	Dielectric Constant 100 KHz	Loss Tangent 100 Hz	Loss Tangent 100 KHz
PEK 62G (#1)	5.94	5.9	0.0062	0.0016
PEK 48G (#1)	5.22	5.17	0.0043	0.0013
Xydar 62G (#2)	6.24	5.94	0.0047	0.0158
Xydar 48G (#2)	7.86	7.62	0.0214	0.0067

The liquid crystal polymer (Xydar®) blends show increasing dielectric constant and losses with decreasing glass levels.

The effects of exposure to moisture of various glass-polymer blends is currently under study. Preliminary results indicate the Xydar-48% glass blends are unaffected by 100-hour treatments at 85°C and 85% relative humidity.

CONCLUSIONS:

Low temperature oxide glasses of the alkali zinc pyrophosphate type with Tgs in the 300° to 350°C range have been blended with a variety of high temperature thermoplastic resins including polyaryletherketones, liquid crystal polyesters and polyetherimides. The resulting materials show uniform fine microstructures some of which are analogous to polymer-polymer blends. Higher-inorganic loadings can be achieved than with any solid fillers. Dielectric properties measurement show low losses despite moderate to high alkali contents in the glass. The key advantages of these materials lie in thermal and dimensional stability, isotropic behavior, high stiffness and flame resistance.

ACKNOWLEDGEMENTS

The authors acknowledge the efforts of J. Dickinson in studies of structure and durability, B. Monahan, W. Bahn, J. Pierson, R. Monahan, and J. Ference in materials preparation, the Experimental Melting Department for supplying quantities of new glasses; members of Analytical Services for providing properties measurements and micrographs, particularly J. Gigliello for electrical data, and the support and encouragement of D. Morse.

REFERENCES

1. C. J. Quinn and G. H. Beall, Low Temperature Glass-Polymer Blends, in Chemical Processing of Advanced Materials, ed. by L. L. Hench and J. K. West, Wiley-Interscience, N.Y., pp. 799-805, 1992.

2. H. S. Katz and J. V. Milewski, Handbook of Fillers for Plastics, Van Nostrand Reinhold, New York, 1987.

3. S. J. Bai, M. Dotrong, E. J. Soloski, and R. C. Evers, Characterization, Processing, and Properties of an Aromatic Poly(ether ketone), Jour. of Polymer Science: Part B: Polymer Physics, Vol. 29, 119-128 (1991).

4. J. C. Seferis, Polyetheretherketone (PEEK): Processing-Structure and Properties Studies for a Matrix in High Performance Composites, *Polymer Composites*, June, Vol. 7, No. 3, pp. 158-169.

5. R. A. Weiss and C. K. Ober, Editors, *Liquid-Crystalline Polymers*, ACS Symposium Series No. 435, the American Chemical Society, 1990.

6. C. J. Quinn, G. H. Beall, and J. E. Dickinson, Alkali Zinc Pyrophosphate Glasses for Polymer Blends, International Congress on Glass, Bull. Spanish Society of Ceramics and Glass, 31-C, v. 4, pp. 79-84, 1992.

7. W. C. Kutyla and A. J. Papa, Flame Retardancy of Polymeric Materials, Marcel Dekker, New York (1978).

ALUMINUM NITRIDE/POLYMER COMPOSITES FOR THERMAL MANAGEMENT

Thomas J. Mroz, Jr. and Eric A. Groat
Advanced Refractory Technologies, Inc.,
Buffalo, NY 14207

ABSTRACT:

Polymer matrix composites are of great potential use for a number of electronic applications. Of particular interest are systems exhibiting high thermal conductivity for thermal management applications. In order to take advantage of the characteristics of the filler within the composite, particle-to-particle contact must be maximized. The effect of filler-particle interaction on thermal conductivity of polymer matrix composites has been examined. Composites are prepared by mixing aluminum nitride powders with the polymer, as well as by infiltrating packed and sintered powder beds. Thermal conductivity of each composite type is measured and related to the preparation method as well as the filler quantity. Thermal conductivity results are compared to an established model.

INTRODUCTION:

The need for high thermal conductivity polymeric materials has lead to much research in the area of filled polymer systems. These composite systems typically use an inert solid filler to modify the thermal properties of the polymer without degrading the other structural and electronic properties of interest. Metal particulate and short fibers have been used successfully to produce thermal and electrically conductive polymer articles.[1-2] However, many of the potential applications require high electrical resistivity, which prohibits metal-filled systems.

Ceramic particles have been used with a degree of success where high thermal conductivity and high electrical resistivity are required.[3-7] Because the thermal transport mechanism in these materials is phonon, rather than electron based, the thermal conductivity is independent of the electronic conductivity. Highly thermally conductive fillers such as alumina, BN, and particularly AlN and diamond have all been examined for this application.[5] Despite the higher inherent thermal conductivities of AlN and diamond over alumina and BN, the thermal conductivity of samples described in the literature appears to be relatively insensitive to filler type. In fact, in some cases, BN filled systems display higher thermal conductivities at lower solids contents than AlN filled systems.[3-4] Thus, it is expected that some other variable has a much larger effect on the composite thermal conductivity than the conductivity of the filler.

Most filled polymer systems are produced by a simple mixing process. In this process, the filler powder and the polymer are mixed together, cast, and cured to form the final part. This production method results in particles that are well dispersed in the polymer matrix, and are thus separated from each other by a thin film of polymer.

Models of composite structure effects on properties have been constructed for 2 dimensional macroscopic systems.[8] These models have also shown applicability in some cases for describing microscopic systems in 2 or 3 dimensions. Because the standard mixed polymer composite system is not interpenetrating, the composite thermal conductivity is best described by the perpendicular plate model:

$$1/K_t = f_f/K_f + f_p/K_p \qquad [1]$$

where K_t represents the thermal conductivity of the composite, K_f and K_p are the thermal conductivities of the filler and the polymer respectively, and f_f and f_p are the volume fractions of the filler and the polymer respectively. The thermal conductivity of the composite increases non-linearly with increased filler content, and is dominated by the conductivity of the least conductive phase.

Should the filler phase become interconnecting as well as the matrix phase, the behavior of the system is best described by the parallel plate model:

$$K_t = f_f \cdot K_f + f_p \cdot K_p \qquad [2]$$

In this case, each component affects the thermal conductivity as a function of its volume concentration, thus the response is linear over the entire range of filler concentration.

When mixing conductive fillers into a fluid polymer, it is expected that the filler particles are wet by the polymer, and therefore are not able to come into contact with other filler particles. This is particularly true in cases where the system is mixed using high shear, and the system is mixed to a constant viscosity. In this case, it is likely that each particle has been completely wetted by the polymer, and therefore, isolated from all other filler particles. This type of structure can be represented by the perpendicular composite model as described in equation [1]. Because the conductivity of the polymer is much lower than that of the filler, it acts as a thermal barrier between filler particles and dominates the properties of the composite.

Evidence of this effect has been shown clearly in the work of Agari and Uno.[6-7] They find that when the filler is an electrical conductor, the conductivity relative to filler concentration can increase rapidly once the percolation threshold is reached. However, the thermal conductivity of the same samples only show a smooth curve relationship with no discontinuity. This suggests that the filler particles are separated from one another by a thin layer of nonconducting polymer. Electrons can tunnel through this layer defeating the resistivity of the polymer. Phonons cannot tunnel through this layer, so they are scattered at the interface. Thus, the perpendicular plate model is the appropriate general model for explaining thermal conductivity in filled systems where the filler phase is dispersed into the polymer phase.

The simple version of the perpendicular plate model does not fully predict the response of actual filler/polymer composites. Several variables, mostly related to the three dimensional influences of particle packing and shape effects must be considered in the model to better represent real systems. There are a great number of proposed models in the literature,[9-10] however, Nielsen's model has been shown to best represent actual experimental data.[11]

Bujard has used Nielsen's model to fit the thermal conductive results obtained from AlN and BN filled polymers.[3-4] However, in his experiments a non-spherical shape factor was required to fit the AlN data despite the use of approximately spherical powders. Additionally, Bujard reports that the thermal conductivity of a composite with a given filler concentration is reduced by additional mixing in this system.[3] He goes on to explain these results in terms of conductive "clusters" of filler particles. These clusters are much more efficient at conducting heat as there is no polymer interface between particles within the cluster. If mixing efficiency is increased, the clusters break down to individual particles, reducing the composite thermal conductivity.

Given this explanation, it is proposed that much higher thermal conductivities can be reached in filled polymer systems in which particle interconnections are maximized. A proposed method of obtaining this structure is to prepare a preform of AlN in which every grain is strongly bonded to all adjacent grains. The porosity can then be backfilled with the polymer. Such a structure would allow AlN grain contacts to persist, and result in unimpeded thermal flow between grains. This structure would then resemble the parallel plate model, characterized by a linear response of the intended characteristic with increased filler content. Presumably, this type of composite would exhibit much higher thermal conductivities than standard mixed systems due to the intimate contact between filler particles.

EXPERIMENTAL:

Commercially available aluminum nitride* powders were used for all composite samples. A low viscosity epoxy system** was selected for the polymeric matrix, as it allowed infiltration of even micron diameter pore channels.

Test samples were prepared using two general methods. The first method, termed "mixed composite system" entailed mixing the AlN powder directly into the polymer. Nominal AlN concentrations ranging from 25 to 55 volume percent were prepared. Samples above 55 volume percent were too viscous to be formed. The mixture was stirred until the powder was completely wetted by the polymer, and the appearance was uniform. Samples were evacuated in a vacuum chamber for 2 hours in order to degas the polymer, then each composition was thermally cured into disk shaped samples approximately 7 cm in diameter by 1.5 cm high.

---

*AlNel A-100 and AlNel A-100 SD, Advanced Refractory Technologies, Inc. Buffalo, NY 14207

**Low Viscosity Spurr Kit, SPI Supplies, Division of Structure Probe, West Chester, PA 19380

The second method, termed "infiltrated preform composite system" utilized a partially sintered AlN preform that was subsequently backfilled with epoxy. The AlN preforms were prepared by pressing AlN powder into bars, cutting the bars to square cross sections of approximate dimensions of 3 cm x 3 cm x 1 cm, then sintering in nitrogen or argon at 1850 °C for 3 hours. A spray dried grade of AlN powder was used for this experiment, as it resulted in better pressing uniformity and greater green strength in the pressed part. In order to obtain samples with varying porosities, parts were pressed at pressures ranging from 35 MPa to 210 MPa.

After sintering the preforms, they were placed into containers and epoxy was added to cover the part. The containers were placed in a vacuum chamber and evacuated for several hours. The containers were then allowed to cycle back to atmospheric pressure, and were then heated to cure the epoxy. The vacuum to atmospheric pressure cycle was sufficient to completely infiltrate the preforms with epoxy.

Attempts to infiltrate as-pressed samples with epoxy were not successful. Parts prepared from spray dried powders would not completely infiltrate, possibly due to interaction with the pressing binders. Parts pressed from raw powders disintegrated during infiltration, resulting in a non-connected AlN network within the epoxy.

Once the samples were cured, they were cut to size using a diamond blade. In the case of the mixed system samples, each test sample was cut from the center of the disk. For infiltrated samples, each test sample was cut entirely from within the boundary of the AlN preform. All samples were then ground to the final test dimensions of 2.5 cm x 2.5 cm x 0.8 cm.

Sample densities were obtained for each composite by geometric measurements on the ground test sample. A ground sample of neat epoxy was also prepared for comparison testing. Actual volume loadings of AlN were calculated from these density results for every sample.

Portions of select composite samples were fractured and mounted for SEM analysis. SEM photomicrographs were obtained using a combination signal from secondary and backscatter electrons. The partial use of backscatter electrons provides some contrast between the AlN and polymer materials.

Ground samples were tested for thermal conductivity using a comparative method***. Reference specimens were selected with thermal conductivities close to the values obtained for the test samples.

***Quickline™ - 10, Anter Laboratories, Inc., Pittsburgh, Pa. 15235

RESULTS/DISCUSSION:

Mixed Composite Results

Figure 1 shows the microstructure of an AlN/polymer composite prepared by the mixed system method. The sample was prepared at a solids loading of 55 volume percent. Note that the composite is well mixed, and each particle is separated from neighboring particles by the polymer matrix.

Figure 1. SEM micrograph of mixed composite

Table 1 shows the composite density and actual AlN loading concentration calculated for each experimental sample. The actual AlN loading compared well with the intended solids content.

Thermal conductivity of each sample is also given in Table 1. The thermal conductivity of the neat polymer was measured at 0.22 W/mK. As expected, the thermal conductivity increases with additions of AlN to a maximum of 1.39 W/mK,

Table 1. Thermal conductivity results, mixed composite system

Sample Type	Composite Density	Actual vol.%	Thermal Conductivity
100% Polymer	1.10 gm/cc	0.0% AlN	0.22 W/mK
25 vol.% AlN	1.68 gm/cc	27 % AlN	0.60 W/mK
35 vol.% AlN	1.87 gm/cc	36 % AlN	0.86 W/mK
45 vol.% AlN	2.05 gm/cc	44 % AlN	1.20 W/mK
55 vol.% AlN	2.24 gm/cc	53 % AlN	1.39 W/mK

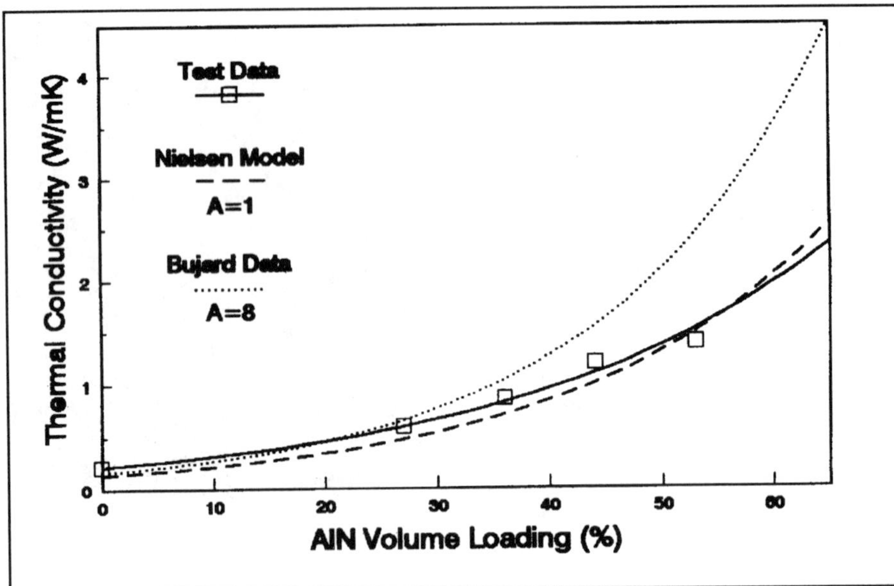

Figure 2. Thermal conductivity as a function of AlN loading in mixed system

corresponding to a 55 volume percent AlN concentration. Thermal conductivity is plotted as a function of AlN loading concentration in Figure 2. The increase in thermal conductivity with concentration is shown to be monotonic with no discontinuities. The data has been fit to Nielsen's model using a packing fraction of 0.61 as determined by the pressed density of the powder pressed at 10 MPa, a particle thermal conductivity of 6 W/mK, and a particle geometric factor (A) of 1. The particle thermal conductivity was estimated by first calculating the effect of 0.4 percent internal oxygen within the particle using the relationship developed by Slack.[12] The value of 0.4 percent internal oxygen was determined in previous work.[13] Then, the remaining oxygen was assumed to be present at each particle surface in the form of a hydrated aluminum oxide with an estimated thermal conductivity of 1 W/mK and the effect was calculated using the perpendicular plate model. The value of 6 W/mK calculated by this method compares extremely well with the value obtained in Bujard's work.

The results were compared to Bujard's earlier work in this system.[3] He found that a particle geometry factor of 8 was required for data fit (data reproduced in Figure 2). Bujard suggested that this indicated AlN particles that were poorly mixed with the polymer, and thus formed conductive clusters. With the current data, the low geometric factor suggests little or no cluster formation, thus the system was well mixed. Bujard also reports that thermal conductivity decreases with increased mixing efficiency (due to the thermal conductivity effect of clusters as predicted by the Nielsen model). This effect explains the lower conductivities obtained with the current samples, compared to Bujard's data.

Infiltrated Preform Composite Results

Figure 3 shows the microstructure of a typical sintered AlN preform. Note the well developed bonding regions between the particles, as well as the size and distribution of porosity. Because sintering aids were intentionally left out of the powder, individual particles necked together and grew via vapor phase transport, but no densification occurred. The pore structure also appears to be interconnected and is generally in the same size range as the grain size. The resulting part exhibited substantial strength, but contained on the order of 30-40 percent porosity which was interconnecting. The porosity concentration was found to correlate with the original pressing pressure used to fabricate the sample.

Figure 3. SEM micrograph of sintered preform

Figure 4 shows the microstructure of a typical polymer infiltrated AlN preform. The polymer completely infiltrates the pores throughout the part. Intergranular neck regions have not been affected by the infiltration; grains remain interconnected.

Table 2 shows the composite density and actual AlN volume percent calculations for the infiltrated preform composites. AlN volume percentages ranged from 58 percent to 75 percent.

The thermal conductivities for infiltrated samples can also be found on Table 2. Thermal conductivities ranged from 22.2 W/mK to 45.1 W/mK. The composite results appear to be insensitive to the sintering condition. The thermal conductivities of these samples are nearly an order of magnitude higher than values previously obtained for any filled polymer system. The continuous nature of the AlN particles allows unimpeded transfer of thermal energy through the preform structure.

Figure 4. SEM micrograph of polymer infiltrated preform

Table 2. Thermal conductivity results , infiltrated composites

Sinter Condition	Composite Density	Actual Vol.%	Thermal Conductivity
1 (N_2)	2.36 gm/cc	58 % AlN	28.6 W/mK
1 (N_2)	2.40 gm/cc	60 % AlN	23.6 W/mK
1 (N_2)	2.73 gm/cc	75 % AlN	27.3 W/mK
2 (Ar)	2.42 gm/cc	61 % AlN	24.0 W/mK
2 (Ar)	2.41 gm/cc	61 % AlN	22.2 W/mK
2 (Ar)	2.50 gm/cc	65 % AlN	26.1 W/mK
2 (Ar)	2.66 gm/cc	72 % AlN	45.1 W/mK

Figure 5 shows the thermal conductivity of infiltrated samples as a function of AlN volume loading. The data is generally consistent with the linear relationship anticipated by the parallel plate mixing model. The single sample which achieved a thermal conductivity greater than 30 W/mK appears to be inconsistent with the results from the other samples.

Extrapolation of the best fit line to 100 percent AlN content provides an estimate for the thermal conductivity phase of approximately 40 W/mK. Oxygen analysis on the preforms was measured to be 0.7 percent. Because the surface area of the preform was below 0.5 m^2/g, it is expected that most of this oxygen is contained within the lattice where it has the greatest effect on thermal conductivity. Thus, a thermal conductivity as low as 40 W/mK for this sintered AlN part is not unusual. This value is considerably higher than the calue calculated for the powder in the mixed system as described previously. This is because the direct contact between grains in the sintered preform eliminates the phonon scattering effect of oxygen at the particle surfaces.

Because the oxygen content is so high, it is expected that the thermal conductivity of the preform, (and likewise the composite) could be substantially increased by utilizing preform sintering techniques that would further reduce the lattice oxygen. In fact, the large effect that lattice oxygen has on the thermal conductivity of the AlN may be the source of the observed spread in data. Small changes in oxygen content and location between samples would be expected to result in noticeable differences in thermal conductivity.

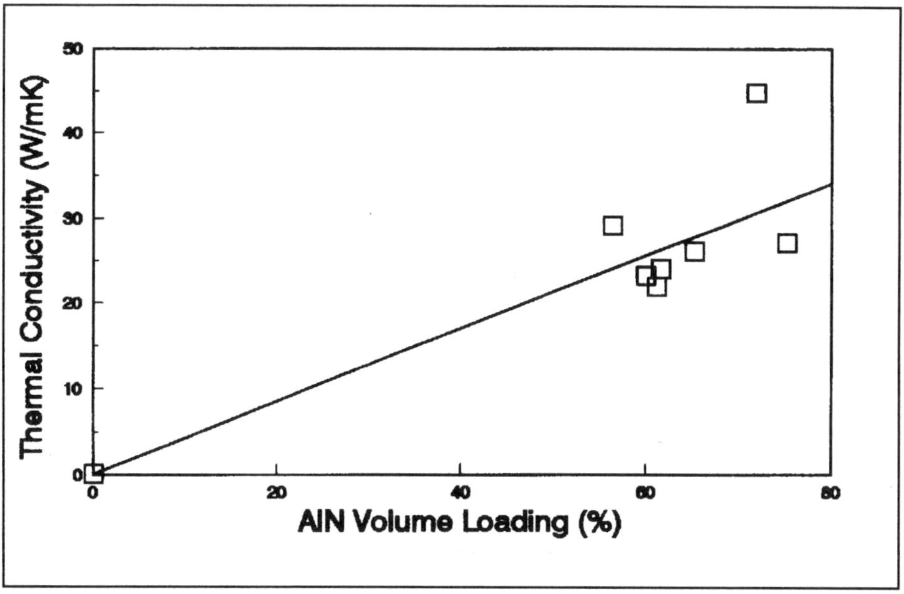

CONCLUSION:

Interaction between particles in AlN filled polymer composites has been shown to have a significant effect on thermal conductivity. Standard mixed samples exhibited thermal conductivity values consistent with a modified perpendicular plate composite model.

Samples prepared with sintered AlN preforms exhibited thermal conductivities an order of magnitude higher than standard mixed samples. Enhanced thermal conductivity has been related to the connectivity of the AlN structure. The thermal conductivity response to increased filler concentration has been shown to fit a parallel plate composite model.

By selectively processing an AlN filled composite system to insure interparticle contact, the thermal conductivity of the system can be significantly modified. While changes to standard processing methods would be required, they would be justified by the significant increase in thermal conductivity compared to conventionally prepared composites.

REFERENCES:

1 D. Bigg, "Mechanical, Thermal, and Electrical Properties of Metal Fiber-Filled Polymer Composties," *Poly. Eng. & Sci.,* **19** [12] 1188-1192 (1979)

2 A. Holbrook, "Aluminum Flake Filled Conductive Plastics for EMI Shielding and Thermal Conductivity," *Int. J. Pwdr. Metal.*, **22** [1] 39-45 (1986)

3 P. Bujard, J. Ansermet, "Thermally Conductive Aluminum Nitride-Filled Epoxy Resin," *Fifth Annual IEEE SEMI-THERM™ Symposium,* IEEE, Piscataway, N.J., 126-130, 1989

4 P. Bujard, "Thermal Conductivity of Boron Nitride Filled Epoxy Resins: Temperature Dependence and Influence of Sample Preparation," *Intersoc. Conf. Therm. Phenom. Fabr. Oper. Electron. Compon.*, IEEE, New York, N.Y., 41-49, 1988

5 N. Tsutsumi, N. Takeuchi, T. Kiyotsukuri, "Measurement of Thermal Diffusivity of Filler-Polyimide Composites by Flash Radiometry," *J. Polym. Sci., Part B, Polymer Physics*, **29** 1085-1093 (1991)

6 Y. Agari, T. Uno, "Thermal Conductivity of Polymer Filled With Carbon Materials: Effect of Conductive Particle Chains on Thermal Conductivity," *J. Appl. Poly. Sci.*, **30** [5] 2226-2235 (1985)

7 Y. Agari, T. Uno, "Estimation on Thermal Conductivities of Filled Polymers," *J. Appl. Poly. Sci.*, **32** [7] 5705-5712 (1986)

8 W. Kingery, H. Bowen, D. Uhlman, *Introduction to Ceramics,* John Wiley & Sons, Inc., New York, 1976

9 R. Progelhof, J. Throne, R. Ruetsch, "Methods for Predicting the Thermal Conductivity of Composite Systems, A Review," *Poly. Eng. Sci.*, **16** [9] 615-625 (1976)

10 S. Hoshi, G. Astarita, "Heat Transfer in Polymer Composites Containing Thermally Active Fillers, Mathematical Modeling," *Polym. Eng. Sci.,* **21** [8] 457-460 (1981)

11 T. Lewis, L. Nielsen, "Dynamic Mechanical Properties of Particulate-Filled Polymers," *J. Appl. Polym. Sci.*, **14** 1449-1471 (1970)

12 G. Slack, "Nonmetallic Crystals with High Thermal Conductivity," *J. Phys. Chem. Solids*, **34** 321-335 (1973)

13 T. Mroz, E. Groat, "Improved AlN Powders For Electronic Applications," *This Proceedings*

Section IV. Characterization Techniques

NIST/NCMS PROGRAM ON ELECTRONIC PACKAGING: FIRST UPDATE

Joseph A. Carpenter, Jr., Brian Dickens, Kenneth G. Kreider, John R. Manning, Richard L. Mattis, Gasper J. Piermarini and David T. Read*
National Institute of Standards and Technology
Gaithersburg, MD and Boulder, CO*

Ronald D. Evans
National Center for Manufacturing Sciences
Ann Arbor, MI

INTRODUCTION

In 1992 the National Institute of Standards and Technology (NIST) expanded its intramural efforts on development of metrology for microelectronics packaging and interconnection. The new intramural projects concentrate on development or refinement of metrology for measuring the properties of the materials as they actually exist in modern packaging and interconnection, and not as measured on idealized or bulk specimens. The metrology is intended for use in research, manufacturing, product quality control and failure analysis. Also, work begun in 1991 accelerated in a major consortium effort led by The National Center for Manufacturing Sciences (NCMS) and jointly supported by the consortium members and the NIST Advanced Technology Program (ATP) to develop new materials and production technology for printed wiring boards. This is an update on the progress of those efforts.

NIST INTRAMURAL EFFORTS

Prior NIST intramural efforts in thermal analyses of packaging, wire bonding, hermeticity, and electromigration have been augmented by new projects in the degree of cure of polymers, solderability, mechanical responses of thin films, residual stress, thermal properties, and materials databases.

Prior Efforts

The Semiconductor Electronics Division and its predecessors have been conducting work relating to packaging since the early 1970's. At that time, Frank Oettinger, David Blackburn, and Sherwin Rubin began studying the thermal properties of discrete power devices. One aspect of the thermal characterization of devices was the effective conduction of heat away from the chip to the package [1]. This became especially important as the chip size and therefore the power consumption increased. The technical output of the thermal studies over the years formed the basis of fifteen EIA-JEDEC

(Electronic Industries Association - Joint Electronic Devices Engineering Council), SEMI (Semiconductor Equipment and Materials International), and military standards [2].

Another aspect of packaging in which NIST has years of experience is wire bonding. In the 1970's these studies focused on problems relating to ultrasonic aluminum wire bonding [3], but later became much broader. An important output of the years of work is the book [4] by George Harman, the leader of the wire bonding effort and an internationally recognized authority on bonding problems. The book covers wire bond testing, intermetallic compounds at the bond-bonding pad interface, mechanical problems, and cleanliness.

More closely related to the package as a whole is work done by the late Stanley Ruthberg on hermeticity. Among other things, Ruthberg took a test method which could detect fine leaks by using helium as a tracer gas and extended the test's applicability to gross leaks [5]. He and Harry Schafft also held workshops concerned with hermeticity and moisture measurement [6]. Another series of workshops began as the VLSI Chip Packaging Workshop in 1982 under joint NIST/IEEE (Institute of Electrical and Electronic Engineers) sponsorship. George Harman plays a leading role in these meetings which now occur twice yearly.

An ongoing activity in electromigration in integrated circuit chips also has application to packaging. Harry Schafft and John Suehle have been studying the movement of metal ions in metal conductors under the influence of high current densities. The work has led to three ASTM standards relating to electromigration testing [7] and an understanding of the mathematical distribution of electromigration failures. The work is becoming applicable to packages as the metal conducting lines become more narrow and carry higher currents.

Degree of Cure of Polymers

Brian Dickens in the NIST Polymers Division is exploring various approaches for determining the degree of cure of polymers in packaging and interconnection. The degree of cure is the determining factor in the performance of the polymers used in packaging as dielectrics, adhesives, or encapsulants. Polyimides, used in electronic packaging to provide low dielectric constant films, which insulate the metal conductors, are being investigated first.

Polyimide films are usually deposited from solvent as precursors containing ionic (acid) or labile (ester) groups. On being heated, the precursor forms new chemical bonds. If this is done in dilute solution, the new bonds are ideally formed intramolecularly. In the more concentrated solutions in which the polyimide precursors are applied in electronic packaging, the new bonds may also form intermolecularly to link previously separate molecules together into much larger and ultimately insoluble molecules. The useful polyimides contain rigid structures, originally because this confers thermal stability. These rigid structures are groups of atoms bonded together in small two-dimensional arrays. The two-dimensional arrays have the potential for preferred orientation such as lying parallel to one another.

The cure of a polyimide thus entails chemical changes, which render a soluble material insoluble and heat resistant, and physical changes in which no new primary chemical bonds are formed but the long-range order in the solid increases. The state of the polyimide governs its dielectric properties, the anisotropy of its thermal expansion coefficients, and its inertness. To relate cure at all realistically to these operational properties, measurements of cure must measure the chemical and physical transformation of specimens a few to tens of micrometers (μm's) thick.

For chemical changes, infrared (IR) spectroscopy has traditionally been invaluable in following which groups appear and disappear. To some extent, IR can also be used to infer the existence of orientation. Depth profiling using IR is possible with attenuated total reflectance, but only to a depth of 1 to 5 μm, depending upon the wavelength of the incident radiation and thus varying across the IR spectrum. Much greater depths, up to 25 μm or so in favorable cases, can be attained with photoacoustic spectroscopy (PAS). Furthermore, degrees of chemical cure and physical ordering may be separated, even as functions of film thickness. PAS is probably not suitable to all cases and some investigation will be necessary to determine the extent to which it will help in determining the degrees of chemical and physical cure in polyimide films. At the very least, it provides a direct measure of changes in chemical groups over depths comparable with film thicknesses used in electronic packaging. It is rapid, potentially nondestructive, and requires minimal sample preparation.

The technique is simple to describe. Radiation with wavelength absorbed by the film is pulsed onto the film. The radiation enters the film, is absorbed, and heat is generated. At some depth in the film, effectively all the radiation will already have been absorbed and no heat will be generated. The heat diffuses in the film and some arrives at the film surface, where it heats the gas next to the film surface. A small shock wave is produced and the resultant pressure is registered by a microphone. Pulsing the incident radiation more rapidly leaves less time for heat to diffuse to the film surface before more heat is generated throughout the film, thus providing a form of depth profiling.

Assembly of the cell and interferometer to perform the PAS measurements has just been completed.

Solderability

John Manning, Carol Handwerker, and Bill Boettinger of the Metallurgy Division are attempting to develop a new understanding of the material parameters affecting solderability, i.e., the ability to attach components via solder. This new project attempts to identify which material parameters other than composition and temperature should be measured and to eventually devise appropriate metrology for improved control of soldering.

When lead-tin solder reacts with copper to form a solder joint, two copper-tin intermetallic compounds often form in the joint. These intermetallics, Cu_6Sn_5 and Cu_3Sn, may affect the solderability as well as the mechanical, electrical, and thermal properties of the joint. For proper analyses of both solderability and solder joint properties, the properties of the intermetallic alloys must be known.

Such measurements have been unavailable in the past because of the inability of usual processing methods to produce suitable bulk samples of the intermetallics. By applying inert gas atomization techniques to rapidly cool and solidify droplets of the alloy into very fine powders, NIST workers were able to produce both homogeneous Cu_6Sn_5 and fine-grained intermetallic bulk specimens. Since the intermetallics in general are brittle, special techniques need to be applied during hot isostatic pressing to produce fully dense bulk samples from the powder [8]. NIST now has produced homogeneous powders and fine-grained bulk samples of both Cu_6Sn_5 and Cu_3Sn intermetallics as well as of the Ni_3Sn_4 intermetallic that forms when nickel is used as a protective barrier layer.

Measurements have been made at NIST of the wettability of Cu_6Sn_5, Cu_3Sn, and pure Cu by eutectic lead-tin solder. Both area-of-spread tests, where the wetting of a substrate by a molten solder droplet is measured, and wetting balance tests, where a sample is dipped into a molten solder bath and measurements are made of the wetting rate, were performed. In each case, the intermetallic samples showed much poorer wettability, which implies much poorer solderability, than did the copper samples [9]. This difference is even larger if intermetallic oxidation is allowed to occur. Measurements of these important intermetallic properties have been in collaboration with the U.S. Army's Harry Diamond Laboratories

[10]. Extensions of this work, especially effects of intermetallics on solderability and composite solders, are continuing.

Further studies are underway at NIST to examine solder-substrate reactions, effects that influence solder joint properties and solderability, and ways to design improved solderability tests. This work is designed to provide guidelines for improved control of solder connections.

Mechanical Response of Thin Films

Adhesion of the micrometer-sized features of modern packaging to one another is of great interest. It is controlled by the mechanical responses of the two features (adherends) being joined and the interface between them. Dave Read in the NIST Materials Reliability Division and Jim Dally of the University of Maryland (College Park) are developing techniques and procedures for measuring the separate mechanical responses of the adherends. Future work is anticipated to address the issues of the mechanical properties of the interface.

Methods for measuring the mechanical properties of thin films have been under development at NIST for about three years [11,12]. A new method of measuring the constitutive properties of thin films has been developed. Miniature arrays of four tensile specimens, each 0.25 mm wide, 1 mm long and 2.2 μm thick, are prepared using deposition, patterning, and etching processes common to the semiconductor industry. An enlargement of a specimen array is shown in Fig. 1. Each array of four specimens is carried on and protected by a 6 by 8 mm rectangular silicon frame similar in size and shape to the dies that carry integrated circuits. Thirty-six such specimens are produced on a single 51-mm-diameter (2-in-diameter) by 0.25-mm-thick (0.010-in-thick) silicon wafer procured with both sides polished.

The first step in specimen fabrication is to heat the wafers in a steam environment to obtain an oxide layer about 0.5 μm thick. The wafers are patterned by photolithography and etched on one side to remove the oxide layer from the gage section (the thinnest part, see Fig. 1) of the tensile specimens. Next, metal films are deposited over the entire surface of the same side of the oxidized silicon substrate using electron-beam evaporation. Multilayer films consisting of a 0.1 μm titanium layer, a 1.98 μm aluminum layer, and another 0.1 μm titanium layer have been deposited thus far. The wafers are then photolithographically patterned on the other side and etched with hydrazine to remove the silicon from beneath the gage sections of the tensile specimens.

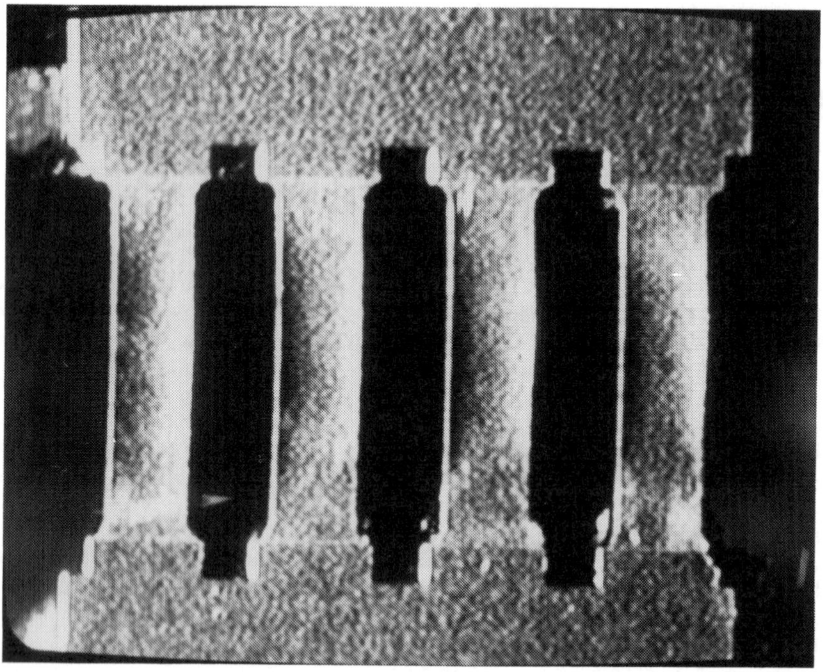

Figure 1. SEM micrograph showing an enlargement of the specimen array. Each of the four tensile members is 0.25 mm wide, 1 mm long, and 2.2 μm thick.

The specimens are subjected to a uniaxial force in a micro-tensile loading system. The specimens with protective frames attached are adhesively bonded to the loading platens. The vertical sides of the protective frame are severed without visually apparent damage to the specimens. The load is applied by micrometers through a special tension spring. The loading system is equipped with eddy current sensors which provide a continuous record of applied force and elongation.

Tensile properties of the titanium-aluminum-titanium films have been determined. The ductility of these films is very low (0.54 to 1.1%) when compared to the ductilities of bulk aluminum or titanium, which are an order of magnitude higher. In seven measurements of the ultimate tensile strength of the titanium-aluminum-titanium films, a mean strength of 170.6 MPa and an estimated standard deviation of 22.3 MPa were observed.

Fatigue life as a function of number of stress cycles has been determined for a set of three-layer titanium-aluminum-titanium thin-film specimens. The specimen preparation was the same as described above for tensile property measurement. The test procedure, adapted from the tensile test procedure, was to bond a specimen assembly to the platens, clamp the sliding platen in place, fracture the frame, release the clamp, inspect to make sure the specimen was undamaged and the frame was completely severed, and then conduct the fatigue test.

The fatigue loads were applied by operating the micrometer by hand. The specimen was first loaded to the preselected fatigue load, then cycled between the fatigue load and a minimum load, which was 10 percent of the fatigue load. Each cycle required approximately one minute. Hysteresis loops were plotted on an xy plotter and recorded in a laboratory computer. Specimen failures, as observed in the microscope, were obvious from the instrument response; once a specimen could no longer sustain the maximum load, the specimen was extended by the spring until complete failure.

For nominal stresses ranging from 50 to 90 percent of the upper-bound ultimate strength, the maximum fatigue life was 76 cycles, which is quite low when compared to normal bulk specimens. The failure mode was crack propagation across the specimens; active cracks were observed from the first few load cycles on. Fatigue life data are always subject to high scatter, and the present data set is no exception. The extremely short, low-cycle fatigue life of the present specimens is attributed to their lack of an extended range of plastic strain before failure. Possible causes of this limited range of plastic strain are the intrinsic natures of thin films, the aggressive specimen preparation procedures, and imperfections in specimen geometry. Imperfections in the specimen geometries were clearly evident. Crack-like features were often observed before the loading was begun. Elucidation of the causes of these low fatigue lives and refinement of the test are the subjects of ongoing research.

The present study is believed to be the first to report the tensile fatigue resistance of thin films.

Residual Stress

Residual stress is undoubtedly present to varying degrees at various locations in the packaging but is difficult to measure with accuracy at micrometer spatial resolutions typical of the size of conductor and dielectric features of the package. The presence of these residual stresses limits the performance characteristics and reliability of the electronic package. Gasper Piermarini of the

NIST Ceramics Division is exploring the use of pressure-dependent shifts of first-order Raman scattering lines for making such measurements of residual stresses in ceramics.

Available methods for measuring residual stresses in ceramic components include x-ray diffraction, micro-infrared absorption, micro-Raman scattering, photoluminescence, and cathodoluminescence techniques. The micro-Raman technique currently under investigation has the required micrometer spatial resolution for the evaluation of miniature electronic packages. In first-order Raman scattering, the energy, relative intensity, and line shape of acoustical phonon peaks are very sensitive to structural properties such as disorder and stress. Furthermore, the micro-Raman technique has potential as a rapid, routine, and relatively simple method suitable for industrial quality control applications for nondestructive evaluation of residual stresses.

The instrument in the present work uses a triple monochromator which spatially filters and frequency disperses the optically collected Raman scattering signal. A diode array operating at -30°C detects the frequency-dispersed signal. The detector can be operated either in the continuous mode or activated with a gate pulse synchronized to the laser pulse for time-resolved spectroscopy.

Alumina, widely used in electronic packaging, has been investigated first. Its characteristic first-order Raman spectrum is simple but relatively weak, consisting mainly of four observable peaks. These weak peaks are sometimes superimposed on an intense broad background, usually attributed to alumina fluorescence. Alumina, therefore, presents some potential detection problems. Initial attempts at obtaining spectra from alumina confirmed these problems. Nevertheless, it was possible to measure and confirm spectra for alumina with peak intensities up to 1500 photon counts above background. This level of scattering is sufficient for conventional Raman measurements on ceramic pieces in the open atmosphere, a result which encourages its use as an industrial tool. However, for the technique to be useful as a test method for the detection and measurement of residual stresses, a calibration of the pressure dependence of the Raman shift with pressure must be made. The maximum strength of alumina crystals is on the order of 1000 MPa, or about 10,000 atm, so the calibration needs to be made at least over this pressure range. To define the calibration curve accurately, however, the calibration probably should be done to 2000 MPa.

The calibration will be done with the aid of a diamond anvil high pressure cell (DAC) specially modified with low fluorescence diamond anvils for micro-Raman scattering measurements coupled with the ruby fluorescence method of pressure measurement. The sharp ruby fluorescence R_1 line is pressure-sensitive and its pressure shift has been calibrated previously against the volume compression of NaCl by x-ray powder diffraction measurements. The volume compression of NaCl is related to pressure through an equation of state. Thus, although the ruby scale is not absolute, it is the best pressure scale available at present.

Initial attempts to measure first-order Raman scattering from alumina samples in a DAC, using water as the pressure-transmitting medium were unsuccessful because of a very intense fluorescence background from the particular diamond anvil used and probably also from the water. The weak Raman peaks from alumina are masked by the high fluorescence levels and do not appear. A high scattering material, sulfur, was then used as the sample in the DAC without using a pressure-transmitting liquid. Spectra were obtained for sulfur both inside and outside the DAC. An 80 per cent loss in intensity was observed by using the DAC. With such a large decrease in intensity, it is clear that weak alumina peaks will not be observed under these conditions.

To reduce the fluorescence background, diamond anvils were examined to select one exhibiting the lowest background. The diamond anvil selected has a background about two orders of magnitude less than the one used previously. This anvil will be mounted in the DAC and used in all future Raman measurements.

Other factors to be explored in the future to improve the measurement capability are the use of longer wavelength exciting lines (e.g., the 514-nm argon ion laser line or the 532-nm frequency-doubled output of a Nd:YAG laser) which produce less intense fluorescence and the possible use of time-resolved spectroscopy.

Thermal Properties

As the features on the microchips become smaller, careful packaging design is needed to prevent overheating. Previous approaches have relied upon mathematical modeling of the heat flow inside the packaging and corroboration by means of thermocouples placed on the outer surfaces of the packaging. Ken Kreider of the NIST Process Measurements Division is exploring the possibility of actually embedding thin-film thermocouples in the bodies of the packaging in order to make direct measurements of the internal temperatures without having to rely on predicted values. The technique being investigated employs the integration of thin-film thermocouples during the fabrication of multiple polyimide spun-on layers. This enables the in-situ evaluation of the polymer layer thermal conductivity and diffusivity.

A test coupon design, a 7 cm square heated on one side with a copper heat block, permits the simulation of the heat transfer through an electronic package of a multilayer polyimide structure on an alumina circuit board. Four thin-film thermocouples incorporated in the fabrication process at the two inner and two outer interfaces of three layers of polyimide enable determination of a thermal profile of the multilayer polymeric package. This design permits one-dimensional heat transfer analysis of the test data and generation of in-situ property analysis of the polymer layers. The test method should be useful for testing various types of polymeric or ceramic layers in the packaging.

Databases

Joe Carpenter of the NIST Ceramics Division, George Harman of the Semiconductor Electronics Division, and John Rumble of the Office of Standard Reference Data are investigating databases for the materials in packaging and interconnection. Plans are being made for NIST efforts to address the most important database needs.

Currently, the only substantial database on packaging and interconnection materials is that available from the Center for Information and Numerical Data Analysis and Synthesis (CINDAS) at Purdue University. That database is intended for proprietary use by members of the Semiconductor Research Corporation (SRC) which supports its development. Discussions are underway with representatives of CINDAS, the Microelectronics and Computer Technology Corporation (MCC), the University of Texas at Austin, and Lehigh University as to the possibility of collaborative work aimed at the eventual development of a nationally available database for packaging and interconnection materials.

NIST and MCC are involved in STEP (STandard for the Exchange of Product Model Data), the global effort led by the International Standards Organization (ISO) aimed at development of standard formats for the exchange of data on products, including packaging and interconnection. One possibility being considered is for the national database for packaging and interconnection materials to adopt the STEP standard. A National Conference on STEP for Materials was held at NIST December 2-3, 1992, to address the possible uses of STEP for packaging and interconnections, as well as other applications.

NIST EXTRAMURAL/NCMS EFFORTS

In mid-1991, a consortium led by the NCMS of Ann Arbor, Michigan, won a NIST ATP award for a major effort to develop new materials and production technology for printed wiring boards (PWB's). The consortium members include AT&T, Texas Instruments, Allied Signal and United Technologies/Hamilton Standard Interconnects. The Sandia National Laboratory in Albuquerque, New Mexico is a subcontractor to the NCMS. Ron Evans is the NCMS Program Manager and Tom Leedy of NIST monitors the program for the NIST ATP.

The (planned) five-year NCMS Printed Wiring Board (PWB) Interconnect Systems Program is structured into four technical teams and a steering group that provides direction, control, and coordination of the technical activities. The four technical teams are Materials, Surface Finishes, Imaging, and Products; each is discussed in more detail below.

The first year of the program focused on identification or development and benchmarking of PWB manufacturing and test technologies, especially those important to the areas of materials, surface finishes, imaging, and chemical processing. The participants believe that millions of dollars in direct cost savings have already been realized, primarily due to the elimination of duplicated R&D projects, test method developments, and resources. The program participants have begun transferring the first year's results to the U.S. PWB supply chain, primarily through timely presentations at IPC (Institute for Interconnecting and Packaging Electronic Circuits) meetings. Additional technology transfer will be accomplished by the program participants individually and collectively through various NCMS technology transfer vehicles.

Materials

During the first year of the program, the Materials Team has been working to identify or to develop methods for measuring material properties. One accomplishment has been the development of a micro/macro-mechanical model to predict thermal expansion in fiber-reinforced laminates. The model uses the properties of resin, glass, ply thickness, yarn counts, crimp variations, fabric weight, and laminate thickness to predict material properties such as the coefficient of thermal expansion. For example, the effect of fabric weight variations on thermal expansion coefficient can be calculated for various ply thicknesses for both the fill and warp directions for various fabric styles. This model is being used on test panels as part of the benchmarking process underway. The first release of user software is in use by the Materials Team.

Surface Finishes

Following the objectives spelled out by the NIST ATP proposal, the Surface Finishes Team has been working to define at least one fabrication test which can be used to estimate solderability performance at the point of use. In order to achieve this objective, the team has been evaluating seven combinations of previously developed and new methods. Most of this work has been done as part of the benchmarking process among the program participants. The seven methods are:

- Sequential Electrochemical Reduction Analysis (SERA)
- Wetting Balance
- Spread Test
- Capillary Action
- Wave Solder
- IPC-S-804 (Dip & Look)
- Energy Dispersive X-Ray (EDX) Analysis

The team completed their evaluation of these solderability test methods during the summer of 1992 and a report covering the findings was released at the fall 1992 IPC meeting.

Imaging

The first year's goals of the Imaging Team were to explore the limits of existing technology while maintaining high production yields. In order to meet this goal, a standard test method needed to be selected so that benchmarking and progress could be documented. The AT&T Bell Laboratories' electrical test method, described in the IPC technical paper IPC-TR-979, was chosen. This electrical test method is used to characterize conductor formation capability. The test method can be used to monitor process performance or to adjust process parameters to improve the quality of the product. The key elements of the electrical test methods are: 1) specially designed test patterns, 2) a two-probe electrical test system, and 3) the analysis of the test data leading to capability metrics. Typical data collected for each feature size in the test pattern include conductor yields and numbers of conductor lines, spaces, opens, shorts, and good and bad conductors. The test method has shown that conductor yields typically decrease with smaller feature sizes. Using this tool, the team has been designing experiments using different process techniques to maximize yields specifically related to 76.2 micrometer (3 mil) lines and spaces.

Products

The original NIST ATP proposal outlined chemical process technology to be worked on by the Chemical Team. At the end of the first year the objectives for the Chemical Team were expanded by the Steering Group to cover more product-oriented features. During the first year of the program, the Chemical Team conducted a survey of the PWB industry that asked questions specific to chemical processing. The survey resulted in useful data related to the manufacturing of PWB's in the U.S. The Chemical Team also published a paper, based on work done at Texas Instruments, entitled "A Comparative Analysis of Thin Film Metallization Methodologies for High Density Multilayer Hybrids." The paper compares evaporation, enhanced ion-plating (EIP), sputtering, and electroplating processes, while measuring adhesion, stress, uniformity and resistivity for copper metallization onto substrates for high density applications.

The Product Team (formerly the Chemical Team) has been given the additional tasks of supervising and coordinating the design and manufacture of integrated test vehicles. These vehicles will be used to provide crosslinking of advanced technology between technical teams and to provide a method of assuring that there will be process compatibility among all four technical teams.

ACKNOWLEDGEMENTS

Thanks go to Grady S. White and Robert I. Scace of NIST for their reviews and critiques of this manuscript.

REFERENCES

1. F. F. Oettinger and R. L. Gladhill, "Thermal Response Measurements for Semiconductor Device Die Attachment Evaluation," Tech. Digest, 1973 International Electron Devices Meeting, Washington, D.C., Dec. 3-4, 1973, 47-50, 1973.

2. F. F. Oettinger and D. L. Blackburn, "Semiconductor Measurement Technology: Thermal Resistance Measurements," NIST Special Publication 400-86, 42 and 67, 1990.

3. G. G. Harman and K. O. Leedy, "An Experimental Model of the Microelectronic Ultrasonic Wire Bonding Mechanism," 10th Annual Proceedings, Reliability Physics 1972, Las Vegas, Nev., April 5-7, 1972, 49-56, 1972.

4. G. G. Harman, Wire Bonding in Microelectronics, International Society for Hybrid Microelectronics, Reston, Virginia, 1989.

5. S. Ruthberg, "A Rapid Cycle Method for Gross Leak Testing with the Helium Leak Detector," IEEE Trans. Components, Hybrids, and Manufacturing Tech., CHMT-3, 564-570, 1980.

6. H. A. Schafft, "Semiconductor Measurement Technology: ARPA/NBS Workshop II. Hermeticity Testing for Integrated Circuits," NBS Special Publication 400-9, 1974; and D. Kane, B. A. Moore, and E. J. Walters, eds., RADC/NBS International Workshop on Moisture Measurement and Control for Microelectronics (IV), NBSIR 87-3588, 1987.

7. H. A. Schafft, "Test Methodology for Electromigration Characterization," Tutorial Notes, 1989 International Reliability Physics Symposium, Phoenix, Ariz., Apr. 10, 1989. Another publication on electromigration is J. S. Suehle and H. A. Schafft, "Current Density Dependence of Electromigration t50 Enhancement Due to Pulsed Operation," Proceedings of the International Reliability Physics Symposium, New Orleans, Louisiana, Mar. 27-29, 1990, 106-110, 1990.

8. R. J. Schaefer, F. S. Biancaniello, and R. D. Jiggetts, "Preparation of Bulk Samples of Copper-Tin Intermetallic Compounds," 175-181, The Metal Science of Joining, M. J. Cieslak et al., eds., The Minerals, Metals and Materials Society, Warrendale, Pennsylvania, 1992.

9. W. J. Boettinger, C. A. Handwerker, and L. C. Smith, "On the Wetting of the Intermetallics Cu_6Sn_5 and Cu_3Sn by Pb-Sn Alloys," 183-189, The Metal Science of Joining, M. J. Cieslak et al., eds., The Minerals, Metals and Materials Society, Warrendale, Pennsylvania, 1992.

10. R. J. Fields, S. R. Low III, and G. K. Lucey, Jr. "Physical and Mechanical Properties of Intermetallic Compounds Commonly Found in Solder Joints," 165-173, The Metal Science of Joining, M.J. Cieslak et al., eds., The Minerals, Metals and Materials Society, Warrendale, Pennsylvania, 1992.

11. D. T. Read, "A Microfabrication Approach to Mechanical Testing of Thin Films," 147-151, Electronic Packaging in Materials Science V, E. D. Lillie et al., eds., Materials Research Society, Pittsburgh, Pennsylvania, 1991.

12. D. T. Read, "In Situ Measurements of Strength and Ductility of Thin Films," 463-468, Advances in Electronic Packaging 1992, Vol. 1, EEP-Vol. 1-1, W. T. Chen and H. Abe', eds., American Society of Mechanical Engineers, New York, NY, 1992.

THE THERMAL CONDUCTIVITY OF THIN CERAMIC FILMS BY THE THERMAL COMPARATOR METHOD

J. C. Lambropoulos[(1)(2)] and S.-S. Hwang[(1)]

(1) Department of Mechanical Engineering, (2) Laboratory for Laser Energetics, University of Rochester, Rochester, NY 14627

ABSTRACT

Various measurements of the thermal conductivity of thin ceramic films are summarized. Such measurements show that the thermal conductivity of thin films with thickness in the micron and sub-micron range may be up to two orders of magnitude lower than the thermal conductivity of the corresponding bulk solid. The applicability is extended of the thermal comparator method to films deposited on substrates of arbitrary thickness, and a correlation is made of the thermal conductivity of ZrO_2 films with thickness in the range 0.15 to 40 μm and deposited on a variety of substrates. The contributions of the film microstructure and of the film/substrate interface to the effective film thermal conductivity are also discussed.

INTRODUCTION

The thermal conductivity of thin films may be much lower than the thermal conductivity of the corresponding bulk solids, and an interfacial thermal resistance may develop along the film/substrate interface [1]. Such reductions in the thin film thermal conductivity have been measured for metallic, ceramic, semiconducting, and superconducting thin films. Therefore, bulk thermal conductivity values are not appropriate when the removal or deposition of heat in thin films is a consideration. A summary of applications of technical importance can be found in [1].

The reduced thermal conductivity of thin films is attributed to scattering of phonons or electrons from microstructural defects such as grain boundaries, vacancies, chemical inhomogeneities, dislocations, or from the film boundaries. Such effects have been examined in [2]. Macroscopically, a reduced thermal conductivity may be also due to the presence of an interfacial thermal resistance along the film/substrate interface [2], or porosity within the film [1]. This report only considers the thermal conductivity of ceramic films.

In 1984, Decker et al. [3] measured the thermal conductivity of evaporated and sputtered films by applying thermocouples to the films. For SiO_2 films (thickness of 0.5-1 μm) they measured 0.17-0.28 W $m^{-1}K^{-1}$ (bulk value for fused silica is 1.4 W $m^{-1}K^{-1}$). For Al_2O_3 films with 1 μm thickness they found 0.04-0.25 W $m^{-1}K^{-1}$ (the bulk value for dense, polycrystalline Al_2O_3 is 36 W $m^{-1}K^{-1}$).

Ristau and Ebert [4] used a complex optical approach involving the heating of the free film surface by a laser beam and the recording of the infrared radiation of the sample. They examined oxide films of thickness 1-4 μm deposited on fused SiO_2 substrates. Their measurements yielded an effective film thermal conductivity k_{eff} with contributions from the film material itself and from the film/substrate interface. For HfO_2, Ta_2O_5, TiO_2 and SiO_2 films they found film thermal conductivities which were 1 to 3 orders of magnitude lower than that of the corresponding bulk solids (1.6, 0.33, 7.4, 1.4 W $m^{-1}K^{-1}$, respectively, for the bulk solids). For Al_2O_3 films they found thermal conductivities similar to that of bulk alumina.

Ogden et al. [5] measured the thermal conductivity of thick anodic oxide Al_2O_3 coatings on Al by placing the sample between two blocks of Cu (held at a higher temperature) and Al (at a lower temperature) and measuring the evolution of transient temperature with thermocouples. They found that the thermal conductivity of thick coatings of Al_2O_3 was less than 1 W $m^{-1}K^{-1}$ (bulk value 36 W $m^{-1}K^{-1}$).

Lambropoulos et al. [6] used the thermal comparator method, originally developed by Powell for the measurement of the thermal conductivity of bulk solids [7], to measure the thermal conductivity of thin films such as oxides (SiO_2, TiO_2, HfO_2, Sc_2O_3, ZrO_2, ThO_2, Al_2O_3 CeO_2), fluorides (AlF_3, HfF4, YF_3, CeF_3, ScF_3, ThF_4, LaF_3, MgF_2) and others (ZnS). The film thickness was in the range 0.06 to 4.4 μm and the substrates were Si (k = 150 W $m^{-1}K^{-1}$ at room temperature), sapphire (k = 46 W $m^{-1}K^{-1}$) or fused SiO_2 (k = 1.4 W $m^{-1}K^{-1}$). They found that the film thermal conductivity could be up to two orders of magnitude lower than that of the corresponding bulk solid.

The technique of Lambropoulos et al. [2, 6] allowed the separation of the film conductivity k_F and the interfacial thermal resistance R_{int} from the effective film thermal conductivity k_{eff}. For thin films of thickness H, this relation is

$$\frac{H}{k_{eff}} = \frac{H}{k_F} + R_{int} \tag{1}$$

Such separation of the film thermal conductivity k_F from the effective conductivity k_{eff} can shed light on the thermal properties and microstructure of the film material and on the structure of the film/substrate interface. In a technical application it is usually the effective film thermal conductivity k_{eff} that controls the film heat flow.

In the thermal comparator technique the measurements yield the film thermal conductivity perpendicular to the film. Extensions to anisotropic film thermal conductivity and to the case of large interfacial thermal resistance across the film/substrate interface have been recently discussed and analyzed by Lambropoulos et al. [2] Todate the thermal comparator technique has been used to find the thermal conductivity of thin films only at room temperature.

The thermal comparator technique was also used by Shaw-Klein et al. [8] to measure the thermal conductivity of sputtered AlN films of thickness 0.05-0.15 μm deposited on sapphire and thickness 0.3-1 μm deposited on fused SiO_2 by magnetron sputtering. For the thinner films the thermal conductivity was more than two orders of magnitude lower than that of bulk AlN (70-180 W $m^{-1}K^{-1}$), but only one order of magnitude lower for the thicker films.

Shaw-Klein et al. [9] also measured the thermal conductivity of $YBa_2Cu_3O_{7-\delta}$ films with thickness in the range 0.25-1 μm and deposited on MgO. It was found that the film thermal conductivity (0.26 W $m^{-1}K^{-1}$) was more than an order of magnitude lower than that of bulk $YBa_2Cu_3O_{7-\delta}$ (8-10 W $m^{-1}K^{-1}$).

A different version of the thermal comparator was used by Henager and Pawlewicz [10] to measure the thermal conductivity k_F of thin films (SiO_2, Al_2O_3, Ta_2O_5, BN, Si_3N_4, SiC) and multilayers (SiAlN, SiAlNO, Al(SiO_2/Si_3N_4), Al(Al_2O_3/AlN). For the single films the thermal conductivity was one to two orders of magnitude lower than the bulk value. In this technique the interfacial thermal resistance of the film/substrate interface could not be extracted.

More recently, Reichling et al. [11] used high frequency photothermal reflectivity and displacement techniques to measure the thermal conductivity of several dielectric thin films coated with a thin Au layer. For 1 μm thin films deposited on BK7 glass substrates (k = 1.1 W $m^{-1}K^{-1}$), they found a significant reduction of the film thermal conductivity as compared to the bulk value. For example, for SiO_2 they found an effective film thermal conductivity k_{eff} of 0.25 W $m^{-1}K^{-1}$; For $Zr0_2$ films they measured 0.05 W $m^{-1}K^{-1}$; For Ta_2O_5 films 0.2 W $m^{-1}K^{-1}$ (the corresponding bulk values are 1.2, 1-2, and 0.33 W $m^{-1}K^{-1}$, respectively). It is interesting to note that for Al_2O_3 films Reichling et al. [11] extracted a film thermal conductivity which was several times greater than single crystal sapphire.

Schafft et al. [12] used resistive heating to measure the effective film thermal conductivity k_{eff} of SiO_2 films. They reported 0.68 and 0.97 W $m^{-1}K^{-1}$ for films of thickness 1.74 and 3.04 W $m^{-1}K^{-1}$, respectively. Brotzen et al. [13] measured the thermal conductivity k_{eff} of thin SiO_2 films deposited by CVD on Si by depositing a thin Al strip on the film. The measured film thermal conductivity was a function of the film thickness: At 100 °C, the film thermal conductivity k_{eff} ranged from 0.045 to 0.22 to 0.40 W $m^{-1}K^{-1}$ as the film thickness varied from 0.1 to 0.7 to 1.4 μm.

It is thus clear that the thermal conductivity of thin ceramic films is usually lower, by as much as two orders of magnitude, than that of the corresponding bulk solid. As the thermal comparator method is an inexpensive, rapid, and non-destructive technique for measuring the thermal conductivity of thin films, we show in this report that this technique may be used to measure the thermal conductivity of thin or thick films deposited on substrates of arbitrary thickness. We use as examples measurements of the thermal conductivity of ZrO_2 films with thickness in the range 0.15 to 40 μm and deposited on substrates of BK7 glass (k = 1.1 W $m^{-1}K^{-1}$), fused silica (k = 1.4 W $m^{-1}K^{-1}$), sapphire (k = 46 W $m^{-1}K^{-1}$) or single crystal Si (k = 150 W $m^{-1}K^{-1}$) with substrate thicknesses in the range 500 μm to 6 mm.

CALCULATION OF THERMAL RESISTANCE

The thermal comparator method has been extended to the measurement of the thermal conductivity of thin films on substrates [1, 2, 6, 8, 9, 10, 14]. In this method, a heated probe tip is pressed into contact with a half-space whose thermal conductivity is to be measured. The probe tip temperature drop is directly related to the apparent thermal conductivity of the half-space. By first measuring the temperature drops for a series of standards of known thermal conductivity a calibration curve is established. The measurement of the probe tip temperature drop for an unknown solid can be converted to an apparent thermal conductivity by using the calibration curve. Here we extend the thermal comparator method to samples consisting of films deposited on thin or thick substrates.

Consider a film bonded to a substrate, see Fig. 1. The film effective thermal conductivity k_{eff} includes that of the film and any interfacial thermal resistance, see eqn (1). The free surface of the film is heated by the influx of the power Q over a circular area whose radius a is the heat flow radius. The steady-state temperatures $T_1(r, z)$ and $T_2(r, z)$ in the film and substrate satisfy the boundary conditions

$$-k_{eff}\frac{\partial T_1}{\partial z} = \frac{Q}{\pi a^2} f\left(\frac{r}{a}\right) \quad \text{at } z = -H$$

$$T_1 = T_2 \quad \text{and} \quad k_{eff}\frac{\partial T_1}{\partial z} = k_{sub}\frac{\partial T_2}{\partial z} \quad \text{at } z = 0 \tag{2}$$

$$k_{sub}\frac{\partial T_2}{\partial z} + \sigma (T_2 - T_a) = 0 \quad \text{at } z = L$$

where σ is the heat transfer coefficient and T_a the ambient temperature at $z = L$. The function $f(r/a)$ gives the power flux profile at the surface of the film $z = -H$.

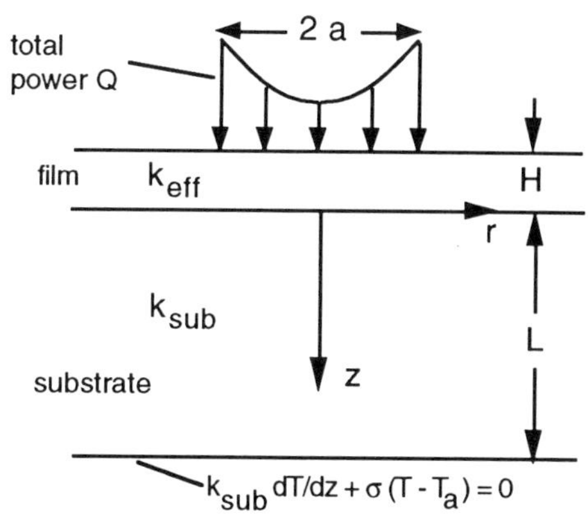

Figure 1: The geometry of a film (thickness H, thermal conductivity k_{eff}) bonded to a substrate (thickness L, thermal conductivity k_{sub}). The power Q enters the surface of the film via a circular area of radius a (the heat flow radius).

The steady-state temperature distribution can be determined by using Hankel transforms of order zero [2]. The thermal resistance R is then defined as [1, 2, 6]

$$R = \frac{\left(\pi\, a^2\right)^{-1} \int_0^a \{T(r, z = -H) - T_a\}\, 2\,\pi\, r\, dr}{Q\left(\pi\, a^2\right)^{-1}} \tag{3}$$

i.e. as the ratio of the average temperature change over the heat flow radius to the average power flux per unit area at the surface of the film.

For the thermal comparator, the power flux distribution in the film surface is adequately described by the shape function $f(r/a) = [1 - r^2/a^2]^{1/2} / 2$ for $r \le a$ [2]. For this power distribution, the thermal resistance of eqn (3) is then calculated as

$$R = \frac{a}{k_{eff}} \int_{\lambda=0}^{\infty} \frac{\sin\lambda}{\lambda} \frac{J_1(\lambda)}{\lambda} K(\lambda, H/a, L/a, k_{eff}/k_{sub}, B)\, d\lambda$$

$$K(\lambda, H/a, L/a, k_{eff}/k_{sub}, B) \equiv \frac{\tanh(\lambda H/a) + \dfrac{k_{eff}}{k_{sub}} \dfrac{\lambda + B \tanh(\lambda L/a)}{B + \lambda \tanh(\lambda L/a)}}{1 + \dfrac{k_{eff}}{k_{sub}} \dfrac{\lambda + B \tanh(\lambda L/a)}{B + \lambda \tanh(\lambda L/a)} \tanh(\lambda H/a)} \tag{4}$$

where J_1 is the Bessel function of the first kind and first order, and the Biot number B is defined as $B \equiv \sigma\, a / k_{sub}$.

When the thermal comparator is used to measure the apparent thermal conductivity k_{app} of the sample shown in Fig. 1, k_{app} is found by comparing the temperature drop of the probe tip of the thermal comparator to the temperature drops of a series of bulk standards of known thermal conductivity. Since the measured standards are bulk half-spaces, the measured apparent thermal conductivity k_{app} corresponds to the thermal resistance R of a half-space of thermal conductivity k whose surface is heated via the influx of the power Q over an area of radius a [2, 6]

$$R \equiv \frac{\pi}{4} \frac{a}{k_{app}} \tag{5}$$

Thus, by equating the thermal resistance as given by eqn (5) to that of eqn (4), the effective thermal conductivity k_{eff} of the film can be calculated when the heat flow radius a, the film thickness H, the substrate thickness L, the substrate conductivity k_{sub}, and the heat transfer coefficient σ are known, and the apparent thermal conductivity k_{app} is measured.

MEASUREMENTS

ZrO_2 single films of thickness 40 μm were deposited on substrates of fused SiO_2 (thickness 500 μm) and single crystal Si (thickness 1,000 μm and oriented along 111) by the Optical Coatings Laboratory (OCLI) using the MetaMode™ process. Five measurements of the apparent thermal conductivity k_{app} were performed on each sample at room temperature by using the thermal comparator. For the ZrO_2 film deposited on fused SiO_2 (k_{sub} = 1.4 W $m^{-1}K^{-1}$) the measured k_{app} was 0.44±0.05 W $m^{-1}K^{-1}$, and for the ZrO_2 film deposited on crystalline Si (k_{sub} = 150 W $m^{-1}K^{-1}$) the measured k_{app} was 0.9±0.3 W $m^{-1}K^{-1}$.

DISCUSSION

Fig. 2 shows the dependence of k_{app} on k_{eff} for the 40 μm ZrO_2 film deposited on 500 μm of fused SiO_2. The heat flow radius used was a = 100 μm [6].

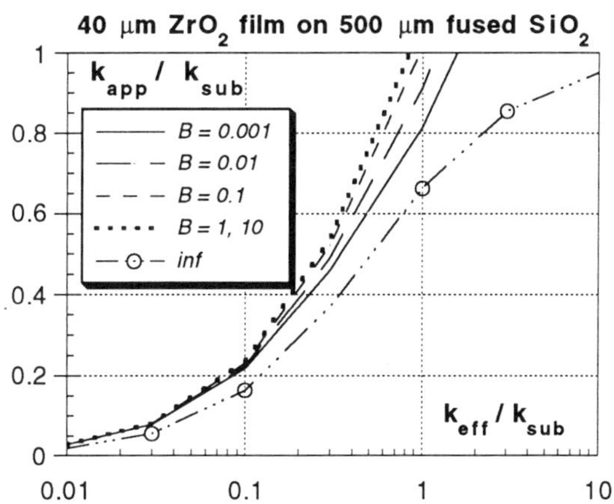

Figure 2: The dependence of the film effective thermal conductivity k_{eff} on the apparent thermal conductivity k_{app} measured by the thermal comparator. The curve labeled "inf" is the prediction of eqn (6), which is valid for a thin film (H<<a) of low thermal conductivity (k_{eff}<<k_{sub}) supported on a semi-infinite substrate.

For comparison, Fig. 2 also shows the dependence of k_{app} on k_{eff} for a thin, thermally insulating film deposited on a semi-infinite substrate [2, 6] where

$$\frac{\pi}{4}\frac{a}{k_{app}} = \frac{H}{k_{eff}} + \frac{\pi}{4}\frac{a}{k_{sub}} \tag{6}$$

This expression previously has been used to convert the measured k_{app} to the film's k_{eff} for a variety of film and substrate combinations when the substrate is on the order of a few mm in thickness [6]. Fig. 2 shows that eqn (6) begins to lose its validity as the film conductivity k_{eff} approaches that of the substrate k_{sub}. In general, for a given measured k_{app}, eqn (6) overestimates the film effective thermal conductivity k_{eff}. It is also seen from Fig. 1 that when $k_{app}/k_{sub} \ll 1$, the extracted k_{eff} is independent of the Biot number B. For larger values of the ratio k_{app}/k_{sub}, there is some dependence on the Biot number for small values of B: When the Biot number $B \geq 1$, the effect of the Biot number on the extracted k_{eff} is still negligible, and we find $k_{eff} = 0.21 \pm 0.04$ W $m^{-1}K^{-1}$. On the other hand, if the Biot number is small (B = 0.001) we find that $k_{eff} = 0.25 \pm 0.05$ W $m^{-1}K^{-1}$. It is concluded that the dependence on the Biot number is not strong.

The ZrO_2 film deposited on Si had $k_{app}/k_{sub} \ll 1$, and had no dependence of k_{eff}/k_{sub} on the Biot number. It was found that $k_{eff} = 0.33 \pm 0.10$ W $m^{-1}K^{-1}$. Table 1 below summarizes the measurements of ZrO_2 thin film thermal conductivity for various film thicknesses and substrates.

The thermal conductivity of bulk ZrO_2 is at least 1.25 W $m^{-1}K^{-1}$ [15]; In its bulk crystalline form the thermal conductivity is in the range 1.8-5.2 W $m^{-1}K^{-1}$ (depending on the stabilizer used, e.g. Y_2O_3 or MgO, and on the crystal structure, e.g. cubic, tetragonal or monoclinic [16]). We conclude that ZrO_2 films possess microstructural features which are not found in the bulk form. These features may be grain boundaries in the film microstructure, voids, or the interfacial thermal resistance associated with the film/substrate interface. Even for the thicker films used in this study, the thermal conductivity is several times smaller than the minimum thermal conductivity of bulk ZrO_2.

Table 1: Thermal conductivity of ZrO_2 films on various substrates.

Substrate	k_{sub} W $m^{-1}K^{-1}$	Film thickness, μm	k_{eff} W $m^{-1}K^{-1}$	Reference
Si (as deposited)	150	0.25 0.50 0.75 1.00	0.015 0.03 0.05 0.06	Shaw-Klein et al. [14]
Si (deposited and annealed)	150	0.25 0.50 0.75 1.00	0.006 0.012 0.017 0.022	Shaw-Klein et al. [14]
Al_2O_3	33	0.15 0.32 0.47	0.04 0.04 0.12	Lambropoulos et al. [6]
BK-7	1.1	1.0	0.05	Reichling [11]
SiO_2	1.4	40	0.24±0.06	This work
Si	150	40	0.33±0.10	This work

ACKNOWLEDGEMENTS

The work was supported by NSF through Grant MSS-8857096 and by the New York State Science and Technology Foundation via the New York State Center for Advanced Optical Technology at the University of Rochester. The authors thank Mr. John Fritschen of OCLI, Santa Rosa, CA for the film samples.

REFERENCES

[1] Lambropoulos, J.C., Jacobs, S.D., Burns, S.J., Shaw-Klein, L., and Hwang, S.-S., 1991, Thermal conductivity of thin films: measurement and microstructural effects, in Thin-film heat transfer: properties and processing, ed. M. K. Alam et al., Am. Soc. Mech. Eng. (ASME), HTD-vol. 203, p. 21, ASME Press.

[2] Lambropoulos, J.C., Jacobs, S.D., Burns, S.J., and Shaw-Klein, L., 1992, Effects of anisotropy, interfacial thermal resistance, microstructure and film thickness on the thermal conductivity of dielectric thin films, in Fundamental issues

in small scale heat transfer, ed. Y. Bayazitoglu and G.P. Peterson, Am. Soc. Mech. Eng. (ASME), HTD-vol. 227, p. 37, ASME Press.

[3] Decker, D.L., Koshigoe, L.G., and Ashley, E.J., 1986, Thermal properties of optical thin film materials, NBS Special Publication 727, Laser Induced Damage in Optical Materials: 1984, p. 291, Government Printing Office, Washington, DC.

[4] Ristau, D. and Ebert, J., 1986, Development of a thermographic laser calorimeter, Applied Optics, vol. 25, p. 4571.

[5] Ogden, T.R., Rathsam, A.D. and Gilchrist, J.T., 1987, Thermal conductivity of thick anodic oxide coatings on Al, Mat. Lett., vol. 5, p. 84.

[6] Lambropoulos, J.C., Jolly, M.R., Amsden, C.A., Sinicropi, M., Diakomihalis, D., and Jacobs, S.D., 1989, Thermal conductivity of dielectric thin films, Journal of Applied Physics, vol. 66, p. 4230.

[7] Powell, R.W., 1969, Thermal conductivity determinations by thermal comparator methods, in Thermal Conductivity, vol. 2, ed. R. P. Tye, Academic Press, New York, p. 276.

[8] Shaw-Klein, L., Burns, S.J., and Jacobs, S.D., 1991, Thermal conductivity of aluminum nitride thin films, in Electronic Packaging Materials Science, ed. E. D. Lillie et al., Mat. Res. Soc. Proc., vol. 203, p. 235, MRS, Pittsburgh.

[9] Shaw-Klein, L.J., Burns, S.J., Kadin, A.M., Jacobs, S.D. and Mallory, D.S., 1992, Anisotropic thermal conductivity of $YBa_2Cu_3O_{7-\delta}$ thin films, Supercond. Sci. Technol., vol. 5, p. 368.

[10] Henager, C.H. Jr. and Pawlewicz, W.T., 1991 (May), Thermal conductivities of thin, sputtered optical films, report PNL-SA-19687, DE92 004486, Pacific Nortwest Laboratory.

[11] Reichling, M., Wu, Z.L., Welsch, E., Schafer, D., and Matthias, E., 1991, High frequency photothermal reflectivity and displacement measurements on thin film samples, presented at the 7th International Topical Meeting on Photoacoustic and Photothermal Phenomena.

[12] Schafft, H.A., Suehle, J.S., and Mirel, P.G.A., 1989, Thermal conductivity measurements of thin film SiO_2, in Proc. IEEE 1989 int. conf. on microelectronic test structures, Vol. 2, No. 1, IEEE, New York, p. 121.

[13] Brotzen, F.R., Loos, P.J., and Brady, D.P., 1992, Thermal conductivity of thin SiO_2 films, Thin Solid Films, vol. 207, p. 197.

[14] Shaw-Klein, L.J., Jacobs, S.D., Burns, S.J., and Lambropoulos, J.C., 1991, Microstructural control of thin film thermal conductivity, in Laser-induced damage in optical materials: 1991, SPIE Proceedings vol. 1624, p. 346.

[15] Cahill, D.G., Watson, S.K. and Pohl R.O., 1992, Lower limit to the thermal conductivity of disordered crystals, to be published.

[16] Hasselman, D.P.H., Johnson, L.F., Bentsen, L.D., Syed, R. and Lee, H.L., 1987, Thermal diffusivity and conductivity of dense polycrystalline ZrO_2 ceramics: A survey, Am. Ceram. Soc. Bull., vol. 66, p. 799.

NEW AND IMPROVED TECHNIQUES TO MEASURE THERMAL PROPERTIES OF THIN FILM ELECTRONIC SUBSTRATES

R. P. Tye[+] and R. L. Gardner*
+ ULVAC Sinku-Riko, Inc. Andover, MA U.S.A.
*ULVAC Technologies, Inc. Andover, MA U.S.A.

ABSTRACT

Reliable values of thermal transport properties of materials available only in small quantities and especially as films and thin sheets are necessary for precise thermal control in electronics packaging. New measurement techniques and modifications of older methods to evaluate the thermal transport and expansion properties appropriate to those applications will be discussed and examples for specific materials or combinations will be included.

INTRODUCTION

Thermal diffusivity, D, (thermal conductivity/specific heat x density, $\lambda/C_p\rho$) and thermal expansion, α, are the prime thermal properties that influence the design and performance of an electronic package. For substrates the former normally necessitates that the thermal conductivity be as high as possible to dissipate generated heat rapidly. However, the latter usually requires that the α values are relatively low for single materials or are similar or exactly matched for layered composites in order to minimize mechanical stresses under the applied thermal loads. In all cases it is necessary that the assigned values are not only reliable and reproducible but also relevant to the particular form and its applications.[1]

The various semiconductor, metal, ceramic, and plastic materials used as substrates can cover a very broad spectrum of thermal diffusivity and thermal expansion. Furthermore, they are used in systems for many different applications involving an increasing broad range of temperature. Many applications using emerging technologies require the materials to be available in new forms including both homogeneous and inhomogeneous wafers, sheets, and films including composite layers.

For some materials the properties of these thin films can differ significantly from those of the bulk solid, while those for the composite cannot be calculated from a knowledge of the properties of the individual constituents.[1] Conventional methods are unsuitable and new and improved more rapid techniques have had to be developed both to measure the properties in these newer configurations and to obtain the required precision.

The present paper discusses several new or improved techniques that have been developed specifically for undertaking reliable, and relevant thermal property measurements on thin specimens used particularly for electronics applications including substrates.

THERMAL DIFFUSIVITY (INCLUDING SPECIFIC HEAT AND DERIVED THERMAL CONDUCTIVITY)

a. Transitory Temperature ("Flash") Method

One of the most significant contributions to the field of measurement of thermal transport properties of homogeneous solids was the development circa 1960 by Parker and colleagues of the flash method.[2] Over the past thirty years the method and its many variations and modifications has evolved into one that has been used for the broadest range of materials over a very wide temperature range.[3]

The principle attractions are:

- relative simplicity in principle and methodology based on a one dimensional analysis with experimental evaluation of heat losses. One surface of a small specimen is subjected to a short-burst of radiant energy. The temperature-time history on the other surface is measured and D for the perpendicular to surface direction is derived from a simple arithmetical relationship involving the square of the thickness and the half time (time to attain 50% of the maximum temperature rise). With suitable modification and calibration using reference materials the specific heat can also be measured on the same specimen.
- small disc specimens usually 6 to 12 mm diameter and 1 to 4 mm thick.
- measurement-times of the order of milliseconds. This avoids potential contamination problems at elevated temperatures and allows much improved volumes of testing in equivalent times to those by steady state methods.
- consequent ability to derive λ from $D=\lambda/\rho C_p$ providing the specimen is homogeneous, representative of the material, not highly anisotropic and that solid conduction is the predominant mode of heat transmission.

While the method has been used to determine absolute transport property values it has proved a most useful and rapid tool for more qualitative and comparative applications relating to the effects of various processing and related parameters on the thermal property. For example, material purity, grain size, additives, orientations, elimination or reduction of imperfections and grain boundaries etc., can all influence thermal conductivity.

A typical example is shown in Figure 1 which illustrates the effects of both oxygen impurity and sintering on the thermal conductivity of aluminum nitride.[4] By increasing the added Y_20_3 to reduce the oxygen content and further sintering in a reducing atmosphere at temperatures in excess of 1900°C it has been possible to remove grain boundary effects and attain values of λ in excess of 250W/m•K (more closer to the theoretical value of 300 W/m•K).

Because of these factors the technique has been utilized predominately in the various phases of research and development of high thermal conductivity substrates. Originally these included aluminium nitride and other ceramics. More recently work has been extended to studies on diamond.

However, materials such as AlN and diamond are now being produced commercially in larger sizes and quantities such as100 and 150 mm round or square wafers. A practical and pressing need is to evaluate the homogeneity and uniformity of property of such wafers non-destructively. Clearly this cannot be undertaken on 10 to 12 mm diameter specimens cut from the wafer.

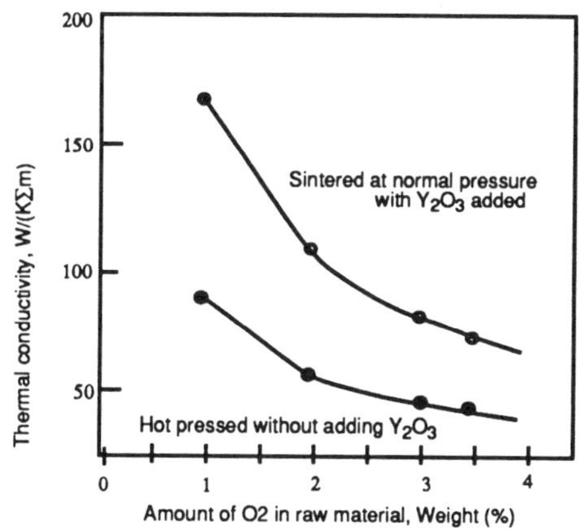

Figure 1. Effects of O_2 Content on Thermal Conductivity of AlN.

One promising approach developed at Sinku-Riko is the modification of the method.[5] This utilizes a two dimensional analysis to provide the spatial distribution of the lateral thermal diffusivity in an infinite thin plate.

The principle and the schematic experimental representation are shown in Figure 2. For incident absorbed energy Q_o (high emittance coating) the temperature history θ at point, r, shown in the figure is obtained from the following equations for an infinite plate and conditions of no heat loss.

$$\theta = \frac{Q_0}{\rho C_p \ell} \left\{1 + 2\sum_{n=1}^{\infty} (-1)^n \exp(-n^2\pi^2 Dt/\ell^2)\right\} * G\left(\frac{r^2}{r_0^2}, \frac{r_0^2}{4Dt}\right) \qquad (1)$$

and

$$G\left(\frac{r^2}{r_0^2}\cdot\frac{r_0^2}{4Dt}\right) = \sum_{n=0}^{\infty}\sum_{k=0}^{n} \frac{(-1)^n\, n!\, (r_0^2/4Dt)^{n+1}}{(n-k)\,!\,(n-k+1)\,!\,k!\,k!} \left(\frac{r^2}{r_0^2}\right)^k \quad (2)$$

$$\theta = \frac{Q_0}{\rho C_p \ell}\, G\left(\frac{r^2}{r_0^2}\cdot\frac{r_0^2}{4Dt}\right) \quad (3)$$

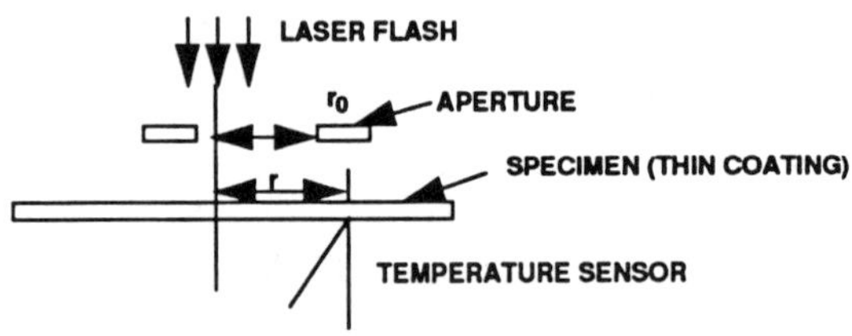

Figure 2. Measurement principle of two dimensional technique.

Providing that the specimen plate thickness, ℓ, is less than 0.1r, eq.1 reduces to eq.3 as illustrated in Figure 3, which shows calculated results based on different values of the parameter r/r_0. Absolute values of thermal diffusivity including anisotropic effects can be obtained from eq.3. but practical comparative values can be determined by using a well characterized reference material for an initial calibration. No significant effect of coated surface condition has been found.

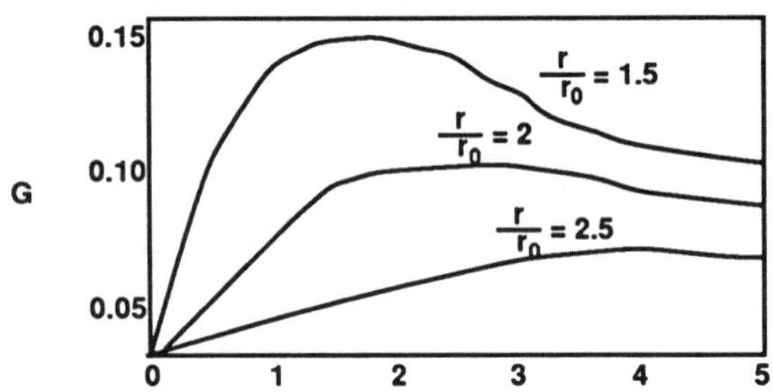

Figure 3. Relation of $G(r^2/r_0^2, r_0^2/4Dt)$ vs. $4Dt/r_0^2$)

The technique has been used at controlled room temperature conditions to study different types of wafers including AlN and Si. For example one AlN wafer some 40mm x 30mm and approximately 0.6mm thick was measured over a matrix of nine positions. The results indicated an average value of λ of 203 ± 2.5 W/m•K. Measurements on a similar Si wafer showed significant variability with an average value of approximately 0.71cm^2/s, some 15% or so lower than the expected one. Subsequent examination of the specimen showed that the lower value could have been caused by the effects of slip planes in the crystal.

This technique has significant impact on NDT and related QC/QA evaluation wafer type specimens of all types of materials. Essentially for the measurement, all that is required is the addition to a flash diffusivity unit of an attachment consisting of a specimen holder and X-Y positioning and measurement table to allow "off center" temperatures to be determined..

b. Modified ac Calorimetry

Ac calorimetry was developed initially as a very sensitive method for measuring specific heat especially on small specimens in the form of wafers, sheets, and films.[6] Overall the sensitivity of the method is some 10^2 to 10^3 greater than that of scanning calorimetry that has been the method of choice for this property when small specimens are used. Essentially one surface of a small rectangular test specimen is irradiated by a halogen lamp chopped at a low frequency (0.01 to 30Hz). The temperature amplitude Tac of the rear surface measured using a very sensitive, stable, lock-in amplifier. The heat capacity is determined either relatively or absolutely (by calibration with known materials) from the heat flow, the frequency and the temperature excursion during one cycle.

The flash diffusivity method is not necessarily suitable for all materials and very thin films. As a result we have developed a modification of ac calorimetry which allows thermal diffusivity (and associated specific heat to be measured on small specimens including layered structures less than 0.35mm thick.[7]

The principle is illustrated schematically in Figure 4 and the basic apparatus is very similar to that for specific heat measurements only. However instead of illuminating the complete surface of the specimen, only part of it is subjected to the periodic thermal energy from the light source. The remainder is masked from the chopped light irradiation. The ac temperature, T_{ac}, at a position on the specimen back surface under the mask is measured using a very fine thermocouple which also supports the specimen. Is has been shown analytically and experimentally[8] that this thermocouple and its attachment do not affect the accuracy of the determination of D. The specially developed lock-in amplifier, developed for measuring accurately the signals obtained at the low frequencies used, detects only the fundamental sine-wave component and is not affected by the periodic deformation.

The thermal diffusivity, D, in the direction parallel to the surface of the specimen is obtained by measuring T_{ac} as a function of distance x between the position of the sensor and the edge of the mask. In the instrument, the mask is moved with respect to a fixed specimen the thickness ℓ.

The relation T_{ac} vs. x is given by:

$$T_{ac} = \frac{Q \exp(-kx)}{2\omega\ Cp} \tag{4}$$

where Q is the amplitude of the energy flux per unit area, ω is the angular frequency of the periodic heating, Cp the heat capacity and k is the thermal decay constant $(\omega/2D)^{1/2}$ or $(\pi f/D)^{1/2}$ where f is $(\omega/2\pi)$ the ac frequency. D can thus be obtained from the decay constant (slope of the $\log T_{ac}$/(frequency) ± v distance relationship). The specific heat is determined separately on the same specimen.

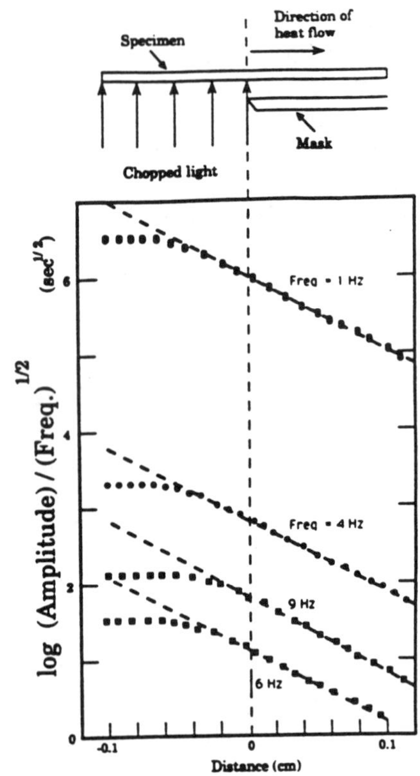

Figure 4. Example of temperature history curve for stainless steel.

Frequently, k is called the temperature wave number and ℓ/k the thermal diffusion length. In order to satisfy the one dimensional system condition it has been determined that the relationship $k\ell < 0.1$ must be satisfied.[9]

In practice, to ensure uniformity of heat flow, the front surface of the specimen is coated with a very thin layer of highly energy absorbent material. It has been demonstrated analytically and experimentally that effects of such layers less than 100nm in thickness can be neglected even on specimens some 10μm thick.[10]

The technique has been used to carry out measurements on a variety of films including layered composites. In the latter case it is possible to use a "method of mixtures" type analysis using the results of separate measurements on the composite and on the substrate alone. Assuming, λ_1, λ_2, and ℓ_1,ℓ_2 are the respective thermal conductivity and thicknesses of the film and substrate experience has shown that reliable measurements can be made providing,

$$\lambda_1 \ell_1 \geq \lambda_2 \ell_2 \quad (5)$$

As $\lambda_1 \ell_1$ approaches 0.1 $\lambda_2 \ell_2$ measurements are still possible although with reduced precision.

Examples of the use of this technique for substrate applications include:

- studies on very thin diamond films to investigate effects of environment, processes, processing parameters, crystal size, impurities etc.[11] Since thermal transport properties are extremely sensitive to structure differences the technique can be used for studying the existence of local variations.
- the thermal properties of superlattice films which are used as replacements of conventional multi-element semiconductors for various. One example[12] was on a AlAs/GaAs superlattice used for the clad layer of a surface oscillation laser diode. The film consisted of stacked layers between 50 and 500 Angstroms thick each built on a substrate to a final thickness of some 10 μm. After removal of the substrates measurements on the different thicknesses were undertaken and indicated that the values for the films were lower than for the bulk alloy and the values decreased as the superlattice layer thickness decreased and the period of the lattices becomes shorter.
- thermal properties of optomagnetic recording films used in discs with shorter access time to replace both floppy and hard discs. Detailed thermal analysis to attain optimum film design is necessary since the recording medium is irradiated with a laser beam of several MHz and the thermophysical properties of material in very thin films can differ considerably from the bulk.

These examples illustrate the flexibility of the technique for thin film applications in electronics. Ongoing studies include changes in properties of ion-implanted materials to improve surface characteristics and performance of various ceramic coatings on metal and composite substrates.

THERMAL EXPANSION

Long term mechanical stability of composite for electronic devices require precise matching of expansion coefficients. As a result the ± 1μm or lower precision of thermal expansion measurements by dilatometry and other thermal analysis techniques is not high enough to design and operate the devices successfully. There is a need to be able to detect changes in length to at least 0.1 μm or better for "matching" coefficients of dissimilar materials is necessary or when only very small specimens lengths are available.

Optical interferometry is one technique that can provide this order of precision. A dual beam automated system based on a modified Michelson interferometer has been developed which provides twice the sensitivity of the single beam system.[13] Furthermore, due to the design for more rugged commercial use the specimen is simple in design and its preparation requires less critical attention.

The method, illustrated schematically in Figure 5 employs a highly stabilized He-Ne gas laser and optical system. A test specimen up to 15mm long is sandwiched between two parallel fused silica mirror having surfaces coated partially with gold. For thin sheets, etc., less than 1mm thick the specimen is sandwiched between two glass distance pieces of known or previously measured property.

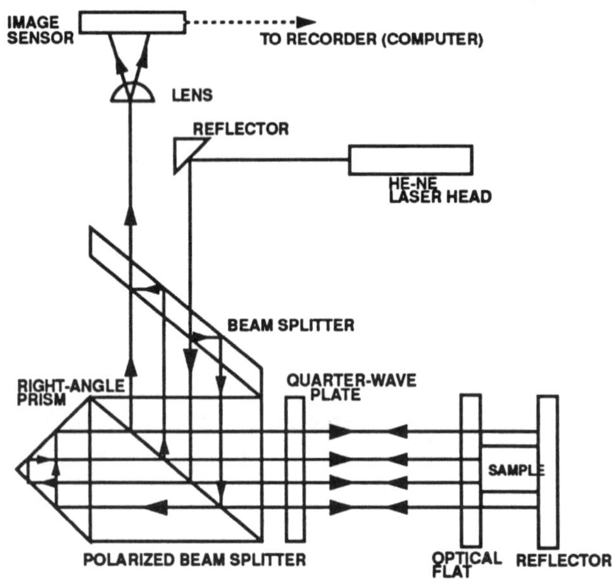

Figure 5. Optical Arrangement of the Dual Beam Interferometer.

The linearly polarized laser beam is split into two by the beam splitter, these go through the quarter waveplate, are reflected at the inner and outer mirrors and return so creating interference fringes as the specimen changes in length. As the specimen is heated and cooled, the fringe pattern is expanded by the negative cylindrical lens and falls on the linear image sensor which detects movement of the multipeak pattern.

A microcomputer analyzes the signal every 0.5 seconds to obtain the position of the peaks A_0, A_1, A_2. . . and calculates the absolute positive or negative length change (ℓ nm) using

$$\ell\ (\text{nm}) = \frac{A_1}{A_1 - A_0} \bullet \frac{632.8\ \text{nm}}{4} \qquad (6)$$

with the wavelength of the He-Ne laser used for calibration. Dimensional changes can be resolved with an absolute accuracy of 0.02nm (200 λ) equivalent to $^1/_{32}$ of the wavelength of the gas laser.

The high performance of interferometry systems has been demonstrated both by measurements on reference materials and by a recent interlaboratory study on two low expansion glasses (Zerodur and ULE) between 100 and 420K.[14] This latter study was organized by the New Glass Forum in Japan in order to ensure that measured values on low expansion functional materials used specifically by the electronics and space industries were comparable to the same order of precision. The results indicated that the reproducibility of measurement of linear coefficients by the present instrument was ±3 and ±3.8 x 10^{-8}/K for the two glasses.

The accurate measuring technique thus provides those associated with the electronics industry, with three valuable attributes precise measurements of very low and negative expansion coefficients or on a “thin” specimens and a QA/QC technique specific “matching” or reproducibility to within precise limits is required.

SUMMARY

Several new or modified techniques for measuring essential thermal properties of thin films being used or considered for special emerging electronics applications have been described together with examples of their use.

REFERENCES

1. Tye, R.P., 1991,”Measurements of thermal conductivity and related properties: a half century of radical change.”, High Temp. - High Pres. vol. 23, p 1.

2. Parker, W.J., Jenkins, R.J., Butler, C.P., and Abbott, G.I., 1961, “Flash method of determining thermal diffusivity, heat capacity and thermal conductivity.”, J. Appl. Phys., vol. 32, p. 1679.

3. Taylor, R.E., and Maglic, K.D., 1984, “Pulse method for thermal diffusivity measurement”, in Compendium of Thermophysical Property Measurement Methods I, Eds. K.D. Maglic, A. Cezairliyan, and V. E. Peletsky, Plenum Press, New York, p. 305.

4. Horiguchi, A., Ueno, F., Kasoni, M., Shinozaki, K., and Tsuge, A., 1986, “Effect of sintering atmosphere on thermal conductivity and miccrostructure of AIN”, Ball. Jpn. Inst. Metals, vol. 25 no. 4, p. 253.

5. Azumi, T., and Kishi, A., 1985, "Measurement of thermal diffusivity and thermal conductivity in surface direction (2-dimensional method)", Proc. 6 Jpn. Symp. Thermophysics Prop., Japan Thermophysical Properties Society, Tokyo, p. 201.

6. Hatta, I., and Ikushima, A.J., 1981, "Studies on phase transitions by ac calorimetry", Jpn. J. Appl. Phys., p. 1985.

7. Hatta, I., Sasuga, Y., Kato, R., and Maesono, A., 1985, "Thermal diffusivity measurement of thin films by means of an ac calorimetric method.", Rev. Sci. Instrum., vol. 56, p. 1643.

8. Hatta, I., Kato, R, and Maesono, A., 1986, "Development of ac calorimeter method for thermal diffusivity measurement I: contribution of thermocouple attachment in a thin sample.", ibid., vol, 25 no. 6, p. 493.

9. Hatta, I., Kato, R., and Maesono, A., 1987, "Development of ac calorimeter method for thermal diffusivity measurement II: sample dimension required for measurement", ibid., vol. 26 no. 3, p. 475.

10. Hatta, I., Hiraiwa, S., and Yao H., 1989, "An ac calorimetric thermal diffusivity measuring method for transparent materials.", Proc. 2nd Asian Thermophysical Prop. Conf., Chinese Physical Society, Beijing, p. 543.

11. Funamoto, H., Nishikawa, A., Moriya, T. and Sangboni, K., 1986, "The influence of methane concentration on the properties of diamond film.", Proc. 7th Jpn. Symp. Thermophysics Prop., Japanese Thermophysical Properties Society, Tokyo, p. 179.

12. Yao, T., 1987, "Thermal properties of AlAs/GaAs superlattices.", Appl. Phys. Lett., vol. 51 no. 22, p. 1798.

13. Takahashi, K., Azumi, T., Amemlya, K., Kato, R., and Sekino, S., 1987, "Laser interferometry type thermal expansion meter.", Proc. 8th Jpn. Symp. Thermophys. Prop. Japanese Thermophysical Properties Society, Tokyo, p.177.

14. Imai, H., Okaji, T., Kishi, T., Sagata, H., Aikawa, H., and Kato, R., 1990, "Measurement of thermal expansivity of low-expansion glasses by interferometric method: result of an interlaboratory comparison.", Int. J. Thermophys. vol. 11, no. 5, p. 937.

NANOINDENTATION OF THIN FILMS USING FINITE ELEMENT ANALYSIS

J. E. Ritter T. J. Lardner, D. T. Madsen, and R. J. Giovinazzo, University of Massachusetts, Mechanical Engineering Department, Amherst, Massachusetts 01003.

ABSTRACT

Design of multi-chip-modules depend on the accurate knowledge of the mechanical properties of the thin films as they exist in electronic packages. Nanoindentation instruments are now available for measuring the indentation load-depth behavior of thin films with resolution in loads of better than 1 mN and in depth of 1 nm. This research explores the use of finite element simulation of the indentation behavior of thin films to determine the properties of these films by "best-fitting" the finite element results to the experimental load-depth data. The thin film systems studied in this research include Al/Si (substrate), Si/Al (substrate), Cu/Si (substrate), Cu/Ti/Si (substrate), and Cu/Cr/Si (substrate). It is found that the strength properties of thin films can be significantly greater than those of the corresponding bulk material.

INTRODUCTION

During the last decade developments within the microelectronic and magnetic disk industries have created a need to predict, control, and understand the mechanical behavior of thin films. One of the most widely used techniques for measuring the elastic and plastic properties of thin films is nanoindentation.[1] However, when the depth of the indentation exceeds 10-25% of the film thickness, the influence of the substrate on the load-depth response of the film must be considered.[1,2,3] Although equations relating the elastic modulus of the film and substrate to the unloading stiffness and equations relating the hardness of the film and substrate to the overall measured hardness have been proposed, there is not good agreement with experiment.[1,2,3] Therefore, this research explores the use of finite element simulation of the indentation behavior of thin films to determine the properties of these films by "best-fitting" the finite element results to experimental load-depth data. Such a finite element investigation is valuable because the deformation properties of very thin films on substrates are thought to be different from those of a free standing film or bulk material.[4-6]

FINITE ELEMENT MODEL

Simulations of elastic/plastic indentations were conducted using the ABAQUS finite element code (Hibbitt, Karlsson, and Sorensen, Inc.). Four-noded axisymmetric elements were used to model the coating and substrate. A rigid conical indenter with a 70.3° half-angle, which gives the same cross-sectional area as the Vickers indenter for a given depth of penetration, was used as the indenter. The indentation procedure was simulated by displacing the indenter downward into the coating and the indenter load was determined during the indentation process by summing the axial reaction forces along the bottom of the substrate. The indenter contact radius was determined directly from the finite element output. The overall dimensions of the model were large enough to be considered remote, i.e. any further increase in the dimensions made no significant change in the load-depth behavior. If the indentation simulation resulted in severe element distortion under the indenter, the remesh/restart feature of the ABAQUS program was implemented. This feature allows regions experiencing extreme distortion to be meshed by new undistorted elements without significantly altering the stress distributions in the model. Accuracy in the numerical results was checked by mesh refinements until no significant changes occurred in the indentation load/depth curve. Non-linear deformations were allowed and both the coating and substrate were modelled as elastic/perfectly plastic materials. Unless otherwise noted, the material properties used in the finite element calculations are given in Table 1.

Table 1. Material Properties Used for Initial Finite Element Simulations with Elastic-Perfectly Plastic Material Behavior

Material	Young's Modulus E (GPa)	Yield Strength Y (MPa)	Poisson's Ratio υ	Reference
Copper	117	69	0.360	7
Titanium	110	483	0.30	7
Chromium	248	452	0.222	7
Alumium	176	485	0.33	8
Silicon	127	4410	0.278	8

The coating/substrate combinations studied were double layer coatings on a silicon substrate consisting of a 3 μm copper top layer with either a titanium or chromium interlayer of thickness 0.02 or 0.20 μm and single layer coatings of 3.02 μm copper and 1.0 μm aluminum on silicon, and 1.0 μm silicon on aluminum. These combinations of aluminum and silicon have been considered with finite element analysis by other researchers[3,8] and represent model systems for a relatively soft film on a hard substrate (Al/Si) and for a hard film on a soft substrate (Si/Al). Also, experimental indentation load-depth is available for an aluminum coating on silicon[9] and a 3 μm copper coating with either a 0.02 μm titanium or chromium interlayer on silicon.[4] Together with investigations of bulk indentations of silicon and aluminum, these film systems will show the capability of finite element analysis in simulating indentation behavior and by "best-fitting" the finite element results to the experimental data, the properties of the film material can be determined.

RESULTS AND DISCUSSION

Indentation Behavior. Finite element simulation showed that "hard" materials with an elastic modulus to yield strength ratio less than 80 tend to sink-in around the indenter and "soft" materials with a ratio greater than 80 tend to pile-up.[2] Thus, bulk silicon (E/Y = 29) exhibits the sink-in effect and bulk aluminum (E/Y = 157) the pile-up effect, see Fig. 1. This behavior has important consequences when considering the relationship between the actual contact area and that predicted from indenter geometry alone using the depth of penetration. From finite element results the actual contact area for silicon was about 25% less than that predicted from indenter geometry and about 30% greater for aluminum. In both cases, the actual contact area was larger than that predicted by the Sneddon elastic solution.[1,3] This sink-in or pile-up behavior also had important consequences in the indentation response of the coatings. An Al coating on Si substrate tended to pile-up around the indenter and plasticity was mainly confined to the soft Al coating; consequently,the load-depth response was not significantly affected until the indenter penetrated into the substrate. On the other hand, a Si coating on an Al substrate tended to sink-in and plasticity was induced primarily in the soft Al substrate with only a small plastic region in the hard Si coating directly under the indenter. In this case, the loading depth response is significantly affected by the substrate. Thus, any modeling of the indentation behavior of coating systems must take into account the fact that soft-on-hard and hard-on-soft coating systems respond in very different ways, e.g. see ref. 10.

Figure 2 compares the finite element indentation simulation to experimental data[9] for a 0.265 μm thick Al coating on Si. The finite element simulations shown are with a specified Al yield strength of 500 and 285 MPa and with all other properties as given in Table 1. It is evident that an Al yield strength of about 550

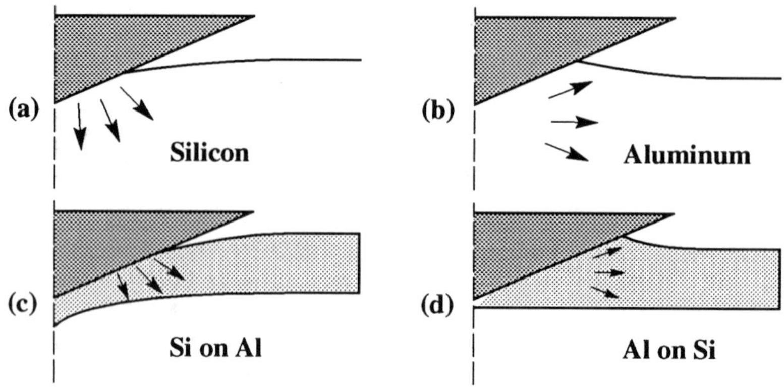

Figure 1. Schematic illustrating flow of material causing the sink-in and pile-up effects.

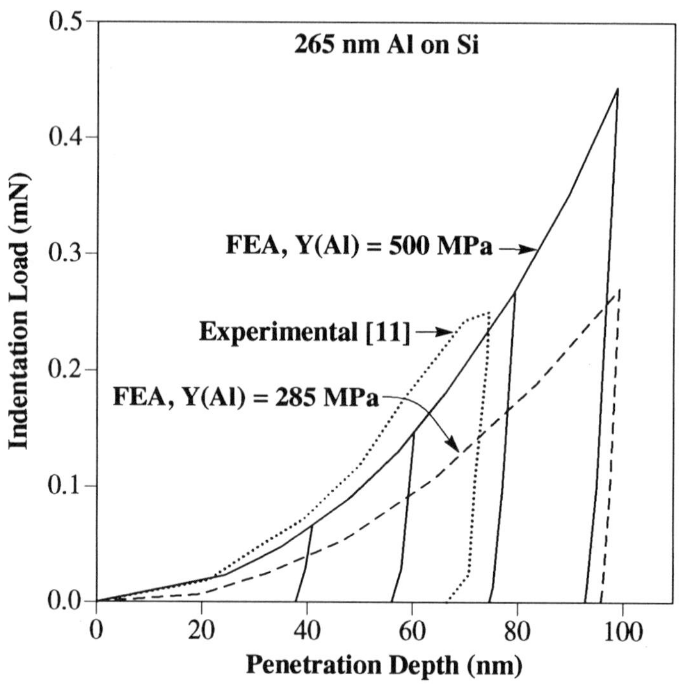

Figure 2. Comparison of experimental indentation behavior of a 0.265 μm Al coating on a Si substrate[9] to finite element simulation.

MPa is required to get agreement between experiment and finite element simulation of the load depth response.

Figure 3 compares the finite element indentation simulation to experimental data[4] for a 3/.02 μm Cu/Ti coating on Si. By increasing the yield strength of Cu to 1000 MPa, it was found that finite element simulation gave a load-depth curve that agreed with experimental data. An important question to answer regarding this indentation behavior is how influential is the interlayer. To answer this question, we ran finite element simulations assuming a thicker (0.2 μm) Ti interlayer in addition to substituting a Cu or Cr interlayer for the Ti interlayer. We found that increasing the Ti interlayer thickness to 0.2 μm had no effect on the indention load-depth behavior. On the other hand, Fig. 4 shows the effect that different interlayers have on the indentation behavior. For these finite element calculations the yield strength of Cu was taken to be 1000 MPa with all other properties as given in Table 1. The response of the Cu/Ti and Cu/Cr coatings are not significantly different while a single 3.02 μm coating (or 3.0/.02 μm Cu/Cu coating) is somewhat stiffer. These results demonstrate that the elastic modulus of the interlayer has far less influence on the indentation behavior of a coating than the yield strength since doubling the elastic modulus of the interlayer (Cr for Ti) had no significant effect but doubling the yield strength of the interlayer (Cu for Ti) made the coating noticeably stiffer.

The adhesion of thin coatings to the substrate can affect the indentation behavior of the coating. Qualitatively, if the adhesion of the coating to a hard substrate is poor, then the constraint imposed by the substrate on the deformation of the coating will be small and this will result in an increase in the depth of penetration for a given indentation load. Such effects have been measured by Lafontaine *et al.*[4] for a Cu/Cr coating on Si where poor adhesion was observed at the Cr/Si interface. Figure 5 compares the finite element indentation simulation to the experimental data assuming that the Cu/Cr coating is fully bonded and debonded to the substrate. The fully bonded and debonded simulation bracket the experimental data. Due to the high compressive stresses underneath the indenter, indentation debonding occurs over a region outside the indenter contact zone. By assuming different sizes for the debonded region, we found that a debonded region between a and 6a, where a is the contact radius, gives finite element load-depth results that agree with the experimental data.

It is important to note that the values of the yield strength used to simulate the experimental data in these Al and Cu thin coatings are quite high and similar to that expected for Al and Cu alloys. This increase in yield strength of thin coatings has been attributed to the small grain size inherent in thin coatings,

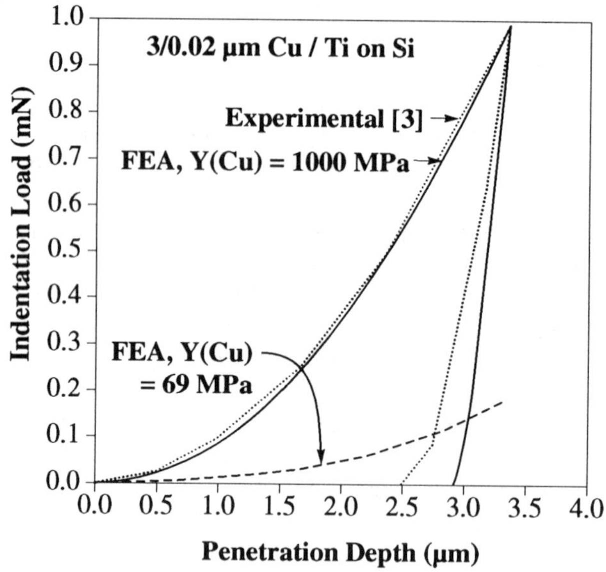

Figure 3. Comparison of experimental indentation behavior of a 3/.02 μm Cu/Ti coating on a Si substrate[4] to finite element simulation.

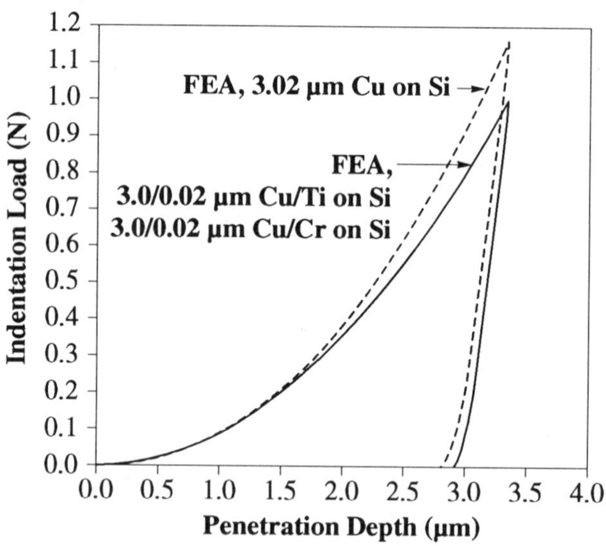

Figure 4. Finite element simulation showing the influence of the interlayer material (Cr, Ti, Cu) on the indentation behavior.

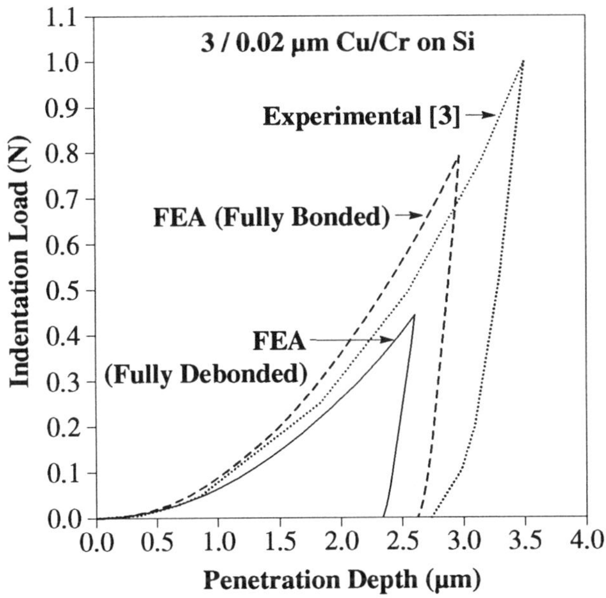

Figure 5. Comparison of experimental indentation behavior of a 3/.02 μm Cu/Cr coating on a Si substrate[4] to finite element simulation assuming the coating is fully bonded and fully debonded to the substrate.

residual stress in the coating, and having the indenter and substrate act as dislocation barriers, and thereby constraining plastic flow in the thin coating.[4-6]

Hardness. Hardness, or a material's resistance to penetration by an indenter, is generally defined as indenter load divided by projected contact area measured upon unloading. Nanoindentation devices however allow accurate measurement of the indenter load as a function of indentation depth and hardness is calculated by first relating the contact area to the depth of penetration. Hardness is calculated by dividing the load by the contact area at that load to give a definition of hardness equivalent to the indentation pressure. However, this calculation does not generally take into account the sink-in or pile-up effect, see Fig. 1, on the contact area. As demonstrated in the previous section, the distinct advantage of finite element simulation of the indentation behavior is that the actual (projected) contact area is available from the calculations at any point of loading and these calculations automatically account for the sink-in or pile-up effects.

The hardnesses determined by finite element simulation for bulk Si and Al, 1 μm Al coating on Si, and 1 μm Si coating on Si are shown in Fig. 6. The hardness

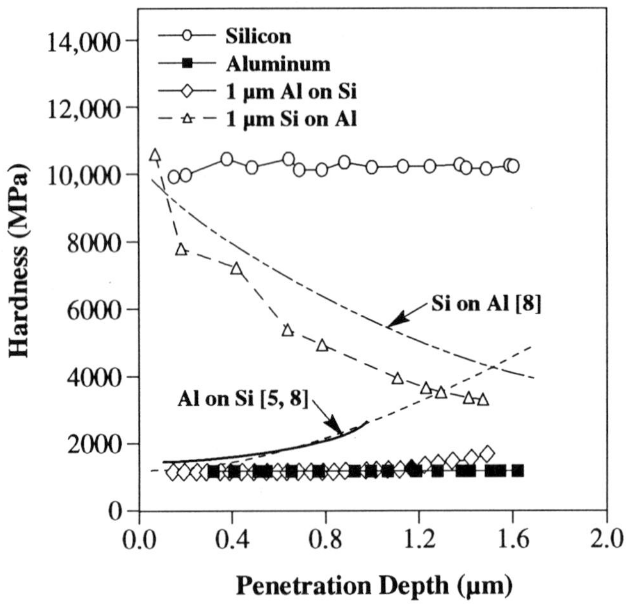

Figure 6. Hardness determined by finite element simulation of bulk silicon and aluminum, 1 μm Al coating on Si, and 1 μm Si coating on Al.

to yield strength ratio for silicon is 2.3 and for aluminum is 2.6. These hardness values agree well with those in ref. 3 and show that the indentation of the bulk materials give hardnesses that are independent of indentation depth, while indentation of the coating systems give hardnesses equal to the coating material at small penetration depths that then diverge toward the substrate hardness at large penetration depths. As expected from the previous discussion of the indentation behavior of these coated systems, the hardness divergence occurs very rapidly for the Si coating on Al but the hardness of the Al coating on Si does not begin to diverge until the indenter penetrates into the substrate and then only slowly because of the pile up effects that causes the coating to support more of the load.[10]

Relations in non-dimensional form for the hardness of a coated system as a function of indenter depth have been developed[5,8] and are plotted in Fig 6. These models do not agree well with the finite element calculations. We believe that finite element simulation should be used to calculate hardness of thin coatings by "best-fitting" the indentation load-depth data.

In the case of the double layer coatings of Fig. 4, we found that the hardness of the 3.02 μm Cu coating was 2.7 GPa while the 3.0/.02 μm Cu/Ti or Cu/Cr coating was 2.4 GPa, independent of depth in both cases before penetration. Consistent with Fig. 4, it is seen that the Ti or Cr interlayer reduces the hardness of the coating. The ratio 2.7 of the hardness of the Cu coating to its yield strength agrees well with the Al data.

Elastic Modulus. On unloading the indenter from the maximum load, the elastic displacements are recovered and an analysis of the initial elastic unloading data can be used to calculate the modulus of the indented material. Specifically, it has been argued that the initial slope of the unloading curve (dP/dδ) at the maximum load is related to the elastic modulus (E) of the indented material by[1,3]

$$E = 0.5\left(\frac{\pi}{A}\right)^{1/2} \frac{dP}{d\delta}(1 - \nu^2) \qquad (1)$$

where A is the projected contact area at the maximum load, υ is Poisson's ratio for the indented material, P is indentation load, and δ is pentration depth. The unloading stiffness (dP/dδ) can be found from the P-δ curve.

Table 2 compares the E calculated from Eq. (1) to that expected from the value assumed for the finite element calculations; for the coated cases, the penetration depth was less then the coating thickness. The agreement between the calculated and expected moduli is not good, contrary to the results in ref. 3. This difference could be due to the differences in the mesh used in the present study to that used in ref. 3 or to approximations inherent in derivation of Eq. (1). This is certainly an area for future research.

Table 2. Comparison of the Calculated Elastic Modulus, Eq. (1), to that Expected from the Elastic Modulus Assumed for the Finite Element Simulation

Material	E (GPa) Calculated	E(GPa) Expected
Si	153	127
Al	107	76
Al Coating on Si	97	76
Cu Coating on Si	112	117
Cu/Ti Coating on Si	106	117

ACKNOWLEDGMENT

This research was supported by NSF Grant DMR-8919374.

REFERENCES

1. G. M Pharr and W. C. Oliver, "Measurement of Thin Film Mechanical Properties Using Nanoindentation," Mater. Res. Soc. Bulletin, XVII (7), 23-33 (1992).
2. D. T. Madsen, R. J. Giovinazzo, J. E. Ritter, and T. J. Lardner, "Finite Element Simulation of Indentation Behavior of Thin Films," pp. 431-36 in Thin Films: Stresses and Mechanical Properties III, eds. W. D. Nix, J. C. Bravman, E. Arzt, and L. B. Freund, Mater. Res. Soc. Symp. Proc., 239, Pittsburgh, PA (1992).
3. T. A. Laursen and J. C. Simo, "A Study of the Mechanics of Microindentation Using Finite Elements," J. Mater. Res., 7 (3), 618-26 (1992).
4. W. R. LaFontaine, B. Yost, and C-Y Li, "Effect of Residual Stress and Adhesion on the Hardness of Copper Films Deposited on Silicon," J. Mater. Res., 5 (4), 776-83 (1990).
5. D. Stone, W. R. LaFontaine, P. Alexopoulos, T-W Wu, and C-Y Li, "An Investigation of Hardness and Adhesion of Sputter-Deposited Aluminum on Silicon Utilizing a Continuous Indentation Test," J. Mater. Res., 3 (1), 141-47 (1988).
6. B. R. Venkatraman and J. C. Bravman, "Separation of Film Thickness and Grain Boundary Strengthening Effects in Al Thin Films on Si," J. Mater. Res., 7 (8), 2040-48 (1992).
7. CRC Handbook of Tables for Applied Engineering Science, 2nd Ed., editors R. E. Bolz and G. L. Tuve, CRC Press, Cleveland, OH (1973).
8. A. K. Bhattacharya and W. D. Nix, "Analysis of Elastic and Plastic Deformation Associated with Indentation Testing of Thin Films on Substrates," Int. J. Solids Structures, 24 (12), 1287-98 (1988).
9. M. F. Doerner and W. D. Nix, "A Method for Interpreting the Data from Depth Sensing Indentation Instruments," J. Mater. Res., 1 (4), 601-09 (1986).
10. J. E. Ritter, D. R. Sioui, and T. J. Lardner, "Indentation Behavior of Polymer Coatings on Glass," Polymer Engr. Sci., 32 (18), 1372-78 (1992).

Section V. Thick Film/Metallizaton Technology

MATERIALS ASPECTS OF THICK FILM RESISTORS

R.J. Bouchard, Dupont Experimental Station, Electronics, P.O.Box 80334, Wilmington, DE 19880-0334

ABSTRACT. Resistor compositions are among the most demanding thick film formulations. The main constituents are conductive oxides and frits, although other additives are critical. Early conductor systems have been displaced in the last 20 years by those containing oxides of ruthenium, either RuO_2 or pyrochlores of the type $A_2Ru_2O_{6-7}$, where A can be Bi, Pb or mixtures of these with rare earths, Cu, etc. The advantages and disadvantages of RuO_2 and pyrochlores are compared. The function and compositions of frits and additives are discussed. Various resistor types and electrical specifications are described, along with materials considerations for each. Newer challenges of the marketplace which place new demands on resistor materials are discussed. Such changes include the trend to smaller part sizes, higher quality and yield, and use in multilayer parts.

Resistor compositions are arguably the most complex of thick film formulations. The most important material variables that impact on the many specs that resistors have to meet are listed below in Table 1 and discussed in detail in the sections that follow.

Table 1. Functional Characteristics of resistor constituents.

CONDUCTIVE PHASE	GLASS BINDER	SUBSTRATE
Resistivity	Softening point	Surface texture
Thermal expansion	Thermal expansion	Thermal expansion
Purity	Purity	Purity
Chemical reactivity	Chemical reactivity	Chemical reactivity
Particle size/SA	Particle size/SA	
TCR	Viscosity	
Thermal stability	Wetting characteristics	

CONDUCTIVE PHASE

Early conductor systems

Early resistor compositions were formulations with mixtures of palladium and silver metals. However, when Pd powder is heated in air, it oxidizes to PdO. The oxide content increases with firing temperature until about 800 C, at which point it evolves oxygen and returns to Pd metal. Since kinetics of this process are influenced by the amount of Ag present, fired films contain mixtures of PdO and Pd/Ag solid solutions. Resistivity is a function of the relative amounts of the conductive alloy and the semiconducting oxide present in the fired film.

Since this is a chemically dynamic system, it is extremely sensitive to processing conditions (firing time, atmosphere, temperature, etc.). Other deficiencies of this early system include resistivity changes in even mildly reducing conditions, limited resistivity range, and high TCR's. Clearly other functional phases were needed, and the various oxides of ruthenium have filled the bill for the last twenty years or more, and have essentially completely displaced PdO/Ag-based systems.

Oxides of Ruthenium

Oxides of ruthenium are unique materials which are critical to modern thick film resistors. Consider the demands: The oxide must be stable in air at 850 C for at least 10 minutes in the presence of a low viscosity, highly corrosive frit. It cannot oxidize or reduce or form a grain boundary phase with grossly different electrical characteristics, unlike most metals which will form a semiconducting or insulating film when fired in air. Finally, it must be highly conductive with the positive TCR's characteristic of metals, so that when diluted with large volumes of electrically insulating frit to adjust resistance, it will still have a TCR that is close to zero.

Ruthenium dioxide, RuO_2, is one of the most conductive oxides known. It derives its conductivity from the strong covalent interaction between Ru+4 and oxygen in the well-known tetragonal rutile structure. In this structure, the metal ions are surrounded by a regular octahedral arrangement of oxygen ions, and orbital overlap among all such octahedra is extensive. This interaction leads to the formation of band states which are continuous throughout the lattice, and because of the electrons provided by the ruthenium ions, the band is only partially filled, thereby leading to conductive, rather than insulating or semiconducting, behavior (1,2).

Other materials containing ruthenium are also conductive, and most flexible of these is the generic $A_2B_2O_7$ pyrochlore family (A=Pb, Bi, Cd, Tl, Gd, etc; B=Ru, Os, Nb, Ta, Sb, Cr, etc), . This cubic crystal structure is best visualized as a combination of A_2O and B_2O_6 sublattices. The basic framework consists of shared BO_6 octahedra, and the main function of the A_2O sublattice is to provide charge balance. Since the O in the A_2O sublattice is not critical (pyrochlores can range from $A_2B_2O_6$ to $A_2B_2O_7$), and the A and B ions can include a very large number of single and mixed ions, there are essentially an infinite number of compositional possibilities.

Pyrochlores based on Ru have the same Ru octahedra present in RuO_2 and are conductive for the same reason. However, because of the difference in crystal structure, conductivity is different than in RuO_2 and depends on the particular A-site cations. The entire spectrum of electrical properties can be found within the pyrochlore family from insulators through semiconductors to excellent metals like $Pb_2Ru_2O_6$ ($PbRuO_3$). Unlike RuO_2, the ability to alter the functional phase via chemical substitution provides a broad latitude for the adjustment of resistivity, TCR and other electrical and physical properties. The most common compositions used in resistor formulations are various chemical variants of bismuth ruthenate ($Bi_2Ru_2O_7$) and lead ruthenate ($Pb_2Ru_2O_6$) (3-7).

Because substitutional possibilities with RuO_2 are extremely limited, the chemical route to modifying intrinsic chemical properties is unavailable, leaving only those properties that can be influenced by the surface area of the oxide and interactions with glasses of different chemical compositions. Nevertheless, each oxide has its advantages and disadvantages which are summarized in Table 2:

Table 2. Advantages and Disadvantages of Pyrochlore vs RuO_2

PYROCHLORE	
ADVANTAGES	**DISADVANTAGES**
Moderate R *vs* loading	High TCE *vs* alumina
Wide range of intrinsic R/TCR	Higher TCR gaps (HTCR-CTCR)
High + TCR ($PbRuO_3$)	Limited to intermediate and high R
High conductivity ($PbRuO_3$)	Decomposed by some frits
Non reactive to organics	Must be milled for high SA

RuO_2

ADVANTAGES	DISADVANTAGES
Wide range of SA	Reactive to organics in high SA
Low TCE *vs* alumina	No chemical variables for R/TCR
Smaller TCR gaps (HTCR-CTCR)	Steep dependence of R *vs* loading
High Conductivity	Forms pyrochlore with some frits
High +TCR	Limited to low and intermediate R

RuO_2 is particularly useful for low and mid range resistors because of its high conductivity, which translates to a lower conductive loading for a given R. This in turn means that the fired composite has higher density, better surface cosmetics, and better ability to withstand shocks like laser trimming. A TCE comparable to alumina helps insure that tension is not created during cooling after firing.

Pyrochlores are more important for the higher resistivities, where they typically give better power handling and TCR control than RuO_2-based formulations, mostly because a higher conductive loading is required for a given resistance than RuO_2.

In addition to the intrinsic properties of the conductive oxides, resistor properties can be influenced by reactions that can occur with the glassy binder phase during firing. For example, pyrochlores like $PbRuO_3$ can partially decompose to give RuO_2 in certain frits where the Pb content is too low. RuO_2 can form pyrochlores with certain other frits with high Pb content.

Because of the particular characteristics of the various conductive types, each is usually best suited for particular resistor values, rather than the whole resistance spectrum. Typically, Pd/Ag alloys, which can be formulated to have essentially zero TCR are required for very low R. For mid range resistors, RuO_2 tends to be best. At the high end, pyrochlores tend to be superior. At the borderline between these categories, mixtures of two conductives are common. The constitution of a typical resistor series could be that shown in Table 3:

Table 3. Conductive type vs resistance

Resistance (ohms)	Materials system
1-10	Pd/Ag alloys
100-10K	RuO_2
1K-10 Meg	$PbRuO_3$

FRITS

The main function of frit is to provide sufficient volume of insulating phase to dilute the conductive properties of the active oxide phase, so that a large range of resistance values can be achieved. It must also bond the resistor to the substrate, and wet oxide particles, so that they can be effectively dispersed into a uniform resistor composite. For most of the resistor systems in use today, lead alumina borosilicates or variants thereof fulfill this function. Their viscosity is low enough under typical firing conditions of 10 min/850 C to form smooth resistor composites over a wide range of conductive oxide loadings.

At low loadings of conductive oxide (high resistance), the constitution of the frit is usually modified with oxides like zirconia, alumina, etc. to raise the viscosity to minimize flow. At high loadings of conductive oxide (low resistance), oxides like PbO and Bi_2O_3 are added to lower viscosity, improve wetting and minimize the amount of frit required to form a dense composite. Depending on the type of conductive and frit, their surface areas, conductor/frit ratio, etc., required for electrical specifications, the fired resistor may have a porous microstructure that will not be durable in use. In such cases, encapsulation by a post-applied, low softening point frit is usually effective in providing durability.

If the resistor must be plated after termination, as in the case of chip resistors, care must be taken that the frit is stable to the corrosive effects of the plating solutions, and additional compositional modifications to frit must be made.

Because the constitution of frits can be varied over essentially an infinite range, a large number of frits are used, depending on the requirements of the particular resistor system. The final frit or mixture of frits in any given resistor formulation is the result of a host of historical, personal, and ad hoc considerations. In many cases, frits of reasonably similar chemistry can substitute for another with little difference in electricals.

Wetting properties of the frit are critical, even though in most cases, these characteristics are not quantified. If the surfaces of the conductive powder are wet well enough and soon enough, they are more likely to be dispersed into the frit volume. If not, they will tend to undergo the same type of sintering and loss of surface area that would occur in the absence of frit. The resistor will then exhibit electricals more characteristic of higher conductive loadings, e.g., lower R and more positive TCR.

It is also well known (8) that many frits can dissolve considerable amounts of alumina from the substrates. This will be a function of its chemical composition, time, and temperature. The resulting alumina gradient can extend many microns up into the resistor, and can have large effects on electricals, so that resistivity will be thickness-dependent. The effect can be minimized by including alumina in the frit.

THE ROLE OF ADDITIVES

Electronic

Often, commercial resistor systems cannot meet all specifications without small amounts of certain chemical additives. TCR "drivers" are a good example. Usually, especially at the lower resistance ranges, the highly conductive oxides have very positive TCR's, even when diluted with the required amount of frit to achieve a given resistivity. It is well known that as little as 1% of oxides like TiO_2 and Nb_2O_5 can reduce TCR by hundreds of ppm/degree (9). In addition, they usually increase resistance, which means that for a given conductive loading, more of the oxide is needed, and this is often the only way to increase the power handling at the higher resistance levels. Occasionally in some systems, an oxide like Cu_2O can have the opposite effect of reducing resistance and driving TCR more positive. There is usually little difference between adding this type of oxide directly to the paste vs introducing it as part of the glassy frit phase.

The exact mechanism by which these additives function is not known. They may change the electronic properties of the glass itself, introducing carriers or traps into the conduction band. They may have more of a physical effect on the microstructure, modifying the equilibrium separation between conductive particles or the wetting characteristics or viscosity of the glass, so that the dispersion of the conductive phase is affected. Microstructural examination shows little difference between samples with and without these additives, even though large changes in electricals result from their incorporation.

Physical

In certain cases, particularly at high conductive loadings of pyrochlore conductives, and with certain relatively high TCE frits, the TCE of the resistor composite will be above the TCE of alumina, resulting in a resistor that is in tension after firing. In addition, although a resistor usually has a composite TCE designed to be less than alumina, if the frit is unstable and tends to precipitate or crystallize a high TCE phase like crystobalite, the actual TCE will be much higher. This can occur with frits of high silica content, for example.

While a fired resistor can withstand some tension because of the reinforcing effect of the oxide particles in the frit, clearly there is some point where long term stability, particularly under high humidity and temperature is compromised. Tension in the resistor manifests itself by the formation of very small hairline cracks, generally at the resistor/termination interface where the stress at the substrate is a maximum. The crack leads to increased resistance and high sigma's in TCR values. Drift of electricals after laser trimming is also common. To compensate an otherwise stable materials system with a TCE slightly higher than desired, it is possible to add low-TCE materials like zircon ($ZrSiO_4$), which at loadings of about 10% tend to act as inert fillers, replacing an equivalent amount of frit.

TYPICAL ELECTRICAL SPECIFICATIONS

Resistivity

Resistivity of thick film resistors is generally expressed in ohms per square, ρ=RW/l, where R=resistivity in ohms/square, l is length, W is width, and R is resistance in ohms. Generally, this is normalized to a standard print thickness of 25 microns and a square resistor configuration so that R is independent of part size. Materials are available with resistivities ranging from 1.5 ohms/square to 10 megohms and above. R can be modified by changing composition of the conductive phase, conductor/frit ratio, etc. or geometry of the resistor. Operationally, resistors are trimmed to final value by vaporizing part of the fired resistor away with a laser.

Temperature Coefficient of Resistivity

TCR is defined as the fractional change in R over a given temperature range, expressed in ppm/deg C and is calculated by $TCR=\Delta R(10^6)/R_{25}\Delta T$ (Figure 1). The referenced temperature is 25 C and two TCR values are usually specified: hot TCR from 25 to 125 C, and cold TCR from 25 to -55 C. Typical resistors today will have TCR values of less than 100 ppm/deg C. As an example a resistor with a TCR of 100 ppm/deg C will change in value by 1% over a range of 100 C.

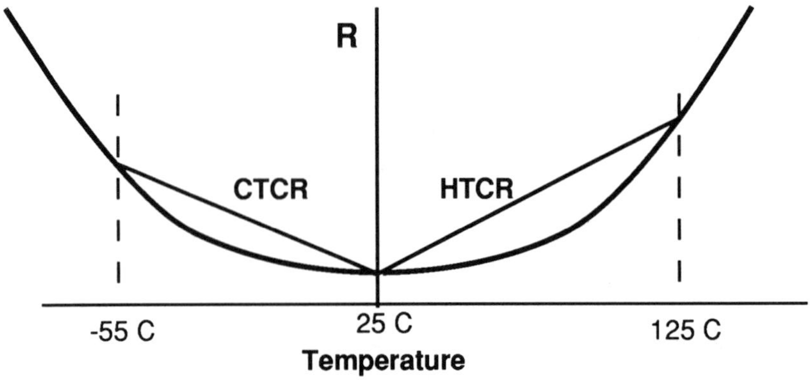

Figure 1. Definition of hot and cold TCR's

Stability and reliability

Thick film resistors are remarkably stable. Both accelerated and long-term performance data have shown that end-of-life tolerances of 1 to 2% are easy to achieve, and 0.5% or less are possible with proper circuit design, resistor design, and processing. Data taken under load of 100 W/in^2 for more than 20,000 hrs shows that, even for early thick film systems, the resistance change for most resistance values is less than 0.1% per decade of time., or <0.5% after 10 years (10). Typical stability after electrostatic discharge is <0.5% change in R after 5 pulses of 5KV/mm.

Process sensitivity

Typical resistors are fired on 96% alumina substrates, over silver or silver/palladium terminations, are about 10 microns thick, and can be as small as 40X40 mils in dimension or smaller. They are fired at 850 C for 10 minutes, with the typical profile from one end of the belt furnace to the other of either 30 to 60 minutes. Typical sensitivity of electricals to firing temperature are $\Delta R<1\%$/minute at temperature and <1%/deg C over the range 825-875 deg C. The corresponding values for TCR are <2 ppm/minute and <2 ppm/deg C.

During firing, a large number of processes occur that are functions of time, temperature, atmosphere, etc.: reduction of Ru-based oxides during burnout (11); tendency of conductive particles to sinter and/or disperse, depending on frit viscosity and wetting (12); dissolution of alumina from the substrate and diffusion into the resistor (8); diffusion of ions from the resistor into the substrate (13);

diffusion of Ru into glass (14,15); diffusion of Ag and other minor constituents from the termination into the resistor (16); crystallization of new phases, decomposition of pyrochlores to form RuO_2, reaction of RuO_2 to form pyrochlores, etc. All of these can influence the final microstructure and attendant electrical properties. All can be controlled to give reproducible resistor characteristics.

Normally, thick film furnaces have carefully controlled profiles, standardized around 10 minutes at 850 C with total cycle time of either 30 or 60 minutes. Nevertheless, some variation is common, and circuits may have to be refired for multilayers. So it is necessary to have microstructure development complete by the time the lowest likely firing temperature is reached and stable enough to be relatively constant at the highest likely temperature. Stability over a range of at least ±25 deg C around the firing temperature is desirable. Refiring the resistor, of course, places extra demands on the stability (17).

TYPICAL RESISTOR CONFIGURATIONS

Thick film resistors may have a number of different configurations, and each has its own special requirements that demand a tailoring of the materials systems.

Hybrid microcircuit resistors

This is the traditional thick film resistor, where the paste is screened onto a substrate which may be as small as 1 in^2 to 35 in^2 or larger. The resistor size can vary over a typical range of 60X60 mils to 200X200, depending on the resistance value and the circuit design. This has been the generic resistor application, used mostly in the US and Europe, and has been the subject of an excellent review (18)

Chip resistors

Chip resistors are screened onto a pre-scored substrate of any size, fired as usual, then broken up into individual resistors which may be as small as 40x40 mils, and is attached to the circuit, usually on PC boards, with pick-and-place equipment. Although more than 100 billion chips are sold annually, mostly in Japan, for <1 cent apiece, the part is surprisingly complex (See Figure 2), and can require nine or ten separate process steps. Because it must be laser trimmed, encapsulated twice, plated twice, and soldered in place, extra demands are placed on the materials system. This requires a materials system that can be printed in very small part sizes without large termination interactions, refired with encapsulants, and dipped in corrosive plating solutions of high ionic strengths.

Chip resistors can be processed at a different set of conditions than the hybrid circuit to which it is later soldered, and can be tested before it becomes part of the circuit. As a result, chips can avoid the high costs of failure associated with conventional circuits, where all components are screened on, and the entire circuit may have to be scrapped if one fails. Chips are the most rapidly growing resistor market, with hundreds of billions of parts produced annually, mostly in Japan.

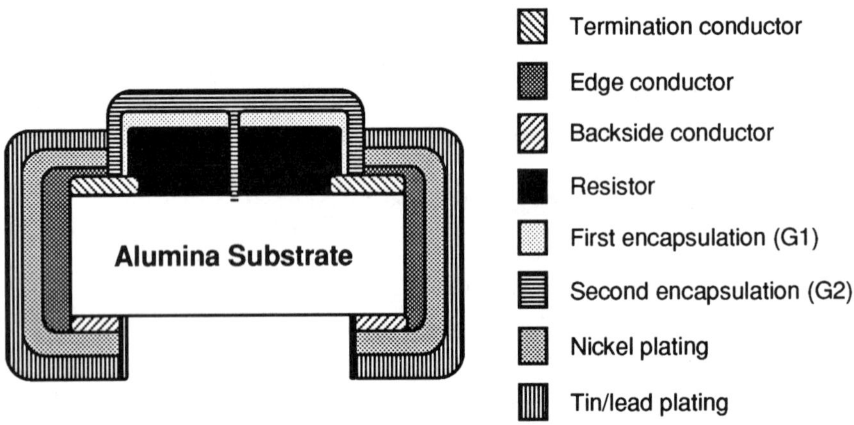

Figure 2. Chip Resistor Configuration

Surge resistors

As the name implies, surge resistors are designed to protect the circuit from sporadic high voltage spikes, like lightning, which can lead to pulses of 1500 volts or more. They are usually very low resistance (~1 ohm/square), and printed in serpentine or spiral patterns with increased path length and resistance to dissipate more power. High concentrations of Pd/Ag alloys which form continuous, highly conductive filaments are used, rather than conductive oxides. Since tight resistance values at high voltages, which can heat the resistor to 500 C or more, are critical, the resistors must be protected from oxidation of the Pd component to PdO by encapsulation.

Variable resistors

Potentiometers and adjustable resistors utilize a relatively long track of resistor material which is wiped by metal contacts to achieve a resistance over a range of values to fine tune a circuit to exact resistance or voltage values. In this application, special morphological requirements of the conducting oxide powder are required to assure that the wiper blade is able to make contact even at higher resistivities, when the conductor loading is very low.

NEW CHALLENGES

Quality

Manufacturers of thick film parts and circuits are becoming much more concerned with quality. They are demanding tighter electrical specifications and reproducibility from lot to lot. Circuits are becoming more demanding, both in design and application. At the same time, competition is forcing more of a focus on yield. Development of higher quality materials systems will require improvements in both the processing of the intermediates constituting thick film formulations, and the formulations themselves. We will need to constantly improve our understanding of resistor technology, the correlation between composition and processing variables, and fitness for use at the customer.

Smaller parts

As the trend toward miniaturization continues unabated, the part size of printed resistors continues to decrease. Part sizes of 40X40 mils are becoming more common and 20x20 mil parts are contemplated over the next few years. Smaller part sizes create a new set of problems. Interactions with the termination, especially diffusion of silver and active bonding oxides that may be included in the termination for adhesion purposes, now will occupy a larger percentage of the volume of the resistor, and can modify electricals dramatically. This may force the use of less reactive Ag/Pd or Ag/Pt alloys in the terminations, or the development of new terminations with much reduced levels of the active oxide constituents normally used to confer adhesion.

In addition, the smaller part sizes tend to print thicker with conventional screens, and this can have adverse influences on TCE mismatch, modifying alumina gradients from the substrate, etc. Thinner terminations will probably be required, and that means that conductivity will be reduced to perhaps below design requirements unless a separate termination and conductor printing step is used, and adhesion can be adversely affected. Clearly this will be a challenge for materials development.

Multilayers

Another clear trend is the adoption of multilayer circuit configurations, where buried resistors are required. In most cases, this will mean that instead of highly crystalline alumina substrates, the resistor will be printed on glass/alumina composites (19). Because of the presence of a large amount of glass phase in the multilayer composite, there is now the possibility of inter-diffusion of glass from the resistor to the dielectric and vice versa. In addition, a different set of chemical reactions will become possible. Degradation of the conductive oxide in the resistor must be prevented, as well as adverse interactions that could effect the performance of the dielectrics. Finally, instead of one interface with which the resistor can interact, there are now two, combined with multiple refires. So it is not surprising that a substantial offset in properties results. Considerable development of both dielectric and resistor thick films will be required. The problem is illustrated schematically in Figure 3.

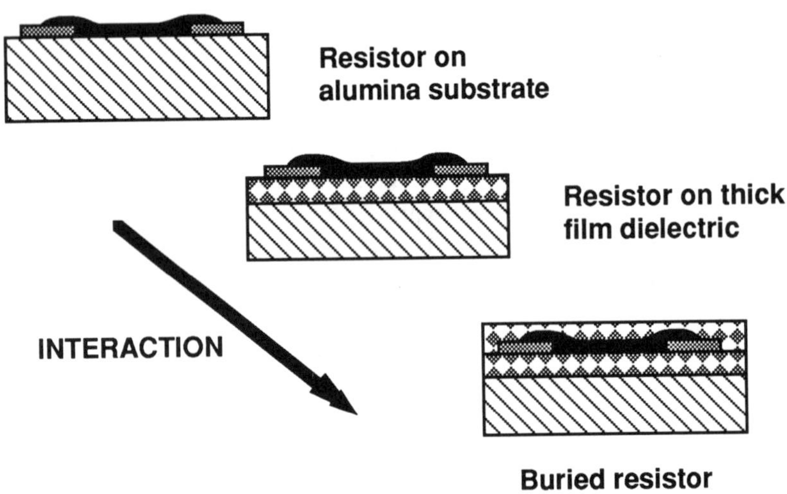

Figure 3. Interaction vs resistor surroundings

Process sensitivity

As the trend continues toward simultaneous goals of better quality and higher yields, it will be necessary to develop more robust materials systems that can be fired under a broad range of furnace loadings, firing temperatures, printing speeds, laser trimmer settings, etc. Manufacturers will want to increase the content of silver in terminations to reduce cost, which tends to increase sensitivity to firing time and temperature.

Environmental Considerations

There is no evidence that fired resistor films result in Pb toxicity problems (20). Existing procedures for powder intermediates containing Pb are well known and widely practiced. However, if there is a continuing trend to limit use of such chemical elements with blanket regulations, Pb in any form may eventually have to be replaced. Then, a complete redesign of resistor materials will be required, incorporating Pb-free conductive oxides and frit. While this may be possible, it will require considerable investment in R&D, and some property penalties may result.

Conclusions

Thick film specifications and processing place demanding requirements on ceramic materials systems. The conductive oxide materials have been dominated for the last twenty years by oxides of ruthenium, namely RuO_2 and various substituted ruthenium pyrochlores. Frit materials are usually lead aluminosilicates or borosilicates, with a wide range of minor substitutions to achieve specific properties. Certain oxide additives are critical to achieve electrical requirements, and the mechanisms by which they function are not well understood. New challenges are constantly forcing continuous improvements in the materials and the understanding of resistor technology.

REFERENCES

1. J. B. Goodenough, Bull. Soc. Chim. France **4**, 1200 (1965).
2. D.B. Rogers et al, "Crystal chemistry of metal dioxides with rutile-related structures", Inorg. Chem 8, 841 (1969).
3. R.J. Bouchard and J.L. Gillson, "A new family of bismuth-precious metal pyrochlores", Mater. Res. Bull. **6**, 669-80 (1971).
4. R.J. Bouchard, "Oxides of cubic crystal structure containing bismuth and at least one of ruthenium and iridium", US patent 3,583,931, 6/8/69.
5. R.J. Bouchard, "Compositions for making electrical elements containing pyrochlore related oxides", US patent 3,681,262, 8/1/72.
6. R.J. Bouchard, "Compositions for making resistors comprising lead-containing polynary oxide", US patent 3,775,347, 11/27/73.
7. A.W. Sleight and R.J. Bouchard, NBS. Special Publication 364, Solid State Chem., Proc. 5th Matls. Res. Symp., July 1972, pp 227-32.
8. M. Prudenziati et al, "Interactions between alumina and high lead glasses for hybrid components", J. Appl. Phys. **65** [1], 146-53 (1989).
9. S. Fu, J. Liang, et al, "Effects of glass modifiers on the electrical properties of thallium oxide thick film resistors", Mater. Res. Bull. **12**, 569 (1977).
10. R. Cote and R.J. Bouchard, "Thick film technology", Electronic Ceramics-Properties, Devices and Applications, Ed. L.M. Levinson, Marcel Dekker, 1987, pp 307-371.
11. J.W. Pierce et al,"The chemistry and stability of ruthenium-based resistors", J. Hybrid Microelectronics **5** [1], 9-17 (1982).
12. R.W. Vest, "A model for sheet resistivity of RuO_2 thick film resistors", IEEE Trans.-CHMT, **14** [2] , 396-406 (1991).
13. Y. Nakamura, "Alumina substrate and glass interactions", Int. J. Hybrid Microelectronics 4 [2], 168-71 (1981)
14. A. Kubovy, "The effect of firing temperature on the properties of model thick film resistors", Silikaty **32** [4], 289-303 (1988)
15. P. Palanisamy et al, "Solubility of ruthenium dioxide in lead borosilicate glasses", J. Am. Ceram. Soc. **72** [9], 1755-6 (1989).
16. A. Cattaneo et al, "Influence of the metal migration from screened and fired terminations of the electrical characteristics of thick film resistors", Electrocomponent Sci. and Tech. **4**, 205-11 (1977).
17. A. Cattaneo and M. Prudenziati, "Effect of refiring processes on electrical and structural properties of thick film resistors", Electrocomponent Sci. and Tech. **6**, 165-71 (1980).
18. W. Borland, "Thick film hybrids", Electronic Materials Handbook, Volume 1:Packaging, ASM International, Materials Park, Ohio, 1989.
19. B. Joshi et al, "The properties of thick film resistors on dielectrics", J. Microelectronics **13** [5], 4-11 (1982)
20. J.N. Boyd, Ceram. Eng. Sci. Proc. **13** [3-4] 160,71 (1992).

MODELING OF THICK FILM RESISTORS

Th. Pfeiffer and R.J. Bouchard, DuPont Experimental Station, Electronics, P.O.Box 80334, Wilmington, DE 19880-0334

ABSTRACT. A simple model of metal and glass conduction predicting TCR's has been applied to thick film resistors. The dependence of resistance and TCR on different conductive/glass ratios between resistors based on Pb-pyrochlore and RuO_2 is described. Activation energies for glass conduction derived from comparison of the model with experiment are related to TCR gap (difference between HTCR and CTCR). Values of the calculated activation energies (1-2meV) are surprisingly low, indicating defect conduction. Finally, the model is used to link microstructure and power handling of thick film resistors.

1. HIERARCY OF INTERACTIONS

In order to understand the different effects contributing to electrical properties of thick film resistors, it is important to consider possible interactions. Because the resistor consists of conductive particles and glass powder, interaction between them obviously has to be considered . The resistor is fired on an Al_2O_3 substrate and is connected to the circuit by the termination, so interactions with both of them are likely, also. The amount of material which can be exchanged between two phases in a given time can be estimated dividing the cross section by the time it takes to impact the whole volume. Therefore, it is interesting to calculate the interaction area A divided by the typical diffusion time t required for an ion to travel through the glass particle or whole resistor, respectively. t is proportional to l_d^2, the typical diffusion length. A/l_d^2 for each interaction gives the hierarchy, assuming that rate constants of possible processes or reactions are similar.

The A can be calculated for a 1 mm by 1 mm resistor, 10 μm fired thickness, a typical conductive loading of 16 vol%, a surface area of 7 m^2/g for the conductive, and 3 m^2/g for the glass. Assuming ld for the conductive/frit interaction is 1μm, (≈ diameter of the glass particles), l_d for the substrate interaction is 10 μm (≈ thickness of the resistor) and l_d for the termination interaction is 500 μm (≈ half of the resistor length), A/l_d^2 can be calculated:

$$A_{cond/frit} = (1mm*1mm*10\mu m)\ (0.16*9g/cm^3*7m^2/g)\ /\ 1\mu m^2 = 10^8$$
$$A_{substrate} = (1mm*1mm)\ /\ 10\ \mu m^2 = 10^4$$
$$A_{termination} = (10um*1mm)\ /\ 0.25\ mm^2 = 4*10^{-2}$$

The result clearly indicates the trivial conclusion, that the resistor is dominated by the interaction between conductive and glass. In addition, interaction of the substrate with the resistor glass, overwhelms interaction of the resistor with the termination. In the following, we will tackle these interactions in sequence.

2. INTERACTION BETWEEN CONDUCTIVE AND GLASS

2.1. Development of Resistance during Firing

Measuring resistance R and thermal coefficient of resistance TCR at different firing temperatures (Fig. 1) reveals that R decreases to a minimum at the softening

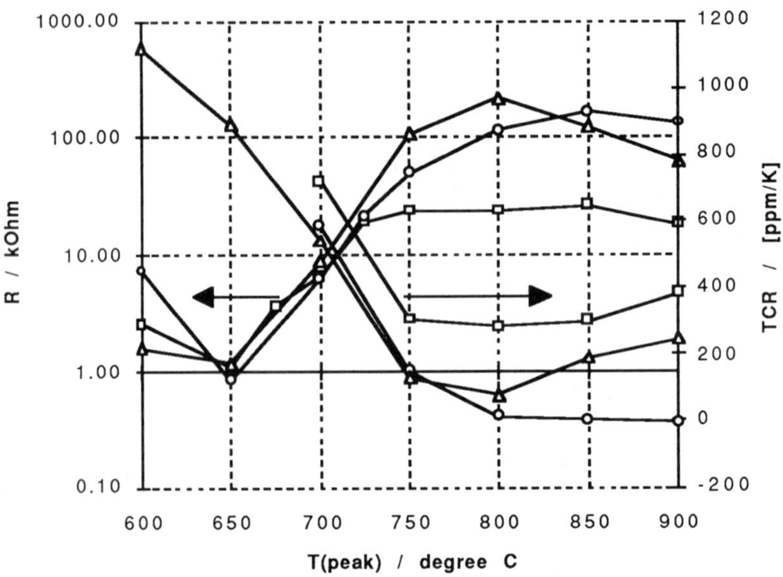

Fig. 1: *R and TCR vs. peak firing temperature for $PbRuO_3$ in several different glasses*

temperature of the glass[1)], then increases with increasing temperature until it hits a maximum. When fired at even higher temperature decreases R again. The TCR behaves just opposite - if R increases TCR decreases, and visa versa.

While temperature is below the softening point, the glass squeezes the smaller conductive particles together, and the conductive sinters. Both effects will decrease R. Above the softening point, the glass starts to disperse the conductive particles, building glass barriers into a conductive chain and stopping sintering. Therefore, R increases. Finally, the glass begins to interact with the substrate, due to its low viscosity. Resulting chemical changes leads to stratification and/or segregation of the conductive, and reduces the effective volume of the resistor. As a result, R decreases again, since the increase of the effective conductive/frit ratio overwhelms the decrease of the total resistor thickness.

2.2. TCR Model

In order to verify the interpretation given above, a TCR model for thick film resistors is proposed, starting from the assumption that the resistor is built of conducting chains. Each chain is constructed of conductive particles and glass barriers[2-6]. For further simplification, the assumption is made that each set of conductive particles and glass barrier has the same resistance. Of course, in the real world, there is a distribution of such resistances. The same assumption with regard to the chains has been made. An average chain has *i* glass barriers and *j* chains connected in parallel to make up the whole resistor (Fig. 2).

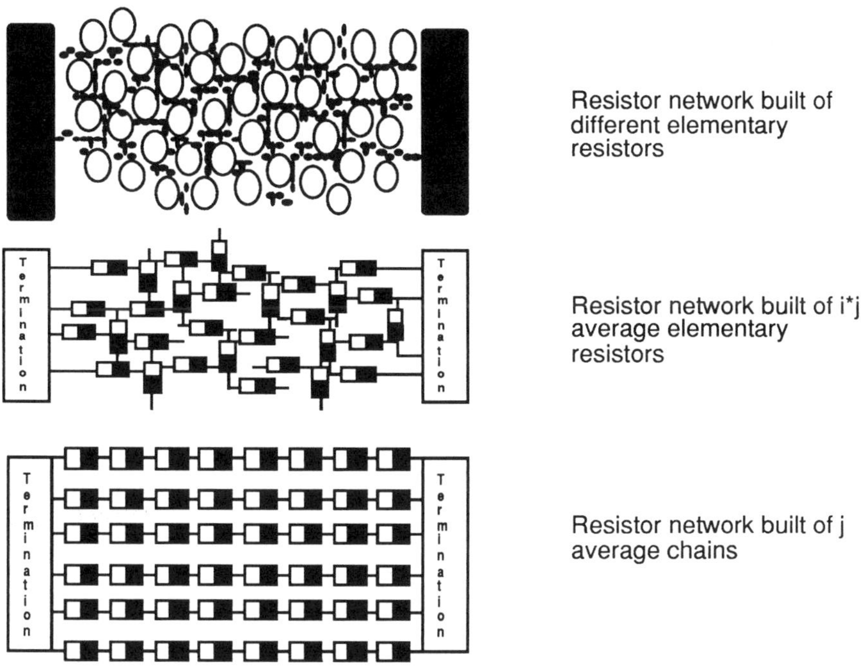

Fig. 2: Assumptions made in order to develop a simple TCR model.

Assuming that the conductive material behaves like a metal with respect to its change of resistance vs. temperature, and that the glass barrier introduces semiconducting resistance behavior, the following equations will describe the elementary resistor r_e:

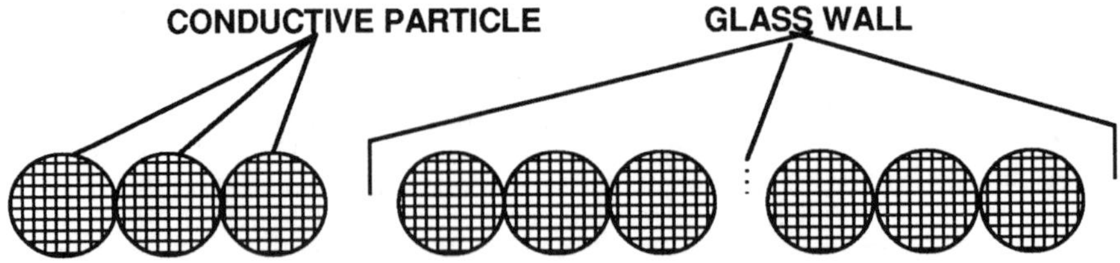

$R_{conductive\ particle} = N_c\, r_c\, (1+a\,(T-T_o))$; $R_{glass\ wall} = r_g \exp(E/kT)$

where N_c is # of conductive particles, having resistance r_c and TCR α; r_g is resistance of a glass wall at temperature T -> oo, E its activation energy, T_o is room temperature, and k is Boltzmann constant.

Applying the assumption that *i* sets of conductive/glass wall make up one chain, and all *j* chains are connected in parallel, the following equations are valid:

$$R_{chain} = i\,\{N_c r_c (1+\alpha[T-T_o]) + r_g \exp(E/kT)\} \quad (1)$$

$$R = i/j\,\{N_c r_c (1+\alpha[T-T_o]) + r_g \exp(E/kT)\} \quad (2)$$

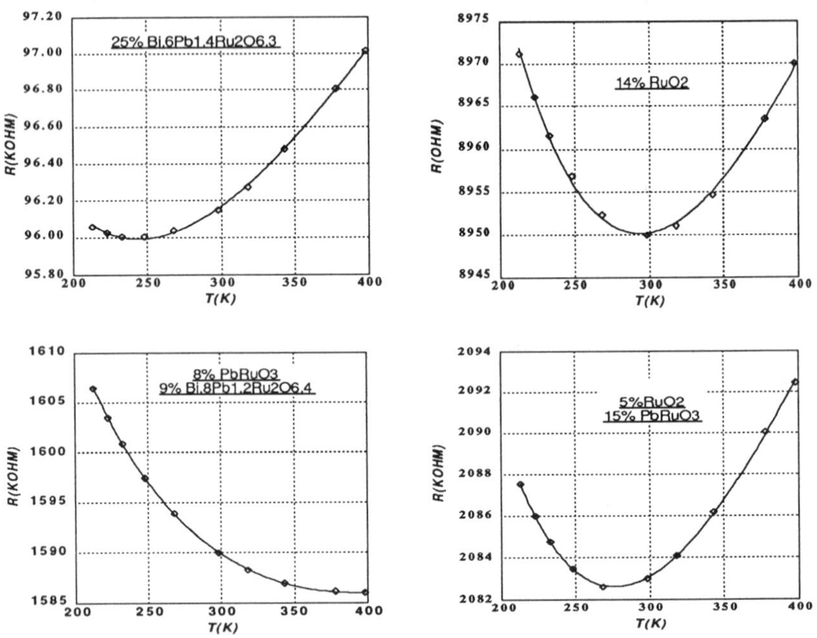

Fig. 3: *R vs. T for resistors based on RuO2, pyrochlore and their mixtures, fitted using equation 2.*

Measured R vs. T data of many thick film resistors have been fitted successfully using equation 2. Examples are shown in Fig. 3 Regardless of conductive or glass, R(T) can be described by equation 2. Mixtures of conductives in different glasses can be handled by the equation as well.

One surprising outcome of the fitting is, that the activation energy E, necessary to describe R(T), is very low, around a few meV, *regardless of glass or conductive*. Literature[7)] typically cites E values for glasses in the range of one eV, 1000 times higher than found by fitting thick film resistors. This small activation energy suggests that the glass barrier is more like an ordered structure rather than a doped amorphous glass.

2.2.1. Resistance

Mathematically, a small E value means that the equation can be simplified. Developing the exponential function into a Taylor sequence reduces equation 2 to:

$$R(T) = (i/j)\,\{N_c r_c(1+\alpha[T-T_o])+r_g(1+E/kT)\} \qquad (3)$$

Analyzing equation (3), some conclusions can be drawn immediately: First, *the resistance of a resistor will not change, if the number of elementary resistors in the average chain changes by the same factor as the number of chains.* An illustration is shown in Fig. 4. One well known consequence of this is the insensitivity of R and TCR to the size of a square resistor.

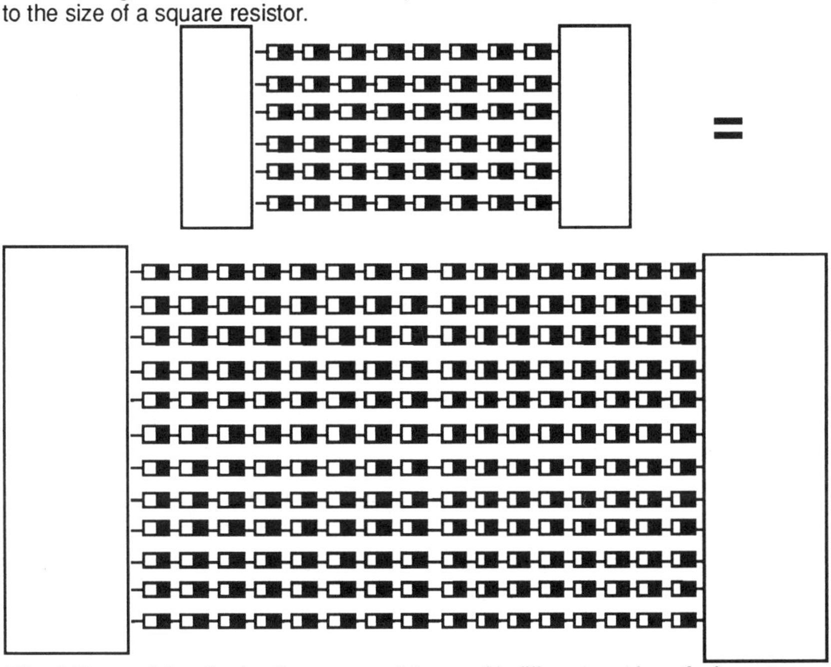

***Fig. 4**: Two resistors having the same resistance with different number of elementary resistors in a chain, but constant ratio of (i/j).*

2.2.2. TCR

The second conclusion becomes clear calculating the derivative of R with respect to T normalized on R , which describes

$$\text{TCR} \rightarrow \frac{(dR/dT)}{R} = \frac{N_c r_c \alpha - r_g E/kT^2}{N_c r_c(1+\alpha[T-T_o]) + r_g[1+E/kT]} \quad (4)$$

TCR is independent of the number of elementary resistors in chain and of the number of chains building the whole resistor. TCR of the whole resistor is solely determined by TCR of the average elementary resistor. While R does mirror the total micro structure via the factor (i/j), TCR is determined by the average set of conductive particles and glass barriers, and therefore by local interaction between glass and conductive particles. Furthermore, equation 4 indicates that TCR in a given glass is determined by the product of resistance times TCR of the conductive material. Therefore, the larger $r_c\alpha$ is, the more positive TCR will be for resistors, made of different conductives at the same loading. Comparing different conductives in a typical lead aluminosilicate illustrates this conclusion (Fig. 5).

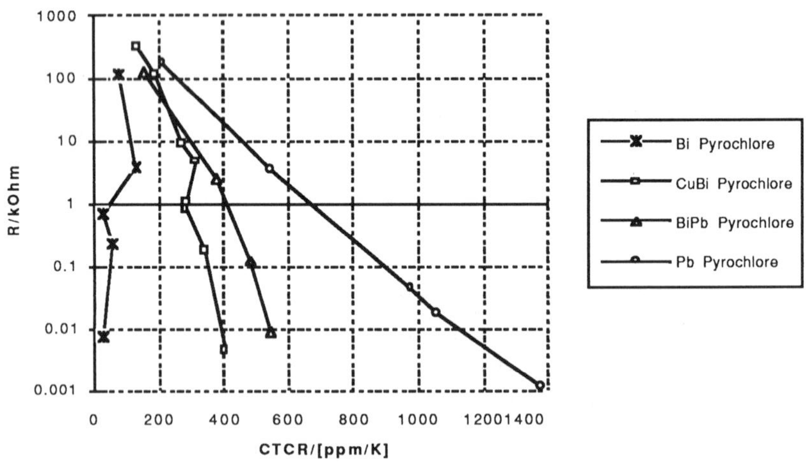

Fig. 5: *R vs. CTCR at different loadings for different pyrochlores in a Lead-aluminosilicate. Bismuth pyrochlore has the lowest (r*α), lead pyrochlore has the highest.*

2.2.3. TCR gap

Calculating the second derivative with respect to T gives TCR-gap, which describes the bending of R(T), or difference between hot (measured from 25C to 125C) and cold TCR (from 25C to -55C) :

$$\text{TCR-gap} \rightarrow \frac{(d^2R/dT^2)}{R} = \frac{2\, r_g E/kT^3}{N_c r_c(1+\alpha[T-T_o])+r_g[1+E/kT]} \qquad (5)$$

It is largely a function of the resistance and activation energy of the average glass barrier. Only normalization of R introduces some dependence on the conductive. The product (r_g*E) describes the TCR-gap completely. *Therefore, the chemistry of glass, and its chemical interaction with conductive, govern the bending of the R(T)-function.*

2.3. Discussion

Next we compare the model with the observed effect of loading on R, TCR and TCR-gap (Fig. 6,7 and 8). While this particular exper-iment illustrates the behavior of $Pb_2Ru_2O_6$ in a specific lead-alumino-silicate, it is representative of most conductives in most glasses.

Fig. 6 shows that log(R) increases nearly linearly with decreas-ing loading. Fig. 7 shows the same dependence with respect to CTCR and finally, Fig. 8 illustrates that TCR-gap increases with decreasing CTCR and decreasing loading.

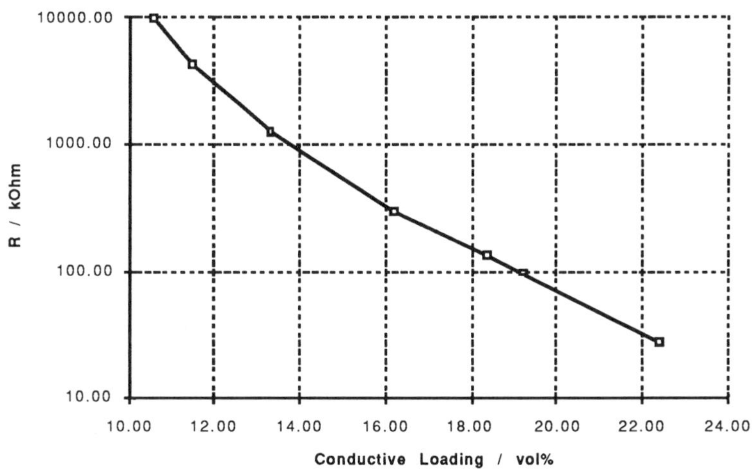

Fig. 6: *R vs. loading of lead pyrochlore in lead aluminosilicate.*

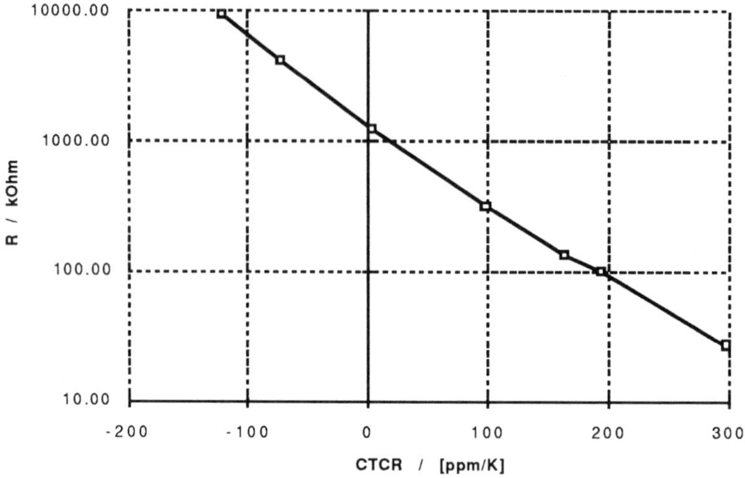

Fig. 7: *R vs. CTCR for different loadings of lead pyrochlore in lead aluminosilicate.*

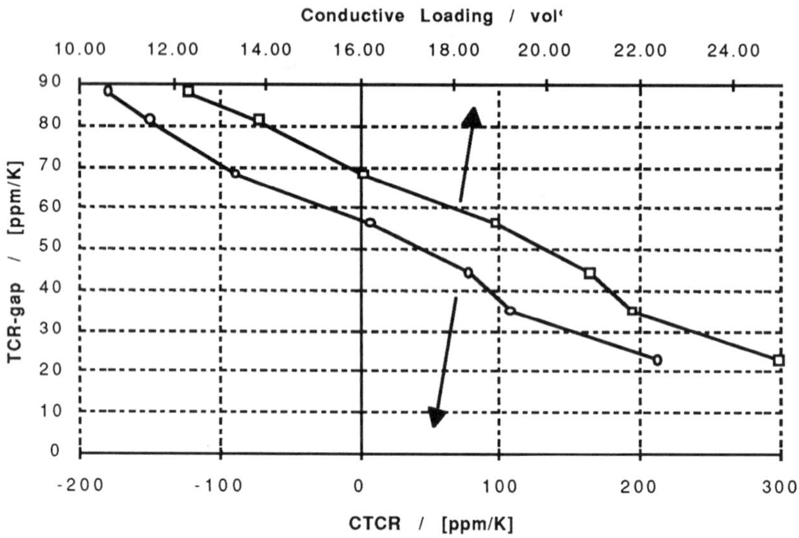

Fig. 8: *TCR-gap as function of CTCR & loading of lead pyrochlore in lead aluminosilicate, 0.75w%TiO_2.*

Comparing TCR-gap dependence on loading with equation 5 leads to the conclusion that N_C, *the number of conductive particles sintered together in an elementary resistor set, decreases with decreasing loading*, because all other parameters have been held constant in the experiment. This explains the decreasing CTCR with decreasing loading as well (see equation 4).

It does not explain the R vs. loading behavior, however, since decreasing N_C would decrease R linearly (equation 3). Therefore, since R increases rather than decreases with decreasing loading, and since $N_C * i * j$ (counting all conductive particles) equals conductive loading, i (number of glass barriers in a chain) increases, while j (number of chains) decreases with decreasing loading.

In summary, three different effects describe the experimental data: *When conductive/frit ratio decreases, the glass is able to separate more conductive particles (N_C goes down), builds more glass barriers into a chain (i goes up) and forms fewer chains (j goes down).* The sum of these effects causes TCR to become more negative, and R, TCR-gap to increase.

Another way of looking at this is, that a certain amount of a particular frit can only wet a certain total surface area of conductive at a given viscosity[8)] (or temperature, assuming viscosity of the glass decreases with increasing temperature). If the conductive/frit ratio decreases, the larger amount of frit can wet a reduced amount of conductive particles at a lower temperature in the firing cycle. Therefore, the competing process of sintering conductive particles together stops earlier, resulting in smaller agglomerates of conductives, i.e. N_C is smaller.

2.4. g/c plot

Equation 3 can be developed further in order to determine the contribution of glass (g) and conductive (c) to electrical properties of the total resistor, regardless of its resistance.

$$R = (i/j)\, R_O\, (g/T + 1 + c(T - T_O)) \quad \text{, with} \qquad (6)$$

$$R_O = N_C r_C + r_g \; ; \; g = (r_g/R_O)\,(E/k) \; ; \; c = (N_C r_C/R_O)\, \alpha$$

$$N_C r_C / r_g = (E/k - g)/g = c/(\alpha - c) \; ;$$

$$g = (E/k) - (E/k)\,(c/\alpha) \qquad (7)$$

Equation 7 says that for one conductive described by α, and a fixed glass chemistry described by E, g plotted against c for different loadings follows a linear relationship. Fig. 9 shows that the linear relationship holds for many systems investigated, regardless of glass and conductive.

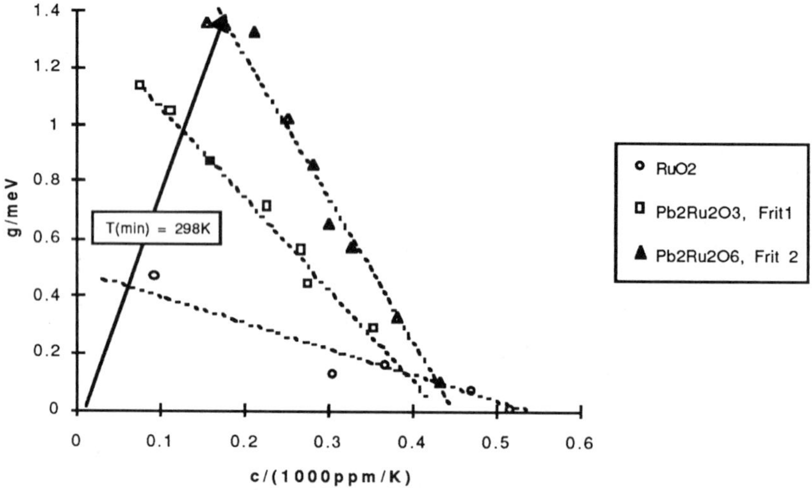

Fig. 9: *g/c-plot at different loadings for a variety of pyrochlore- and RuO_2-systems. Intersect with T(min) line gives loading at which TCR=0ppm/K at room temperature.*

Fig. 9 clearly shows, that with decreasing conductive loading (lower c), the glass contribution (g) to resistance increases. Furthermore, different glasses have different E values, as described by different intersects between loading curves and the g-axis. Setting TCR=0 in equation 4 results in the condition which describes formulations having TCR of 0 ppm/K at temperature $T_{min} = \sqrt{(g/c)}$. A line is added in Fig 9 showing formulations which have TCR of 0 ppm/K at room temperature, like most commercial resistor inks. The difference between two formulations with 0 ppm/K at the <u>same</u> temperature but different g values, is that the formulation having a <u>larger</u> g has a <u>larger</u> TCR gap, i.e. the R vs. T curve is bent more strongly.

2.5. Power handling

One of the most important features of thick film resistors is the ability to handle power without changing resistance. Experience shows, however, that the resistance of a resistor does decrease if it has been exposed to a voltage pulse above some critical value. Assuming that change of resistance is introduced by breakdown of elementary resistors, the model gives some insights why some resistors can survive voltage better than others. Since chains are connect in parallel, each chain sees the same voltage ($V = V_{chain}$). The power, P_{chain}, a chain sees is therefore:

$$P_{chain} = V_{chain}\, I_{chain} = V_{chain}^2 / R_{chain} = V^2 / (i\, r_e) \quad (8)$$

If the power one elementary resistor has to handle becomes larger than a critical value, the glass wall of this elementary resistor breaks down, and its resistance decreases. Consequently, R_{chain} decreases, and the power this chain has to handle increases. This will cause another elementary resistor in that chain to break down as well, etc., etc.. So one breakdown in a chain causes a chain reaction, and a breakdown of the total chain, triggered by the weakest link. In order to maximize power handling of the total resistor, the power P_e an elementary resistor can survive has to be maximized:

$$P_e = P_{chain} / i = V^2 / (i^2 r_e) \qquad (9)$$

Equation 9 suggests two ways to increase the voltage a resistor can handle without damage:

The first way is to increase resistance of an elementary resistor. In commercial systems, this is accomplished by additives like TiO_2 or Nb_2O_5 which increase r_g. Due to the increase of r_g, R increases as well, which is compensated by more conductive to maintain the same resistance. Since increasing resistance of the glass barrier normally brings along an undesirable increase of E, and therefore of TCR gap, this approach has limitations.

The second approach is to increase the number of elementary resistors making up a chain. A good illustration is the improved power stability of larger resistors. In addition, the two resistors shown in Fig. 4 have different power handling capability, although they have the same R and TCR. The second resistor will exhibit much better power handling than the first because *i* is much larger. This is consistent with experimental experience that resistors with higher conductive loading at a given R are much more robust against voltage.

2.6. Microstructure

The power stability of resistors can be used to investigate microstructure development of resistors at different conductive loadings. Fig. 10 shows voltage V needed to change resistance by 1% vs. R of thick film inks made up of one glass and different conductive loadings (higher R, smaller loading). Using a log log plot, constant power P is represented by diagonal lines. While voltage necessary to change R by 1% increases, P_{crit} decreases with increasing resistance or decreasing loading. If for each glass barrier built into a chain, one chain would disappear, while decreasing loading ($N_c ij$), resistors would fail at constant P_{crit} regardless of R because

$$P_{crit}/P_e = i^2 r_e / R \quad \text{demands } ij = P_{crit}/P_e \text{ or } ij = \text{const} \qquad (10)$$

Therefore, since P_{crit} decreases with decreasing loading, for each glass barrier built into a chain, more than one chain is eliminated. This suggests that *R vs. loading is dominated by the number of chainsbuilding the total resistor up, rather than the length of a chain.* This means R vs. loading behaves like one over loading.

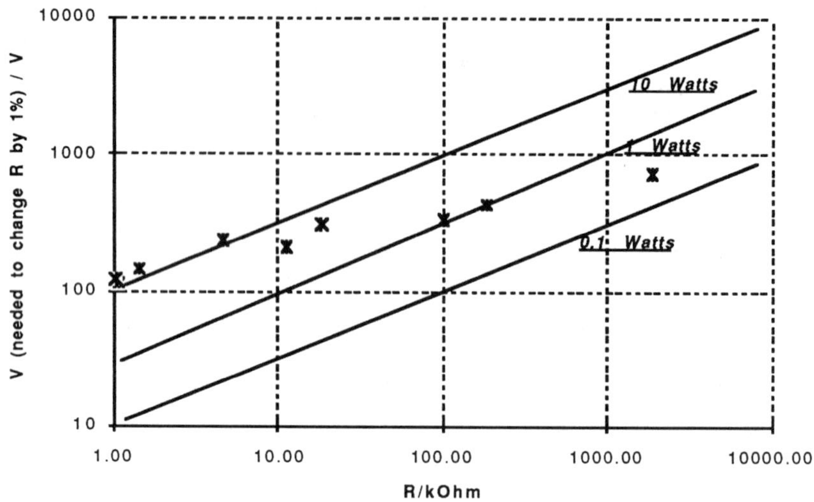

Fig. 10: *Voltage necessary to change R by 1% plotted vs. R for a pyrochlore system at different loadings.*

3. SUBSTRATE INTERACTION

As mentioned earlier, resistance decreases if resistors are fired at increasing temperature. At the same time, a glassy layer develops at the interface between substrate and resistor in some resistor formulations. Diffusion of Al_2O_3 from the substrate into the resistor glass and diffusion of glass modifiers from the resistor into the substrate has been observed[9,10]. Both effects cause a change of conductive/frit ratio. R vs. loading behavior can be described by:

$$R = \frac{\rho_o}{h\,(v_c - v_p)^n} \qquad (11)$$

ρ_o=resistivity of conductive; h=resistor thickness; v_c=volume fraction of conductive; v_p=percolation limit; n=fit parameter.

Assuming that part of glass represented by thickness h_s is free of conductive or has sunken into the substrate, the effective volume fraction of conductive v_c' is calculated by $v_c' = v_c / (1-h_s/h)$. Combining with equation 11 and rearranging

$$R = \frac{\rho_o\,(1-h_s/h)^{n-1}}{h\,\{v_c - v_p(1-h_s/h)\}^n} \qquad (12)$$

Equation 12 describes R vs. thickness h of the resistor. R of stratified resistors is always smaller than of unstratified resistors and if h_s increases, R decreases, as observed at higher peak firing temperature or lower viscosity. Surprisingly, this function has a maximum with respect to h, meaning that the resistance for thin resistors increases with increasing thickness, contrary to intuition. Real world resistors with low conductive loading show this inverse thickness effect at thickness of 8 to 12 um.

4. TERMINATION INTERACTION

Normally, the resistor is terminated with silver bearing conductors. It is well known that silver from the termination diffuses into glass of the resistor[11)]. A typical diffusion length for silver is 0.3 mm. It is known that silver decreases the viscosity of the glass. Therefore, if a resistor element in the neighborhood of the termination is observed alone, its resistance development vs. peak firing temperature exhibits a maximum of R and a minimum of TCR at a lower temperature than the rest of the resistor. Because R and TCR are given by:

$$R = \int (\partial R/\partial l)\, dl \qquad (13)$$

$$TCR = \frac{\int (\partial R/\partial l)\,(\partial R/\partial T)\, dl}{\int (\partial R/\partial l)\, dl} \qquad (14)$$

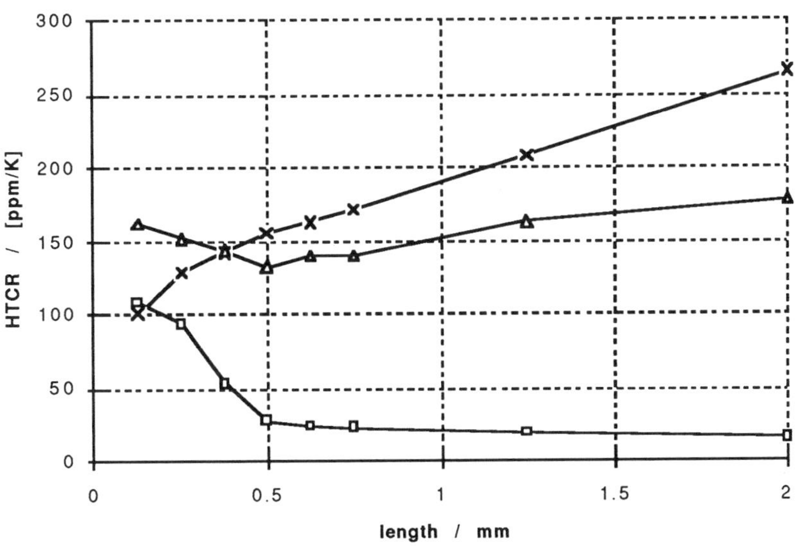

Fig. 11: TCR vs. length of a resistor for 3 different glasses.

R vs. length l does not behave linearly for small resistors, and depending on the formulation, TCR can decrease, increase, or show a minimum with increasing length (Fig. 11).

The effect on TCR is seen even on long resistors, because regions having high R are weighed more heavily than regions having lower R (equation 14). If the viscosity of Ag-rich parts is such that R is maximal or still increasing, TCR increases with length, since Ag-free regions have lower R, but higher TCR. If R of the Ag-rich region is lower than the maximum, but on the low viscosity branch of R vs. viscosity, TCR decreases or exhibits a minimum, depending on whether R of Ag-free region is maximal or still increasing. In summary, the length effect of resistors is a result of two competing interactions (conductive/glass and glass/substrate), triggered by Ag diffusion.

5. References

1. O. Abe, et al, "Glass viscosity and electrical properties of thick film resistors", Denki Kagaku Oyobi Kogyo, Butsuri Kagaku, **56** [2], 22-7(1988)

2. R.M. Scarisbrick, "Electrically conducting mixtures", J.Phys. D, Appl. Phys., **6**,2098-2110 (1973)

3. A. Kusy, "On the structure and conduction mechanism of thick resistive films", Thin Solid Films, **37**,281-302 (1976)

4. D.P.H. Smith and J.C. Anderson, "Electrical conduction in thick film paste resistors", Thin Solid Films, **71**, 79-89 (1980)

5. R. Kuzel, et al, "X-ray and microscopic investigations of resistors containing CdO and RuO_2", IEEE Trans. Components,Hybrids, and Manufacturing Technology, CHMT-4-3, 245-49 (1981)

6. O. Abe, et al, "Conduction mechanism in RuO_2 thick film resistors", Seisankogakubu Hokoku **22** [2], 117-23 (1989)

7. Handbook of Chemistry and Physics, E-105 (1980)

8. Jau-Ho Jean, "Liquid-phase sintering in the glass-cordierite system: particle size effect", T.K. Gupta, J.Mater.Sci. **27**, 4967-73 (1992)

9. M. Prudenziati, et al, "Interactions between alumina and high lead glasses for hybrid components",J. Appl. Phys. **65** [1] 146- 53 (1989)

10. Y.Nakamura, 'Alumina substrate and glass interactions",Int J. Hybrid Microelectronics **4** [2], 168-71(1981)

11. K. Yata and T Yamaguchi, "Effect of additives on Ostwald ripening of silver in glass",J. Am. Ceram. Soc. **75** [10], 2910-14 (1992)

COPPER THICK FILM CONDUCTOR ADHESION TO CERAMIC SUBSTRATES - BONDING MECHANISMS

Raju A. Perecherla* and Relva C. Buchanan*,‡
Department of Materials Science and Engineering
University of Illinois at Urbana-Champaign
Urbana, Il-61801

ABSTRACT:

Copper thick film adhesion and its failure modes on different ceramic substrates (glass ceramic, Al_2O_3 and AlN) were investigated as a function of aging time at 150°C in the presence of Pb/Sn solders. With the glass ceramic and Al_2O_3 substrates, direct bonding of the copper thick film to the substrate occurred through the formation of a ternary eutectic liquid and interfacial reaction compounds, for ink formulations with <2.50 wt% glass frit. At higher frit contents, bonding occurred primarily through an interfacial glassy layer. Both bonding mechanisms resulted in high unaged adhesion strengths, but poor aged adhesion resulted from the higher frit contents (>2.50 wt%) when aged at 150°C for >100 hours, with the aged failures occurring primarily at the copper/glass interface, due to Sn diffusion to the interface. In contrast, aged adhesion strengths for the direct bonded copper films exceeded 8.90 N/0.002m x 0.002m (2 Lbs/80 mil x 80 mil) square, even after 1000 hours of aging at 150°C, with the adhesion failures occurring predominantly along the Cu/Sn intermetallic and solder interface. With AlN substrates no direct bonding was achieved even for low concentrations of glass frit. Both unaged and aged failures took place at the substrate/thick film interface and the corresponding adhesion strengths were also lower.

INTRODUCTION:

Copper as a thick film conductor material was originally investigated because of the potential cost savings, but the activities in this area are currently

---

Based in part on the thesis submitted by R.A. Perecherla for the Ph.D. degree in Ceramic Engineering, University of Illinois, Urbana-Champaign, IL, 1992.
*Member, American Ceramic Society
‡Present address: Department of Materials Science and Engineering, University of Cincinnati, Cincinnati, OH-45221

being driven by the improved performance of copper conductors. However, adhesion of copper thick films to various ceramic substrates is often problematic. Adhesion and other characteristics such as resistivity and solderability depend on composition, and the content of the ink ingredients, substrate composition and the thick film processing conditions.

There are three primary methods used to achieve adhesion between a thick film conductor and the ceramic substrate. These are reactive bonding, flux bonding, and frit bonding.[1] The most common method of achieving adhesion is frit bonding, which involves the addition of 2 to 10% glass powder relative to metal. During firing glass wets the substrate and penetrates to some extent into the metal network binding the thick film to the substrate.

Thick film conductor adhesion on ceramic substrates has been investigated by many researchers.[2-10] A major finding from these investigations is that the formation of "interlocking" or "reentrant" glass microstructures at the metal conductor/substrate interface is essential for good conductor adhesion in frit bonded systems. Firing of thick film conductors at higher temperatures or for prolonged periods destroys this favorable glass structure at the interface and yields lower adhesion strengths. Adhesion strength therefore, is not significantly affected if a chemical bond is also achieved at the metal-glass interface. In the reactively-bonded metal conductor-alumina substrate systems, the reaction rate of solid alumina and copper oxide can be dramatically increased by the presence of any liquid or semiliquid phase which has solubility for either species.[3]

Good bondability of thick films to the substrate is ensured when some reaction between the glass frit and substrate occurs, but severe reactions can result in deeper reaction zones and changes in the substrate structure which have been correlated with lower adhesion strengths. The severity of the reaction between the thick film and the substrate depends on the firing temperature, type of the glass frit, the ink additives and the substrate binder.[8]

Adhesion of soldered thick film conductors tend to degrade with time on exposure to elevated temperatures (100°-150°C). While this is seldom catastrophic of itself, the weakened conductor becomes progressively more sensitive to thermal cycling and fluctuating mechanical stresses and the combined effects can lead to conductor adhesion failures at the substrate/thick film interface.[11-17] This loss of adhesion has been attributed to solid state inter-diffusion of the conductor metal and the Sn in the Pb/Sn solder. This inter-diffusion at elevated temperatures (>125°C) results in the formation of intermetallic compounds, which are inherently brittle, and exhibit lattice structure and properties substantially different from the parent metals. The formation of these intermetallic compounds is often accompanied by Kirkendall voiding as a result of different diffusion rates, and a weak interface may be created.[18] The mismatch in the physical properties between the conductor metal, solder and the intermetallic phases creates internal stresses which eventually leads to loss of adhesion of the thick film to the ceramic substrates.

Formation of brittle intermetallics due to the reaction between the thick film conductor metal and Sn at elevated temperatures and their deleterious effect on the aged thick film conductor adhesion have been studied for many years. Several investigators[19-27] have observed that when copper is coated with or joined with tin containing solder alloys, intermetallic compounds of copper and tin (Cu_3Sn and Cu_6Sn_5) are formed at the interface between the two materials. The formation of the intermetallic compounds is exponentially dependent on temperature, and the intermetallic layer thickness at a given temperature is proportional to the square root of time.

Systematic evaluation of thick film bonding mechanisms, failure modes and thick film/substrate interfacial microstructures are important to completely understand the thick film adhesion to ceramic substrates. In this investigation, an attempt was made to evaluate copper thick film conductor adhesion on RCA type glass ceramic, Al_2O_3 and AlN substrates which represent a good diversity of ceramic substrates used in microelectronics industry.

EXPERIMENTAL:

Copper thick film inks were formulated from appropriate amounts of copper powder, 2 - 5μm in size, finely ground powdered glass frit and an ink additive, CuO along with 14 wt% of screening agent. Three types of glass frits (one alkali and two non-alkali glass frits[28,29]) were used in this investigation. The alkali glass frit contained Na_2O as an oxide component of the frit composition. Of the non-alkali glass frits, one was a Barium-Boro Silicate glass and the other was a Calcium-Boro Silicate glass containing ~20 wt% alkali earth content. The mixture was ground in a steel roller mill until the particle size distribution of all the ink constituents was below 12.5μm. Ink formulations used in this investigation contained glass frit in the amount of 1.00 - 4.50 wt%.

The inks were then screen printed onto the ceramic substrates (RCA type porcelainized steel glass ceramic, Al_2O_3, and AlN) in a test pattern consisting of twelve 0.002 m x 0.002 m (80 mil x 80 mil) square pads using a 200 mesh screen with a wire diameter of 5.33×10^{-5}m (2.1 mil), and fired in both conventional and IR furnaces in a N_2 ambient containing less than 10 ppm of O_2 at a peak firing temperature of 900°C. Composition of the insulation layer of the glass ceramic substrate is given in Table 1. The fired thickness of all the copper thick films was ~25 μm. The test pads were then soldered with eutectic Sn/Pb solder, cured over night and peel tested using a modified Dupont peel test procedure in which soldering was done by a reflow process.

Table 1. The Composition of the Insulating Layer of the Glass Ceramic Substrate

Oxide	Wt%
SiO_2	14.3 ± 1.0
B_2O_3	24.8 ± 2.0
MgO	39.7 ± 1.5
BaO	18.2 ± 1.0
SnO_2	1.5 ± 0.5
P_2O_5	1.5 ± 0.5

Thick film samples for interfacial microstructure evaluation were sectioned and polished in steps to 1 μm finish. All samples for microscopic examination were ultrasonically cleaned, first in a bath of acetone for 10 minutes followed by ethyl alcohol for another 10 minutes. Electron Microscope samples were mounted on appropriate holders and the sample surfaces were carbon coated with evaporated carbon using a vacuum evaporator.

Structural characterization of the samples was carried out using an X-ray Diffractometer. Thermogravimetric Analyzer (TGA) was used to analyze CuO powder in nitrogen ambient and a Thermomechanical Analyzer (TMA) was used to measure coefficient of thermal expansion and softening point of glasses.

An optical microscope equipped with polarized light and photomicrographic system was used to examine the reaction zones within the substrates. A high resolution scanning electron microscope (SEM) with a LINK system EDS analyzer was used for microstructural examination of the sample surfaces and interfaces. Back Scattered electron (BSE) imaging, X-ray dot mapping and quantitative elemental line scanning was carried out using a fully automated, computer driven Electron Microprobe with a WDS analyzer.

RESULTS AND DISCUSSION:

Effect of Glass Frit Composition on Thick Film Adhesion:

Figure 1 compares copper thick film adhesion strength on the glass ceramic substrate as a function of aging time at 150°C, for two ink formulations containing alkali and non-alkali (Ba) glass frits. Both films show a decrease in adhesion strength with aging time but the overall strength is significantly lower for the alkali containing system. Analysis of the substrate/thick film cross sections of unaged samples of the alkali and non-alkali glass frit systems indicated deep reaction zones within the substrate, for the alkali ink system.[28] Deep reaction zones are correlated with lower adhesion strengths.

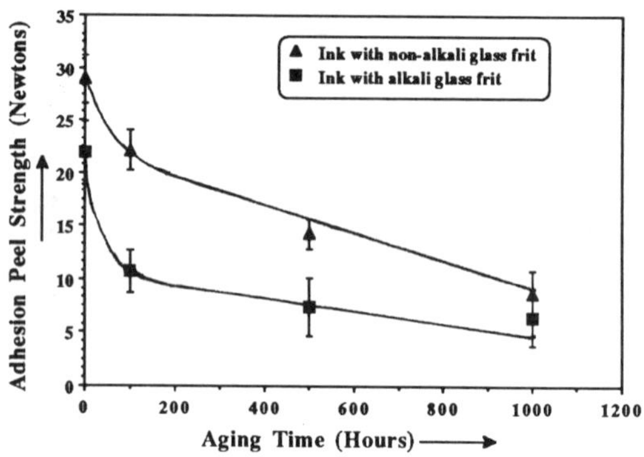

Fig. 1. Dependence of the copper thick film adhesion on aging (150°C) for the glass ceramic substrate systems which are fired in a conventional furnace for two ink formulations containing 4.00 wt% of alkali and non-alkali (Ba) glass frits.

Effect of Glass Frit Concentration on Thick Film Adhesion:

Figure 2 (A), (B) and (C) show the dependence of copper thick film adhesion strength on aging for ink systems containing different amounts of non-alkali (Ca) glass frit on glass ceramic, Al_2O_3 and AlN substrates, respectively.

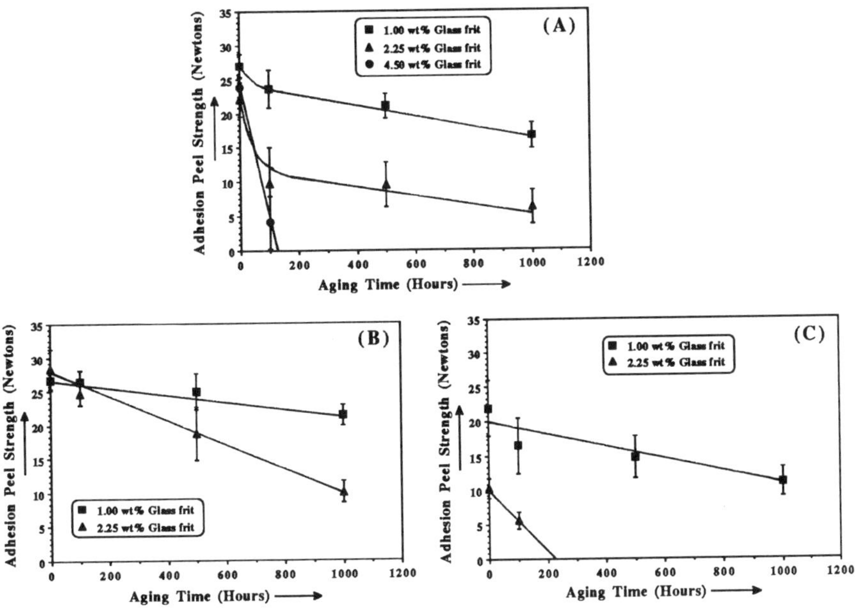

Fig. 2. Dependence of the copper thick film adhesion on aging (150°C) for ink formulations containing varying amounts of the non-alkali (Ca) glass frit, on: (A) glass ceramic substrates, fired in a conventional furnace; (B) Al_2O_3 and (C) AlN substrates, fired in an IR furnace.

The data indicate that aged adhesion strength decreases significantly with the increase in the concentration of the glass frit in all the substrate systems. In the glass ceramic substrate systems, there is a steep drop in the thick film adhesion strength, after a relatively short period (100 hours) of aging for the ink systems containing high amounts of glass frit. For the Al_2O_3 substrate systems the drop in aged adhesion strength is more shallow and adhesion strength degradation is linear, compared to the glass ceramic substrate systems. However, unaged adhesion strength in both the substrate systems was not significantly affected by increase in the glass frit concentration. In contrast, for the AlN substrate systems, both unaged and aged adhesion strengths decreased for high glass frit concentrations.

For Low Glass Frit Concentrations (1.00 - 2.25 wt%):

The adhesion data for the conventional and IR fired thick films on the glass ceramic substrates for the ink system containing 1.00 wt% of the non-alkali glass frit is graphically depicted in Figure 3 which indicates that the aged adhesion strength of copper thick films, fired in IR furnace is comparable or even better than those fired in conventional furnace.

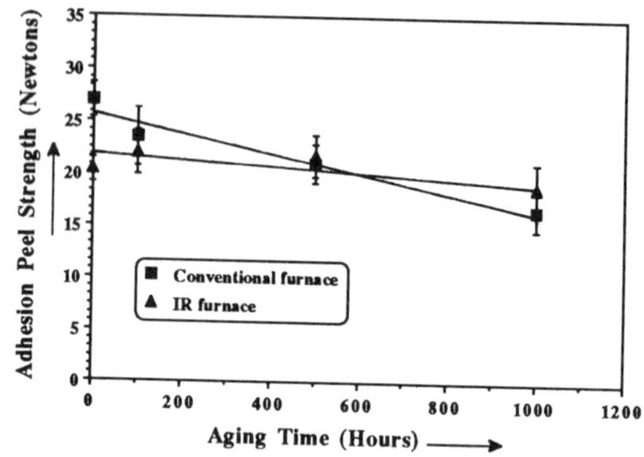

Fig. 3. Dependence of the copper thick film adhesion on aging (150°C) for the glass ceramic substrate/copper thick film systems containing 1.00 wt% of the non-alkali (Ca) glass frit which are fired in conventional and IR furnaces.

Back Scattered Electron (BSE) images of the glass ceramic substrate /thick film/solder cross sectional surfaces of unaged and aged samples made from the ink formulation containing 1.00 wt% of the glass frit are shown in Figure 4. This indicates the progressive development of Cu-Sn intermetallics at the

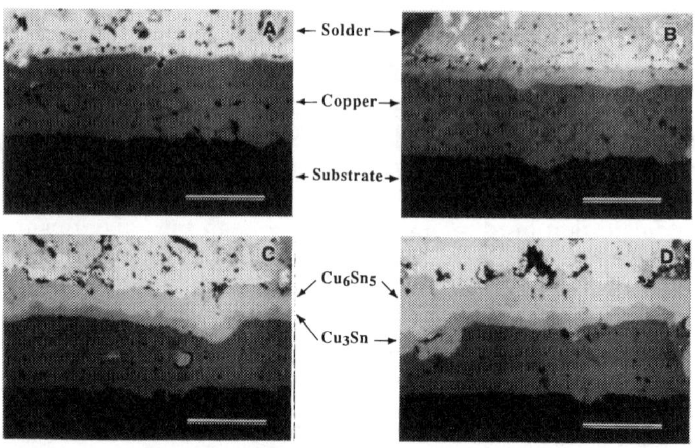

Fig. 4. SEM micrographs of the BSE images of the glass ceramic substrate/copper thick film/solder cross sectional surfaces for the ink system containing 1.00 wt% of the non-alkali (Ca) glass frit, fired in a conventional furnace, showing the progressive development of Cu/Sn intermetallics with aging (150°C): (A) unaged; (B) 100 hrs aged; (C) 500 hrs aged and (D) 1000 hrs aged. Bar = 20 μm.

thick film/solder interface with aging. Figure 5 shows diffusion profiles for the elements Mg, Ca, Sn and Cu across the corresponding surfaces for the unaged and 1000 hours aged samples. Mg represents the substrate, Ca represents the glass frit,

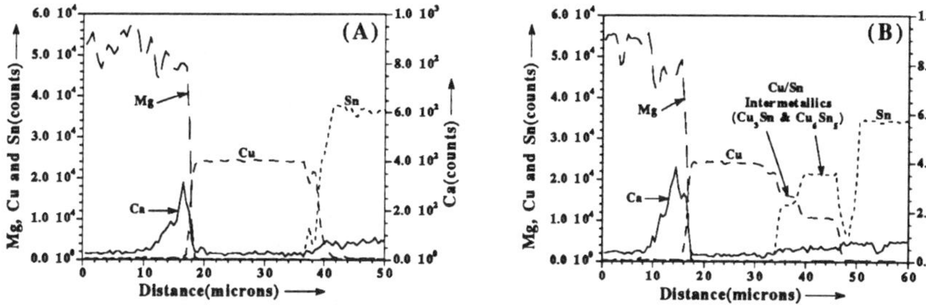

Fig. 5. Elemental diffusion profiles across the glass ceramic substrate/copper thick film/solder cross sectional surfaces of the unaged and the aged (150°) samples for the ink system containing 1.00 wt% of the non-alkali (Ca) glass frit, fired in a conventional furnace: (A) Unaged and (B) 1000 hrs aged.

Cu and Sn represent the thick film and the elemental Sn in the Pb/Sn solder, respectively. The results indicate that virtually all the glass added (1.00 wt%) to the thick film ink in this particular system reacted with the substrate with no visible accumulation of glass at the substrate/thick film interface, and that the thickness of the Cu/Sn intermetallics increases with aging. The depth of the reaction zone, indicated by the Ca diffusion profile within the substrate, differed little with aging, which indicates that for a given thick film ink composition and substrate type, the reaction zone is well defined by the firing process, and is not significantly affected by aging. Also, with aging two Cu/Sn intermetallic phases were formed at the thick film/solder interface, and a distinct Sn depleted region was progressively developed at the intermetallic/solder interface. As observed by many investigators, Sn rich intermetallic, Cu_6Sn_5, is formed next to the solder matrix and Cu rich intermetallic, Cu_3Sn, is formed next to the copper thick film.

Figure 6 shows BSE images for the glass ceramic substrate/thick film/solder cross sectional surfaces of unaged and aged samples made from the ink formulation containing 1.00 wt% of the glass frit and fired in IR furnace. Thick film microstructure developed in these systems is highly porous and has several connected pores. The intermetallic growth rate in these systems is over three orders of magnitude higher than the intermetallic growth rate observed in similar systems fired in conventional furnace[30] and with aging, the intermetallics in these systems even reached the substrate/thick film interface at many areas.

Lower unaged thick film adhesion strengths observed in IR furnace fired systems can be attributed to the lower substrate/thick film interfacial area available in these systems due to the high porosity developed in the thick films. The plateau nature of the dependence of adhesion strength on aging, in these systems can be attributed to the penetration of solder into the highly porous thick film microstructure, mechanically keying the thick film with the solder body and also to

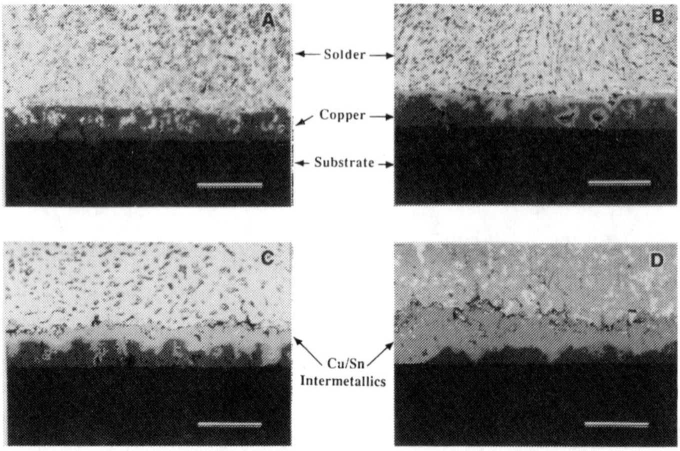

Fig. 6. SEM micrographs of the BSE images of the glass ceramic substrate/copper thick film/solder cross sectional surfaces for the ink system containing 1.00 wt% of the non-alkali (Ca) glass frit, fired in an IR furnace, showing the progressive development of Cu/Sn intermetallics with aging (150°C): (A) unaged; (B) 100 hrs aged; (C) 500 hrs aged and (D) 1000 hrs aged. Bar = 50 µm.

the development of highly irregular Cu/Sn intermetallic and solder interface. Although, the Cu/Sn intermetallics reached the substrate/thick film interface at several points along the interface, high aged adhesion strengths obtained for these systems therefore, suggest an absence of any glass at the interface.

SEM micrographs in Figure 7 show the thick film test pad areas on the glass ceramic substrate after the peel test for unaged and aged samples made from the ink formulation containing 1.00 wt% of the glass frit which are fired in a conventional furnace. Detailed analysis of the corresponding microstructures indicated that the unaged failures in this system took place predominantly within the substrate reaction zone. With aging, the failure mode was progressively shifted toward the metallic solder region and the aged failures primarily took place in the Sn depleted (Pb rich) Cu/Sn intermetallic/solder interfacial region.[31] Figures 8 (A) and (B) show SEM micrograph of the the glass ceramic substrate/thick film interfacial area and X-ray dot maps of the corresponding area for the elements Mg, O, Ca and Cu, respectively indicate that the copper particles at the interface are fused to the substrate. Similar interfacial microstructure is also observed in the corresponding systems which are fired in an IR furnace.[32]

Thermodynamic data and structural analysis of the interface indicate that an eutectic liquid (CaO-CuO-Cu_2O) and an interfacial reaction compound, $BaCuO_2$ are formed in these systems.[32] High thick film adhesion strengths observed in these systems can be attributed to the direct bonding established between the thick films and the substrate through the interfacial reaction compound.

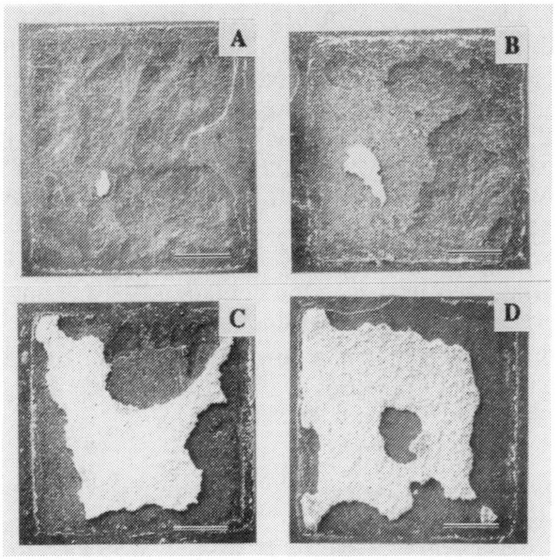

Fig. 7. SEM micrographs of the test pad areas on the substrate after the peel test for the copper thick film/glass ceramic substrate system containing 1.00 wt% of the non-alkali (Ca) glass frit, fired in a conventional furnace, showing the modes of failure with aging (150°C): (A) unaged; (B) 100 hrs aged; (C) 500 hrs aged and (D) 1000 hrs aged. Bar = 500 μm.

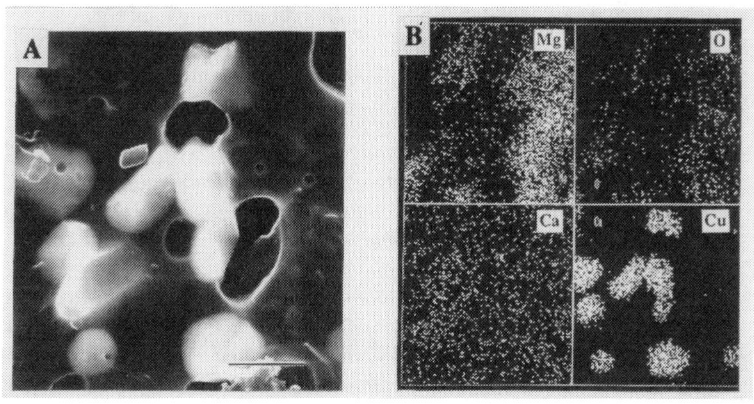

Fig. 8. (A) SEM micrograph of the glass ceramic substrate/copper thick film interface of the unaged sample for the ink system containing 1.00 wt% of the non-alkali (Ca) glass frit, fired in a conventional furnace and (B) X-ray dot maps of the corresponding area. Bar = 2 μm.

SEM micrographs of the substrate/thick film interfacial area for unaged samples made from the ink formulation containing 1.00 wt% of the glass frit for the

Al_2O_3 and AlN substrates which are fired in an IR furnace are shown in Figures 9 (A) and (B), respectively. The interfacial microstructure for the Al_2O_3 substrate system is similar to that observed for the glass ceramic substrate system in which copper particles are fused to the substrate. A detailed analysis of the interface indicated the presence of $CuAl_2O_4$, a spinel interfacial reaction compound and a thin, but discontinuous layer of glass. As in the case of the corresponding glass ceramic substrates systems, high thick film adhesion strengths observed in these systems can also be attributed to the establishment of direct bonding between the thick films and Al_2O_3 substrate at several points along the substrate/thick film interface through the interfacial reaction compound.[32] In contrast, the interfacial analysis in the case of AlN substrate system indicates the presence of a continuous layer of glass at the interface even for low concentrations of the glass frit.[32] Low thick film adhesion strengths obtained for these systems are due to the lack of direct bonding between the thick films and the substrate.

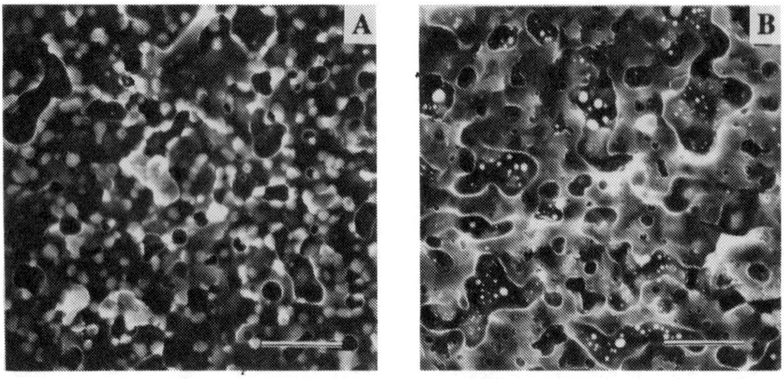

Fig. 9. SEM micrographs of the substrate/copper thick film interfaces of the unaged samples for the ink system containing 1.00 wt% of the non-alkali (Ca) glass frit which are fired in an IR furnace, for: (A) Al_2O_3 and (B) AlN substrate systems. Bar = 10 μm.

Contrary to the observations made in the case of glass ceramic substrate system containing 1.00 wt% of the glass frit which is processed in IR furnace, corresponding cases in Al_2O_3 and AlN substrate systems, appear to have relatively dense thick film structure with no connected pores.[30] This would indicate that the thick film microstructure not only depends upon the firing process but also on the compositions of the substrate and thick film ink systems as well as on the reactions which occur at the substrate/thick film interface during firing.

Unaged failures in Al_2O_3 substrate systems predominantly occur at the substrate/thick film interface whereas, in similar glass ceramic substrate systems the failures occur primarily in the substrate reaction zone. With aging, failure modes in both the substrate systems are progressively shifted to the Sn depleted Cu/Sn intermetallic and solder interfacial region.[31] In contrast, for the AlN substrate system both the unaged and aged failures occur within the glass layer accumulated at the substrate/thick film interface.[31]

Figures 10 (A), (B) and (C) show diffusion profiles across the substrate/thick film cross sectional surfaces of unaged samples containing 1.00 wt% of the glass frit in the ink formulation, for the glass ceramic, Al_2O_3 and AlN substrates, respectively. In the case of the glass ceramic substrates, all the added glass reacted with the substrate with virtually no accumulation of glass at the interface. With the Al_2O_3 substrate system, only part of the added glass is seen to have reacted with the substrate and with AlN substrate no reaction between the substrate and the constituents of the thick film ink is observed. This is attributed to high stability of the AlN substrates at the thick film processing conditions.

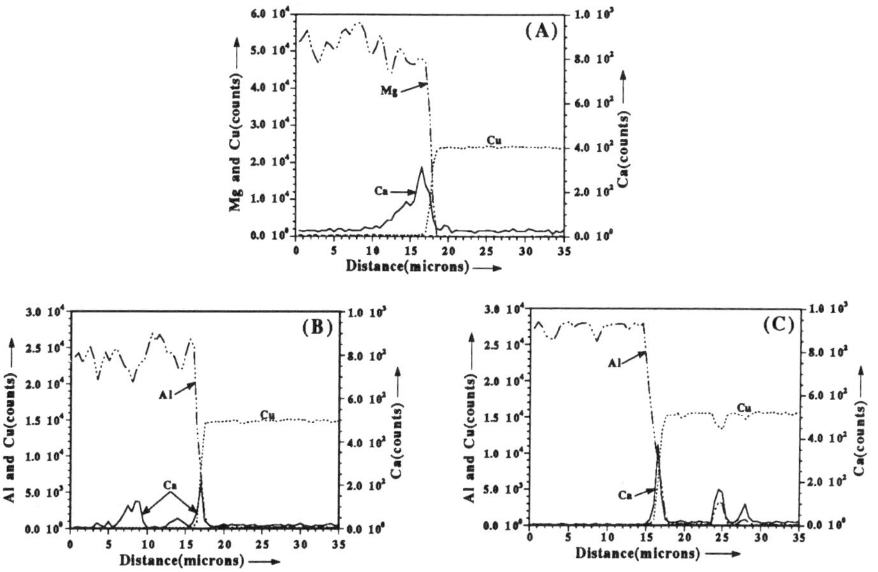

Fig. 10. Elemental diffusion profiles across the substrate/thick film cross sectional surfaces of the unaged samples made from the ink formulation containing 1.00 wt% of the non-alkali (Ca) glass frit, for: (A) glass ceramic; (B) Al_2O_3 and (C) AlN substrates.

For High Glass Frit Concentrations (2.25 - 4.50 wt%):

BSE images for the glass ceramic substrate/thick film/solder cross sectional surfaces of unaged and 100 hours aged samples made from the ink formulation containing 4.50 wt% of the glass frit are shown in Figure 11 (A) and (B), respectively indicating connected porosity in the copper thick films and the progressive development of Cu/Sn intermetallics with aging. Analysis of the substrate/thick film interface for the system indicated a high concentration and accumulation of glass at the interface.[32] With the increase in the glass frit

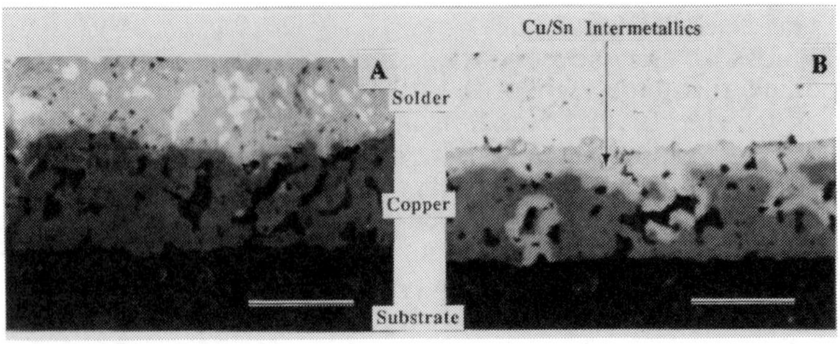

Fig. 11. SEM micrographs of the BSE images of the glass ceramic substrate/copper thick film/solder cross sectional surfaces for the ink system containing 4.50 wt% of the non-alkali (Ca) glass frit, fired in a conventional furnace, showing high thick film porosity and progressive development of Cu/Sn intermetallics, with aging (150°C): (A) unaged and (B) 100 hrs aged. Bar = 20 μm.

concentration in the thick film ink, part of the excess glass accumulates at the interface and the remainder is retained within the thick film. High porosity and the development of connected pores in the thick film microstructure results. High porosity translates into high surface area and, therefore, the surface area available for Sn in the solder to react with the copper thick film during high temperature aging is high. This results in the formation of high amounts of Cu/Sn intermetallic compounds.

Figures 12 (A) and (B) show SEM micrographs of the test pad areas on the substrate after the peel test for unaged and 100 hours aged samples, respectively of

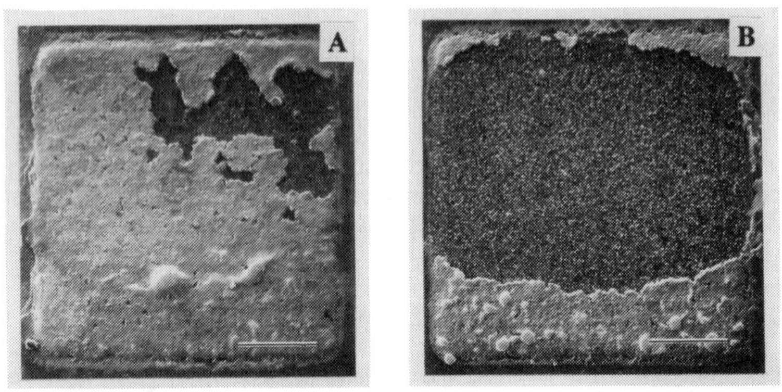

Fig. 12. SEM micrographs of the test pad areas on the substrate after the peel test of the copper thick film/glass ceramic substrate system containing 4.50 wt% of the non-alkali (Ca) glass frit, fired in a conventional furnace, showing the modes of failure with aging (150°C): (A) unaged and (B) 100 hrs aged. Bar = 500 μm.

the same system. X-ray dot map analysis of the corresponding test pad areas[31] indicated that unaged failures in this system took place predominantly within the thick film metallic region, attributed to the inherently weak porous thick film microstructure developed in these systems. But the aged failures in the system primarily took place at the thick film/glass interface due to the diffusion of Sn to this interface through the connected pores in the thick film in a relatively short aging time (<100 hours) breaking the initial bonds established between the thick film and the interfacial glass layer.[31]

Figures 13 (A) and (B) show the diffusion profiles across the substrate/ thick film/solder cross sectional surfaces of unaged samples made from the ink formulation containing 2.25 wt% of the glass frit for the Al_2O_3 and AlN substrates, respectively. The results indicate that in both the substrate systems, with the increase in the glass frit concentration, its concentration at the substrate/thick film interface as well as in the overall thick film body, increases.

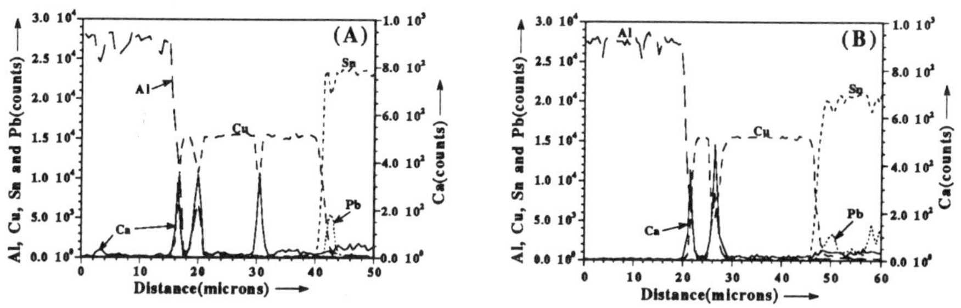

Fig. 13. The elemental diffusion profiles across the substrate/copper thick film cross sectional surfaces of the unaged samples made from the ink systems containing 2.25 wt% of the non-alkali (Ca) glass frit which are fired in an IR furnace, for: (A) Al_2O_3 and (B) AlN substrates.

CONCLUSIONS:

(1) The alkali glass frits are found to reduce the thick film adhesion strength.

(2) For low amounts of non-alkali glass frit, direct bonding is established between the thick film and the substrate both in the glass ceramic and Al_2O_3 substrate systems.

(3) For high concentrations of glass frit, thick films are bonded to the substrate through an interfacial layer of glass. The unaged thick film adhesion strength do not significantly differ with the increase in the glass frit concentration in these substrate systems but, the aged thick film adhesion decreases significantly.

(4) On the contrary, no direct bonding is established between the thick film and the substrate in AlN substrate systems even for low concentrations of glass frit. Unlike in the other two substrate systems, in AlN substrate systems the unaged as well as the aged thick film adhesion strengths decrease with the increase in the glass frit concentration.

(5) The thick film microstructure dependent on several factors: (a) the concentration of glass frit in the thick film ink; (b) the thick film firing process; (c) the substrate/thick film interfacial reactions which occur during thick film firing.

(6) Cu/Sn intermetallic growth rate is higher in thick films with high surface porosity and connected pores. With aging, irrespective of the substrate type, a Sn depleted region develops at the Cu/Sn intermetallic and the solder interface and the interface gets progressively weaker.

(7) The residual thick film aged adhesion strengths retained for the optimum thick film ink formulations in the case of glass ceramic and Al_2O_3 substrate systems, after 1000 hours of aging at 150°C are well above the minimum acceptable value of 8.90 N/0.002m x 0.002m (2 Lbs/80 mil x 80 mil) square. In contrast, the aged adhesion strength achieved for the a similar AlN substrate system is not quite acceptable.

(8) For low amounts of glass frit the unaged failures in the glass ceramic substrate system took place primarily within the substrate reaction zone whereas, the failures in the Al_2O_3 substrate system took place at the substrate/thick film interface. But with aging, the failure modes were progressively shifted to the Sn depleted, Cu/Sn intermetallic and solder interfacial region in both the substrate systems.

(9) For high concentrations of glass frit in the glass ceramic substrate systems, a continuous layer of glass was accumulated at the substrate/thick film interface. The unaged failures in these systems took place predominantly in the metallic thick film and the aged failures were shifted to the thick film/glass interface in a relatively short time.

(10) In AlN substrate systems, no chemical bonding was established between the thick film and the substrate, and a continuous glass layer was accumulated at the substrate/thick film interface even for low concentrations of glass frit. This resulted in the formation of a very weak substrate/thick film interface and both the unaged and aged failures in this case took place within the glass layer at this interface.

ACKNOWLEDGMENTS:

The authors are very grateful to the CTS Corporation's Research and Development Center at Elkhart, IN for its financial support, and would especially like to thank Mr. Terry Bloom and Ms. Evelyn Rice of CTS for their technical help. Use of the electron microscope facility at the Center for Microanalysis of Materials of the Materials Research Laboratory and the X-ray diffraction analysis facility at the Center for Cement Composite Materials, both at the University of Illinois, Urbana-Champaign, and the electron microprobe facility at the Department of Geophysical Sciences at the University of Chicago, are gratefully acknowledged.

REFERENCES:

[1]R.W. Vest, "Materials Science of Thick Film Technology," Am. Ceram. Soc. Bull.,**65** [4] 631-36 (1986).

[2]D.E. Riemer, "Optimized Adhesion Mechanism for Thick-Film Multilayer Gold," pp.418-23, 35th Electronic Components Conference, Washington D.C., May (1985)

[3]T.T. Hitch, "Adhesion, Phase Morphology, and Bondability of Reactively-Bonded and Frit-Bonded Gold and Silver Thick-Film Conductors," J. Electron. Mat., **3** [2] 553-77 (1974).

[4]T.T. Hitch, "Phase Morphology and Adhesion in Thick-Film Conductor Inks," ISHM Proceedings, pp. 7-7-1 - 11 (1971).

[5]P.F. Becher and J.S. Murday, "Factors affecting Thick Film Adhesion," ISHM Proceedings, pp. 235-38 (1976).

[6]D.R. Johnson and F. Uribe, "The Effect of Multiple Refirings on the Adhesion of Thick Film Conductors," ISHM Proceedings, pp. 1-7 (1977).

[7]P.F. Becher and W.L. Newell, "Adherence-Fracture Energy of a Glass-Bonded Thick-Film Conductor: Effect of Firing Conditions," J. Mater. Sci., **12**, 90-96 (1977).

[8]Y. Nakamura, "Alumina Substrate and Glass Interfacial Reactions," ISHM Proceedings, pp. 168-71 (1981).

[9]C.Y. Kuo, "Adhesion of Thick Film Copper Conductors," ISHM Proceedings, pp. 70-78 (1981).

[10]M. Monneraye, H. Baudry and D. Bricout, "The Adhesion of Thick-Film Copper Conductors," 4th. European Hybrid Microelectronics Conference, Copenhagen, Denmark, pp. 240-49 (1983).

[11]W.A. Crossland and L. Hailes, "Thick Film Conductor Adhesion Reliability," Solid State Technol., **14** [2] 42-47 (1971).

[12]S.S. Leven, "Screening procedure for Adhesion Degradation Due to Solder Leaching in Thick-Film Hybrid Microcircuits," IEEE Electronic Components Conference, pp. 285-91 (1976).

[13]D.H.R. Sarma, D.G. Weaver and P.S. Palanisamy, "An Accelerated Lot-Acceptance Test for Adhesion Degradation of Soldered Copper Thick Films in Temperature Cycling," ISHM Proceedings, pp. 554-61 (1987).

[14]A.A. Milgram, "Influence of Metallic Diffusion on the Adhesion of Screen Printed Silver Films," Metallurgical Transactions, **1** [3] 695-700 (1970).

[15]J.R. Larry and R.B. Amin, "High Adhesion, Thick Film Palladium-Silver Conductors," IEEE Electronic Components Conference, pp. 161-70 (1973).

[16]G.P. Wirtz, "Adhesive Strength of Thick Film Pd-Ag Conductors," Proceedings of 21st IEEE Electronic Components Conference, pp. 480-86 (1971).

[17]C.J. Peckinpaugh, "Adhesion Loss in Solderable Thick Film Conductor Systems As a Result of Thermal Shock," Insulation/Circuits, **23** [3] 35-38 (1977).

[18]G.J. Ewell and J.A. Roth, "Metallurgical Criteria for Selection of Solders for Microelectronic Uses," ISHM Proceedings, pp. 239-48 (1976).

[19]L. Zakraysek, "Intermetallic Characterization," 11th Annu. IEEE Conf. on Reliability Physics, pp. 6-9, Los Vegas, NV, (1973).

[20]L. Revay, "Interdiffusion and Formation of Intermetallic Compounds in Tin-Copper Alloy Surface Coatings," Surf. Technol, **5** [1] 57-63 (1977).

[21]B.G. LeFevre and R.A. Barczykowski, "Intermetallic Compound Growth on Tin and Solder Platings on Cu Alloys," Wire J. Int., **18** [1] 66-71 (1985).

[22]E.W. Brothers, "Intermetallic Compound Formation in Soft Solders," Western Electric Engineer, **25**, pp. 49-63 (1981).

[23]P.E. Davis, M.E. Warwick and P.J. Kay, "Intermetallic Compound Growth and Solderability," Plat. Surf. Finish., **69** [9] 72-76 (1982).

[24]P.E. Davis, M.E. Warwick and S.J. Muckett, "Intermetallic Compound Growth and Solderability of Reflowed Tin and Tin-Lead Coatings," Plat. Surf. Finish., **70** [8] 49-53 (1983).

[25]P.J. Kay and C.A. Mackay, "The growth of Intermetallic Compounds on Common Basis Materials Coated with Tin and Tin-Lead Alloys," Trans. Inst. Met. Finish., **54** pt. 1, 68-74 (1976).

[26]A.K. Bandyopadhyay and S.K. Sen, "A Study of Intermetallic Compound Formation in a Copper-Tin Bimetallic Couple," J. Appl. Phys., **67** [8] 3681-88 (1990).

[27]M.A. Howes and Z.P. Saperstein, "The Reaction of Lead-Tin Solders with Copper Alloys," Welding Journal Research Supplement, **48** [2] 80s-85s (1969).

[28]A. Perecherla and R.C. Buchanan, "Copper Thick Film Adhesion on Glass Ceramic Substrate," pp. 439-454 in Ceramic Transactions on Ceramic Thin and Thick Films, Vol. 11, ed. B.V. Hiremath, American Ceramic Society, Westerville, OH, 1990.

[29]R.A. Perecherla and R.C. Buchanan, "Role of Glass Frit in (Cu) Thick Film Adhesion," pp. 433-446 in Ceramic Transactions on Glasses for Electronic Applications, Vol. 20, ed. K.M. Nair, American Ceramic Society, Westerville, OH, 1991.

[30]R.A. Perecherla and R.C. Buchanan, "Aged Adhesion Degradation of Copper Thick Film Conductors on Ceramic Substrates," Unpublished work.

[31]R.A. Perecherla and R.C. Buchanan, "Adhesion Failure Modes of Copper Thick Film Conductors on Ceramic Substrates," Unpublished work.

[32]R.A. Perecherla and R.C. Buchanan, "Bonding Mechanisms and Interfacial Microstructure Development in Copper Thick Film Conductor/Ceramic Substrate Systems," Unpublished work.

Cosintering of Gold/LTCC Substrates: Development of Camber and Via Defects

Robert C. Sutterlin, Guo-Quan Lu, and Tapan K. Gupta
Alcoa Technical Center, 100 Technical Drive, Alcoa Center, PA 15069

Abstract

Linear shrinkage profiles of a low temperature cofireable glass ceramic (LTCC) green tape and unconstrained gold thick film circuit and via fill materials were measured using a noncontact optical technique. The sintering kinetics of the gold thick film materials were found to differ significantly from that of the LTCC tape. These differences in the sintering kinetics of the gold materials from the LTCC lead to various types of defects in the gold/LTCC substrate. Mismatched sintering kinetics between the circuit material and the LTCC can lead to the development of camber (warpage) during cosintering of gold/LTCC substrate. Mismatched sintering kinetics between the gold via fill material and the LTCC can lead to defects such as excessive bumping at the via, separations at the via/LTCC wall, and voids in the via fill material.

Introduction

One of the most important steps in manufacturing multilayer ceramic packages is the cosintering (cofiring) of the metal/ceramic multilayer structure. The cofiring process of copper and a glass-ceramic system[1] was acknowledged to be a crucial technical hurdle in the development of the recently announced IBM low temperature cofired packaging technology. Two types of metal thick film materials are used to form the wiring in a ceramic package. Circuit materials are used to route the wiring in the x-y plane. Via fill materials are used to make the vertical connections between the various layers in the multilayer ceramic package. Mismatched sintering kinetics between the via fill material and the LTCC can lead to defects such as via bumps, via slumps, via voids, and separations at the via fill/LTCC interface. Mismatched sintering kinetics between the circuit metal and ceramic can lead to significant camber or warpage in the product[1,2], a defect which substantially lowers production yield. Typical solutions for minimizing camber have often relied upon engineering approaches, such as weighting, varying material compositions, and varying the cofiring temperature-time profiles.

An understanding of the sintering kinetics of the materials used in multilayer ceramic packages has been hampered by difficulties in measuring the linear shrinkage profiles of screen printed thick films. An optical apparatus, described elsewhere[3], is used to determine the x-y shrinkage profiles of unconstrained ceramic tapes and screen printed metal films to study the sintering kinetics of the materials used in a proprietary gold/LTCC packaging system. Knowledge of the sintering kinetics of the various materials are used to analyze defects associated with the cosintering of the gold/LTCC structures. Circuit gold/LTCC structures are cofired and the development of camber is monitored in real-time, validating the analysis of the stresses generated during the cosintering of the circuit gold and LTCC. Defects observed in the microstructure of gold via fill/LTCC structures are related to the shrinkage kinetics of the gold via fill and LTCC materials.

Experiment

The optical system exploits the contrast between the reflective surface of the setter and the rough surface of the sample to measure the dimensions of the sample during sintering. It is estimated that a minimum linear shrinkage as low as 0.5% can be detected by the technique given the step resolution of the translation stage and the size of the sample. This technique requires free standing samples. Tape samples were made by cutting 2 mm x 2 mm squares from a three layer laminate. The gold via fill and circuit pastes were screen printed and dried on mylar sheets, producing 2.5 mm x 2.5 mm samples approximately 0.20 mm in thickness. The thick films were subsequently lifted off the mylar sheets, and placed on polished alumina setters for sintering.

Multilayer circuit metal/ceramic structures were cofired in a horizontal quartz tube furnace to observe in real-time the development of camber during the sintering process. The furnace is equipped with a video camera to laterally monitor the sample during the entire firing cycle. The tops of three layer laminates, approximately 25 mm x 25 mm x 0.6 mm, were completely covered with a pure gold thick film circuit material to produce the gold circuit/LTCC structures for this experiment.

Samples for the microstructural observation of the gold via fill/LTCC structures were made with various gold via fill materials by filling holes mechanically punched in individual sheets of the LTCC tape. Five via filled layers were stacked and laminated to produce the gold via fill/LTCC structures. The via fill and the circuit paste samples were preheated at 450°C for four hours to remove the organic binder and then heated to a sintering temperature of 925°C at a constant rate typically between 1 and 10°C/min.

Results

Temperature and linear shrinkage as a function of sintering time for the pure gold circuit material, the via fill materials, and the LTCC material are plotted in Figure 1. The lines drawn through the data points are smooth-fitted. A clear mismatch in the sintering behaviors of the materials is evident in the figure. The unconstrained circuit material has nearly completed its densification before the LTCC begins to densify. The total shrinkage of the via fill materials can be modified by altering the composition of the paste.

During the cofiring process the initially flat gold circuit/LTCC structure first bends upward towards the metal side of the structure, reaching a maximum upward camber 85 minutes into the cofiring cycle (775°C). Figure 2a depicts the warpage of the laterally viewed sample upward away from the setter plate. Later in the firing cycle, the sample bends back towards the setter plate, eventually developing downward camber. The maximum downward camber, shown in Figure 2b, occurs 180 minutes into the cofiring cycle (925°C, 60 minute dwell). The time in the cofiring cycle at which the upward camber became noticeable corresponds to when the gold was found to be rapidly shrinking in Figure 1. The upward camber reached its maximum at a time when the metal has nearly completed its densification and the ceramic has begun to densify as indicated by the linear shrinkage data in Figure 1. The time when the coupon started to camber downward corresponds to the time when the LTCC was rapidly shrinking.

Scanning electron micrographs (SEM) depicting the microstructure of the various gold via fill materials/LTCC system are shown in Figures 3 and 4. Figure 3 shows the microstructure of an unaltered gold via fill material A. This via fill material shrinks more than the LTCC, producing the separation of the via fill material from the via wall. Via voids or separations with in the via fill material are also evident in the microstructure depicted in Figure 3. Figure 4 shows the microstructure of via fill material B whose shrinkage has been altered to produce little or no shrinkage as shown in Figure 1. Note the distortion of the ceramic in the region of the via (via bump) related to the disparity in the total shrinkage of the via fill and LTCC materials.

Discussion

Circuit Material and Camber

Little or no chemical interaction occurs between the pure gold thick film material and the LTCC due to the chemical inertness of gold. Any adhesion between the gold and the LTCC results from mechanical interlock at the interface. The adhesion from the mechanical interlock is sufficient to bond the gold circuit materials to the LTCC despite the disparity in the timing of the sintering kinetics of the two materials. The gold circuit materials begins to sinter well ahead of the LTCC, completing its densification before the LTCC starts to sinter. This large disparity in the timing of sintering kinetics of the circuit material and the LTCC suggests that the two materials constrain each other from sintering during the cosintering process. In the early stages of the cosintering process, the circuit material is constrained from sintering by the nonsintering LTCC, producing the upward camber shown in Figure 2a. In the later stages of the cosintering process the LTCC is constrained from sintering by the densified gold circuit material, producing the downward camber shown in Figure 2b. The mechanics of the composite structure, governed by viscous constitutive relations between the stresses and strains in the sintering materials, has been examined in order to understand the development of camber during the cofiring process[3]. Scherer's[4] viscous analogy has been used to analyze the observed development of camber in the gold/LTCC composite structure based on the measured linear shrinkage data of the unconstrained gold and LTCC. The in-plane stress that develops as a result of the mismatched sintering kinetics has been found using the viscous analogy to be related to the densification rate by:

$$\sigma = 2\,G_p\left[\frac{\dot{\rho}}{\rho}\right]_C \qquad (1)$$

where s is the in plane stress, G_p is the shear viscosity, and $[\dot{\rho}/\rho]_C = -\dot{\varepsilon}_z$ is densification rate in the constrained body. It is this in-plane stress developed when the material is constrained from sintering that leads to the development of camber in the composite structure[3].

Since the stress in equation (1) depends on the densification rate of the constrained material, knowledge of how the densification kinetics of the constrained material differ from those of the unconstrained material are required. An equation has been developed using Skorokhod's model for the shear and bulk viscosity[5] that relates the density of the constrained ($[\bar{\rho}]_C$) to that of the unconstrained or free ($[\bar{\rho}]_f$) sintering body[3].

$$\frac{[\bar{\rho}]_C}{(1-[\bar{\rho}]_C)} = \frac{\bar{\rho_0}}{(1-[\bar{\rho}]_f)} \qquad (2)$$

This equation allows the densification curve of the gold under constrained sintering to be calculated based on the linear shrinkage profile of the free gold circuit material.

The calculated relative density vs. time in the constrained gold film along with the curve converted from the linear shrinkage profile measured in the free gold are shown in Figure 5. Also plotted in Figure 5 is the relative density vs. time curve for the free LTCC calculated from its linear shrinkage profile, assuming that the LTCC sinters isotropically. The initial relative density of the LTCC tape after binder burn-out was measured to be about 0.6. Some overlap between the retarded densification curve of the constrained gold (by a rigid substrate) and that of the LTCC which is not quite rigid occurs in Figure 5. Nevertheless, the two curves are still significantly mismatched. In the real system, the densification curve of the gold circuit material is expected to lie somewhere between the free and the constrained (by a rigid substrate) curves.

The development of camber in the circuit gold/LTCC composite structure during the earlier stage of the cofiring process is treated using the analogy of a plate bonded together by two dissimilar elastic materials[6]. When one of the materials is considerably thinner than the other, the elastic bending curvature, κ, of the structure as a result of in-plane residual stresses in the thinner material is found to be[7]:

$$\kappa = \frac{6\,(1-\nu)}{E\,D^2}\,\sigma\,t \qquad (3)$$

where D, E, and ν are respectively the thickness, Young's modulus and elastic Poisson's ratio of the thicker material, and σ and t are respectively the residual stress in and thickness of the thinner material. The porous gold and LTCC do not behave elastically, instead, their responses to stresses are better described by the viscous analogy. The following relation for the rate of bending curvature ($\dot{\kappa}$) in the gold/LTCC structure resulting from the sintering stress in the gold has been derived[3]:

$$\dot{\kappa} = \frac{6\,(1 - \nu_p)}{E_p\, D_{LTCC}^2}\, \sigma_{Au}\, t_{Au} \tag{4}$$

where D_{LTCC}, E_p, and ν_p are respectively the thickness, uniaxial viscosity and Poisson's ratio of the porous LTCC tape, and σ_{Au} and t_{Au} are respectively the residual stress in and thickness of the sintering gold.

Curvature as a function of time can be calculated and is plotted in Figure 6. The corresponding camber values in microns per mm are given on the right vertical axis. The curvature (κ) and camber (c) are approximately related by $c = 0.125\kappa$. It can clearly be seen from the plot that the curvature (or camber) of the part increases with time as the gold is constrained from sintering by the LTCC. This result is in qualitative agreement with the observed early stage of the camber development shown in Figure 3. At onset of the LTCC densification, the curvature (camber) was found to be about $5\ m^{-1}$ ($0.6\ \mu m/mm$), which is in surprisingly good agreement with the observed maximum camber toward the gold as estimated from Figure 5. This result may be fortuitous given the many assumptions that had to be made for the calculation. Again for better comparisons with the experimental results, more accurate values of the material properties for both the gold and the LTCC are needed.

Gold Via Fill Material and Defects Associated with Vias

The cylindrical geometry of the filled via compounds problems associated with the asynchronous sintering of the via fill material and the LTCC material. The earlier onset of sintering and the larger total shrinkage of the pure unaltered gold via fill material leads to open spaces with in the via fill material termed via voids and separation of the via fill material from the wall producing a vertical interconnect free to rattle around in the via hole as shown in Figure 3. Consequently, the densification kinetics of the via fill material need to be retarded in order to produce a defect free via. The onset temperature of sintering needs to be pushed to higher temperatures and the total shrinkage needs to correspond to the total shrinkage of the LTCC material. If the total shrinkage of the via fill material is less than the LTCC, the via will distort the LTCC material in the region of the via producing a via bump as shown in Figure 4. Excessive via bumps deleteriously impact the planarity of the surface of the ceramic package, decreasing yields. Consequently it is desirable to match the total shrinkage of the via fill material to that of the LTCC in order to produce defect free gold via/LTCC structure.

One method of inhibiting the sintering kinetics of the gold via fill material is to decrease the gold particle to particle contact during sintering, retarding the solid state diffusion kinetics. Treatment of the gold powders with organic materials which burn off later in the sintering cycle inhibits sintering during the early to middle stages of the sintering cycle. The addition of more refractory metals and the inclusion of refractory oxides to the via fill material decreases particle to particle contact between the gold particles during the entire sintering cycle.

Conclusion

A noncontact optical-scanning system has been used to determine the linear shrinkage profiles of a free standing gold thick film materials and LTCC tape material. Significant mismatch in their sintering profiles was found. The gold circuit material was found to sinter much faster and at considerably lower temperatures than the LTCC. Substantial cambering (warping) of the gold/LTCC composite structure has been observed in real-time during the cofiring process. The development of camber was discussed based on the constrained sintering of the two materials in the composite structure resulting from their mismatched sintering kinetics in the free states. The sintering kinetics of the gold via fill material were alterable by the addition of materials that decrease the gold particle to particle contact during sintering.

Acknowledgment

We are grateful to R. Lytle, M. Bruno, Jr., A. Wertz, F. Krivda, L. Martin and W. Krafick for their technical assistance. We are also grateful to B. Mogilevsky, G. R. Love, J-H Jean, D. A. Weirauch, Jr., J. F. Kelso, and S. F. Horvath for their valuable comments.

References

[1] A. H. Kumar and R. R. Tummala, "State-of-the-Art, Glass-Ceramic/Copper, Multilayer Substrate for High Performance Computers," *Int. J. for Hybrid Microelectronics*, **14** [4] 137-150 (1991).

[2] W. D. Kingery, H. K. Bowen, and D. R. Uhlmann, Introduction to Ceramics, 2nd edition; pp. 448-515. John Wiley & Sons, New York, 1976.

[3] G.-Q. Lu, R. C. Sutterlin, and T. K. Gupta, "Effect of Mismatched Sintering Kinetics on Camber in a Low Temperature Cofired Ceramic Package," to be published *J. of Amer. Ceram. Soc.*

[4] G.W. Scherer, "Sintering Inhomogeneous Glasses: Application to Optical Waveguides," *Non-Crys. Solids*, **34**, 239-56 (1979).

[5] R. K. Bordia and G. W. Scherer, "On Constrained Sintering-II. Comparison of Constitutive Models," *Acta Metall.*, **36** [7] 2399-2409 (1988)

[6] S. P. Timoshenko and J. N. Goodier, Theory of Elasticity; McGraw-Hill, New York, 1970

[7] R. W. Hoffman, "The Mechanical Properties of Thin Condensed Films"; pp. 211-217 in Physics of Thin Films. Edited by G. Hass and R. E. Thun. Academic Press, New York, 1966.

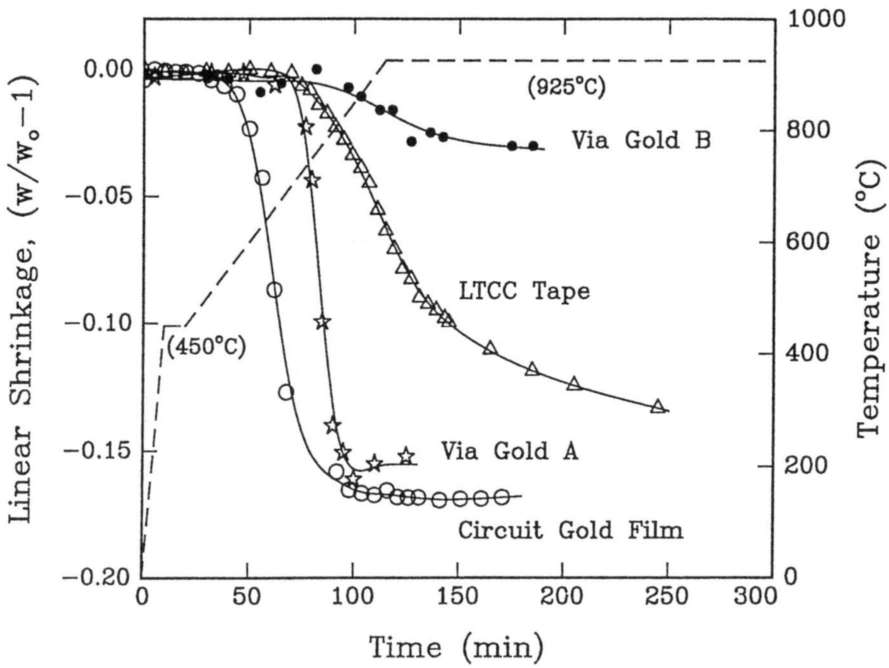

Figure 1: Linear shrinkage profiles of the unconstrained LTCC tape and gold thick film materials. Curves through the data points are smooth-fitted. The sintering temperature is plotted on the right-hand axis.

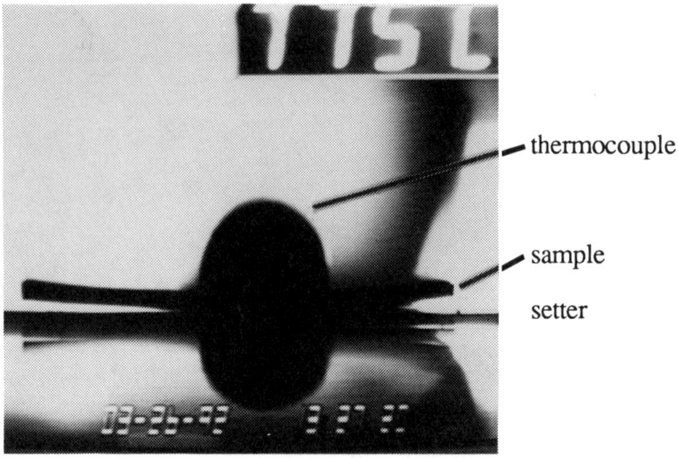

a.)

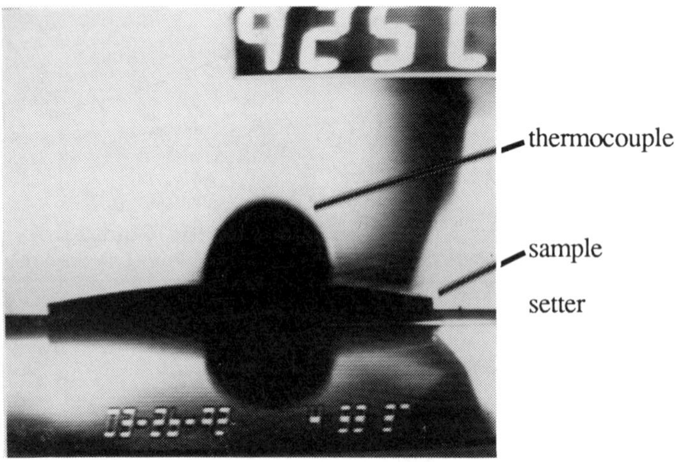

b.)

Figure 2: Photographs of the gold circuit/LTCC structure showing the development of camber in a gold/LTCC composite structure during cofiring; a.) maximum upward camber 85 min. into the cosintering cycle (T = 775°C), b.) maximum downward camber, 180 min. into the cosintering cycle (T = 925°C, 60 minute dwell).

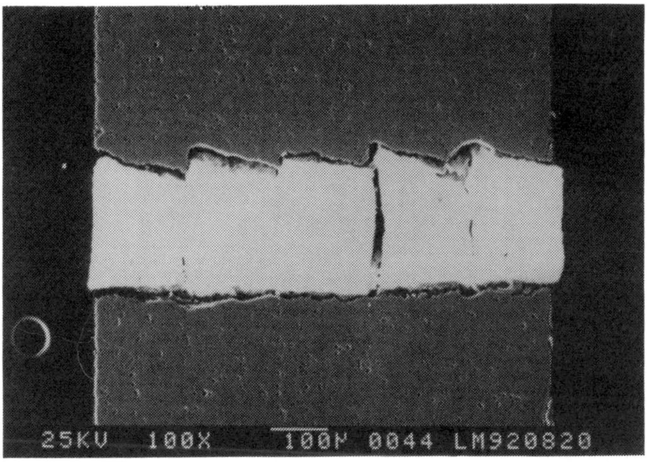

Figure 3: Scanning electron micrograph of the via fill A/LTCC structure showing separation at the via fill/via wall interface and voids with in the via fill material.

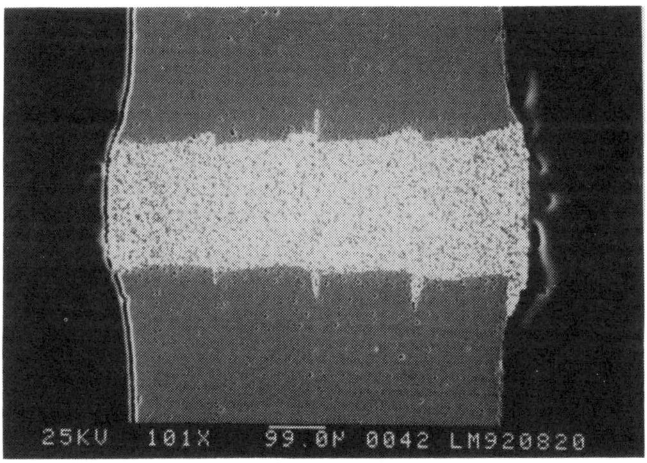

Figure 4: Scanning electron micrograph of the via fill B/LTCC structure showing distortion of the LTCC (via bump) in the region of the via.

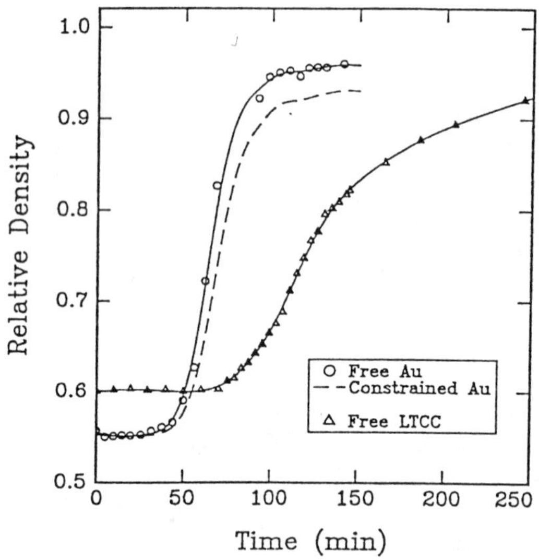

Figure 5: Densification curves of the unconstrained and constrained gold circuit material and the densification curve for the free LTCC tape.

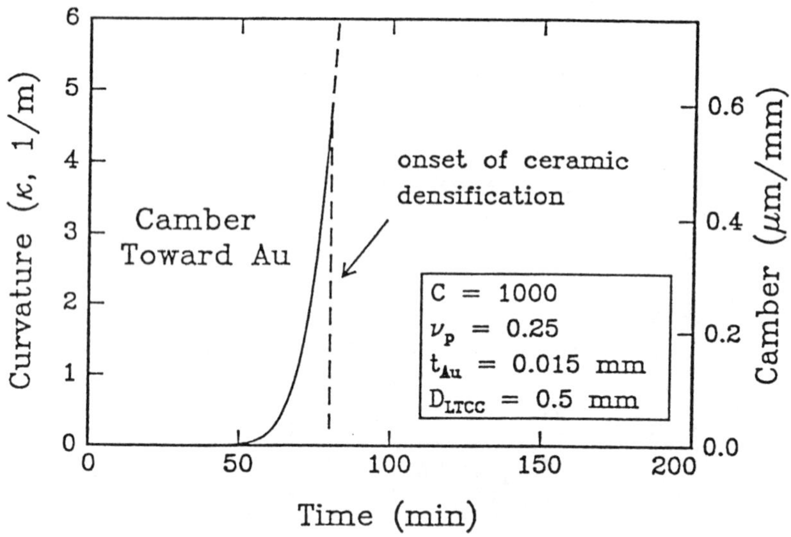

Figure 6: Bending curvature development of the gold/LTCC composite structure.

NEW CHIP-ON-GLASS MOUNTING TECHNOLOGY OF LSI

Yoshihiro Bessho, Yasuhiko Horio, Toru Ishida and Wataru Sakurai
Matsushita Electric Industrial Co,. Ltd.
1006, Kadoma, Kadoma-shi, Osaka, 571 Japan

ABSTRACT

LCDs are now widely use in various display terminals; as a result, there is great demand for an LCD with improved display quality, high resolution and large capacity. For driving such an LCD having a large number of pixels, many drive lines must be connected to driver LSIs. For this reason, we have developed new COG mounting technology, using called Stud-bump type bonding technique[1], which can bond driver LSI chips directly to the glass substrate. Stud-bump is formed on electrode pad of LSI chip by a wire-bonding apparatus. Each Stud-bump has two-stepped construction and bonded with a conductive adhesive to an ITO electrode terminal formed on the glass substrate. The conductive adhesive is very flexible and strong in bond, thus resisting thermal and mechanical stresses. Moreover, repair technique has been developed. With this new COG technology, 50 microns pitch electrode pads can be bonded. The electrical contact resistance is less than 1 Ω. Several reliability tests were carried out. The COG bonding portion was very stable in all tests. High density and reliability in bonding have been attained by the Stud-bump type mounting technology. This report describes the COG mounting technology applied to a small, high-resolution and high-density color LCD panel for the color electronic viewfinder of a home video camera.

INTRODUCTION

LCDs (Liquid Crystal Displays) have been widely used as information terminal devices. There has been a great demand for an LCD with improved display quality, high resolution and large capacity. To realize such LCDs, COG mounting technology is indispensable whereby driver LSI chips are bonded directly onto a glass substrate.

We have developed the Stud-bump type mounting technique which uses Au bumps with two-stepped construction and a conductive adhesive, to enable driver LSI chips to be mounted densely on the ITO (Indium Tin Oxide) electrode terminals of an LCD panel.

The following paragraphs describe the COG mounting technique used to realize a high-density color LCD panel for the color electronic viewfinder of home video cameras.

MOUNTING TECHNOLOGY FOR LCD PANELS

A challenge in the mounting technology for an LCD panel is to enable connection of multiple driver LSIs to the large number of electrode terminals coming out of the panel, compactly, reliably and economically.

The LCD mounting techniques in use today are divided roughly into two categories: one is connects a printed board having driver LSIs thereon to an LCD panel, and the other utilizes what is called COG mounting technique to connect driver LSI chips directly onto an LCD panel.[2]

Today, LCD panels are expected to have high definition and high resolution equivalent to those of a CRT. Accordingly, LCD panels have a trend toward greater number and smaller size of pixels. In addition, electrode terminals are decreasing in width and pitch, and increasing in number.

To comply with these trends, the TAB (Tape Automated Bonding) mounting method has prevailed rapidly for mounting driver LSI chips onto an LCD panel. However, this method has the following disadvantages:

1. It involves anisotropic conductive film, and has a limit in fine-pitch connection.[2]
2. It requires a large mounting area due to the carrier.
3. It incurs a high cost for materials and process.

Thus, it is not practical to apply the TAB mounting method to the LCD panel in an electronic viewfinder which is required to be small and to provide a high picture quality.

The COG mounting technique, in which driver LSIs are mounted directly onto a glass substrate of an LCD panel, would appear to be the optimal option to meet this requirement.

STUD-BUMP TYPE COG MOUNTING TECHNIQUE

Background of the development

The basic concept of the COG mounting technique is to connect driver LSI chips electrically to a large number of fine electrode terminals on an LCD panel glass substrate, rationally and accurately.

The conventional methods of mounting LSI chips onto a glass substrate are the face-up bonding using wires and face-down bonding using solder bumps.

These conventional methods are limited in fine-pitch connection and require surface treatment of the ITO electrode terminals on the glass substrate, leaving problems in rationalization and cost of the mounting process.

With this background, as an optimum method of connecting LSIs onto a glass substrate, we have developed the Stud-bump type COG mounting technique in which LSI chips are bonded with a conductive adhesive directly to the ITO electrode terminals.

Structure

Fig. 1 shows the structure of the bonding portion by the Stud-bump type COG mounting technique using a conductive adhesive and Au bumps.

The Au bumps are formed by a newly developed technique[1] which uses the wire-bonding method in place of the conventional plating method. The Au bump obtained by this method has two-stepped construction, so that conductive adhesive can be transferred collectively only to the bump top. This restricts spread of the conductive adhesive when LSI chips are mounted on the ITO electrode terminals, thus permitting ultra-fine pitch connection.

If the conventional one-stepped structure plated bumps are bonded to the ITO electrode terminals with conductive adhesive, adhesive spread cannot be restricted and short-circuit between adjacent terminals cannot be prevented. This is not suitable for fine pitch connection such as those several tens μ m pitch.

The gap between each LSI chip and the glass substrate is filled with silicone sealing resin, to ensure higher structural stability and weather resistance of the COG bonding portion.

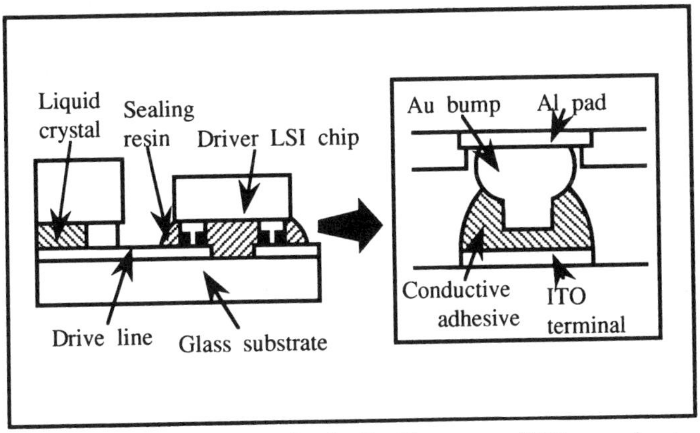

Fig. 1 Structure of the bonding portion by Stud-bump type COG mounting technique

Bump formation method

In Stud-bump formation method, which is different from the conventional plated bump formation method, the adhesive layer and diffusion prevention layer on the aluminum electrode terminal pads of LSI chips are unnecessary. In addition plating and washing processes are also unnecessary.

These advantages of Stud-bump formation method lead to low cost and high reliability.

Fig. 2 shows the formation process of Stud-bump.

First, Au ball formed on a tip of Au wire is fixed on an aluminum terminal pad on an LSI chip to form the bottom of Stud-bump, as well as ordinary wire-bonding technique.

Then the Au wire is cut off to form the tip on the upper surface of the bump bottom.

Thus the two-stepped construction is formed with good reproducibility and a stable shape.

The diameter and height of the Stud-bump are determined by the Au wire diameter. For example, if using Au wire 25 μ m in diameter, Stud-bump approx. 80 μ m in diameter and approx. 50 μ m in height is obtained. As the Au wire becomes thinner, a finer bump can be formed.

Formation speed of a full automatic type wire-bonding apparatus is 0.12 seconds/bump.

Fig. 3 is an SEM photograph of Stud-bump formed on driver LSI chip. This photograph shows Stud-bumps 80 μ m in diameter and 50 μ m in height formed on an aluminum electrode pad 100 by 100 μ m square at pitches of 70 μ m.

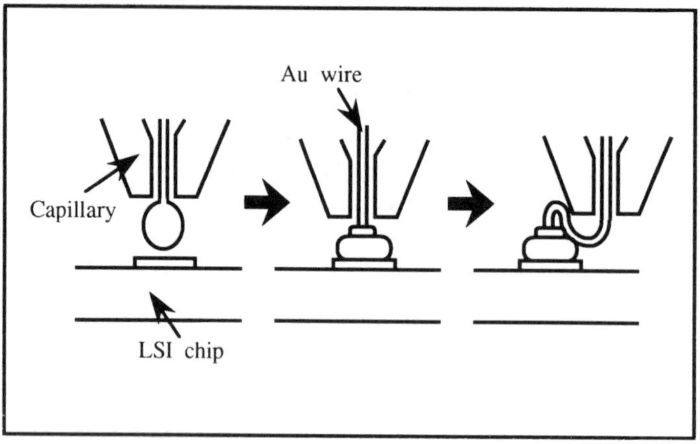

Fig. 2 Formation process of Stud-bump

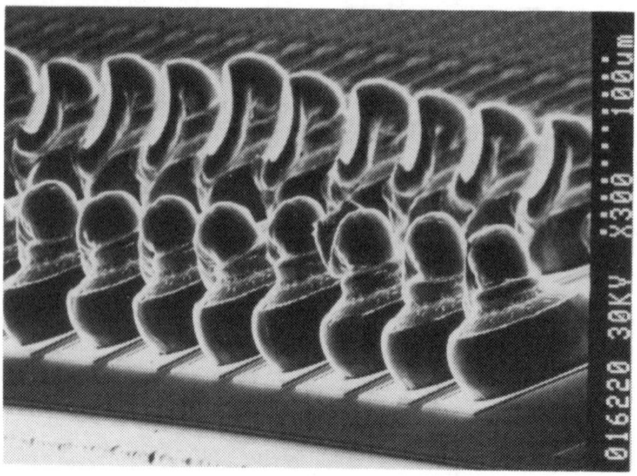

Fig. 3 SEM photograph of Stud-bump formed on driver LSI chip

Conductive adhesive

New conductive adhesive was developed for the COG mounting technique. The quantity of this newly developed conductive adhesive transferred onto each Stud-bump is controlled at a high accuracy by both viscosity and thixotropy, thus enabling the adhesive to be transferred collectively onto bumps.

Table 1 specifies characteristics of the newly developed conductive adhesive.

An epoxy resin flexible over a wide range of temperatures is used as binder, and scaly AgPd alloy particulate 0.8 μ m in mean particle size is used as a conductive filler.

This binder has a high heat shock resistance, permitting LSI chips to be bonded directly onto ITO electrode terminals.

And this conductive filler has a high migration resistance. With 20% Pd content in AgPd alloy, Ag migration speed can be reduced to 1/100 than that of Ag alone.

Curing condition is as short as 120℃ × 2 hour by taking into account damage on a liquid crystal and so on of the LCD panel. Volume resistivity is as low as approx. 10^{-5} Ω · cm with extremely high conductivity so that low contact resistance can be obtained in the COG bonding. To restrict impurity concentration at a low level, the binder and conductive filler are substantially purified.

Table 1 Characteristics of the conductive adhesive

Binder	Special epoxy resin
Conductive filler	AgPd alloy
Solvent	2-Ethanol & 2-Ethyl Acetate
Viscosity @25℃	68 Pa · s
Curing condition	120 ℃ × 2 h
Glass transition point	105 ℃
Volume resistivity	3.5 × 10^{-5} Ω · cm
Expansion coefficient	3.8 × 10^{-5} / ℃

Direct bonding to ITO terminal

Rationalization in particle size and content of the conductive adhesive can be realized to bond directly onto an ITO electrode terminals.

If the conductive filler content is low, despite the increase in bonding strength, bulk resistance of the adhesive increases and contact resistance also increases. Conversely, if there is too much conductive filler content, adhesive flexibility deteriorates. And stress cannot be relaxed, resulting in contact resistance increase.

The relation between the conductive filler content and COG contact resistance is shown in Fig. 4.

By rationalizing the composition ratio of the binder to the conductive filler, low contact resistance and excellent thermal shock resistance have been attained. Consequently, direct connection to the ITO electrode terminals is satisfactorily obtained.

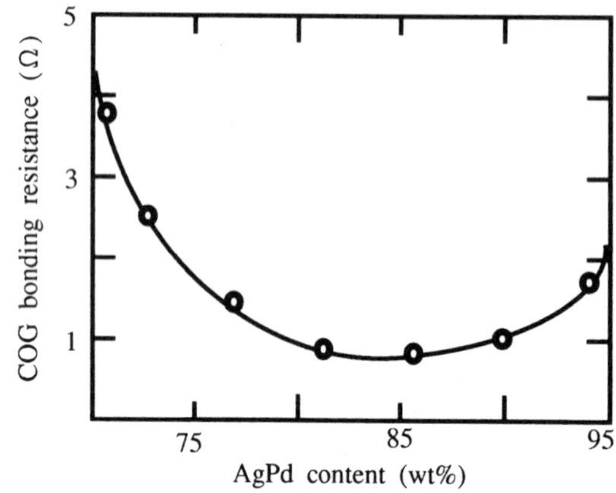

Fig. 4 Relation between the conductive filler content and COG contact resistance

Sealing resin

With Stud-bump type COG mounting technique, silicone resin is inserted to seal the gap between each LSI chip and the glass substrate, to enhance the stability of the bonding portion in mechanical construction and weather resistance.

Table 2 specifies characteristics of the sealing resin.

This silicone resin is of a self-curing type, so it cures at a low temperature and effects a strong adhesion force between LSI chips and a glass substrate. In an actual operating temperature range (-40 ℃ to 120 ℃) of an LCD panel, this resin remains flexible with a hardness of 15 (Durometer-A), and maintains shrinkage stress in the gaps, thus securing the stability of the COG bonding portions.

Table 2 Characteristics of the sealing resin

Resin	Silicone
Viscosity @25℃	2.6 Pa · s
Curing condition	120 ℃ × 2 h
Glass transition point	-120 ℃
Hardness (Durometer-A)	15
Volume resistivity	1.0×10^{16} Ω · cm
Expansion coefficient	2.5×10^{-5} / ℃
Thermal conductivity	0.84 W/m · K
Dielectric constant	3.5

Testing and repair technique

In the general mounting process of bare LSI chips including the COG mounting technique, the most important task is testing and repair after mounting. Establishment of an appropriate repair technique is essential particularly for the COG mounting technique in which a large number of LSI chips are connected directly onto an LCD panel.

With the Stud-bump type COG mounting technique, it is possible to electrically test the driver LSI chips mounted on the LCD panel before conductive adhesive is cured; the conductive adhesive has been designed to provide a high conductivity even before curing.

Fig. 5 shows the testing and repair technique.

Specifically,

1. Two types of solvent with different boiling points are used in the conductive adhesive so that the adhesive is cured in two steps.
2. Electrical testing is carried out with the conductive adhesive in the temporary curing state. If the testing result is acceptable, the solvent with higher boiling point is volatilized to allow the adhesive to cure completely.
3. If defective LSI chips are found in the electrical testing, they are repaired with the conductive adhesive in the temporary curing state.
4. Since the conductive adhesive has not been cured completely at this stage, the defective LSI chips can easily be removed at ordinary temperature, and can be replaced with new driver LSI chips without necessity of cleaning the glass substrate.

These testing and repair techniques have permitted a remarkable increase in the yield of LSI chip bonding by the COG mounting technique.

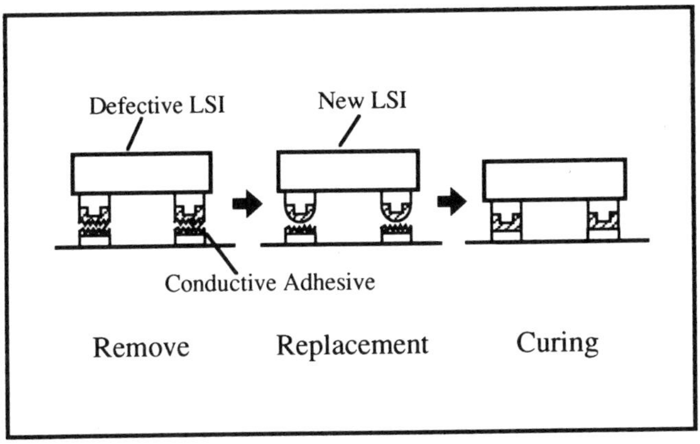

Fig. 5 Repair technique of the Stud-bump type COG mounting technique

COG mounting process

Fig. 6 shows a schematic flow chart of the Stud-bump type COG mounting process. First, a two-stepped Stud-bump is formed from Au wire 25 μm in diameter on each aluminum electrode terminal pad on a driver LSI chip, using the wire-bonding method. Stud-bumps about 80 μm in diameter are thus formed stably on each driver LSI chip.

Next, a flat surface is pressed against the Stud-bumps to level the bump height. The pushing load for the leveling is about 50 g/bump. By this operation, the Stud-bumps have a uniform height of 45 $\pm$ 1 μm.

Then, only the tips of the two-stepped Stud-bumps are dipped in a layer of conductive adhesive, to collectively transfer the conductive adhesive onto the top of each bump. Fig. 7 shows a SEM photograph of Stud-bumps onto which the adhesive has been transferred.

Driver LSI chips thus obtained are bonded onto the ITO electrode terminals on an LCD panel, and are subjected to functional tests with the conductive adhesive in the temporary curing state. After complete curing of the conductive adhesive, the LSI chips on the LCD panel undergo final testing, to complete the mounting process.

The mounting load for driver LSI chips is 100 g/chip, causing no damage on the chips.

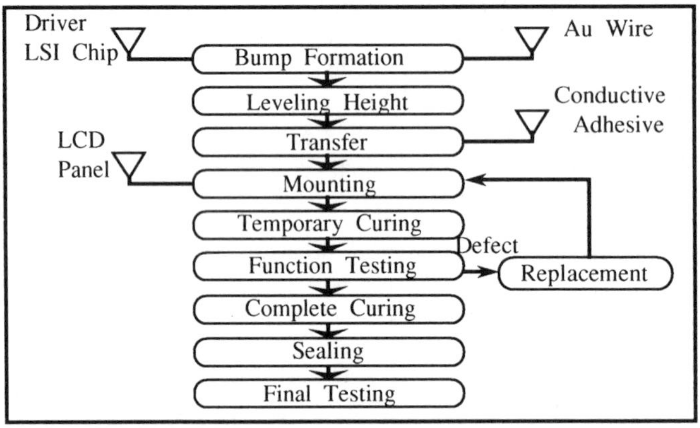

Fig. 6 Flow chart of the Stud-bump type COG mounting process

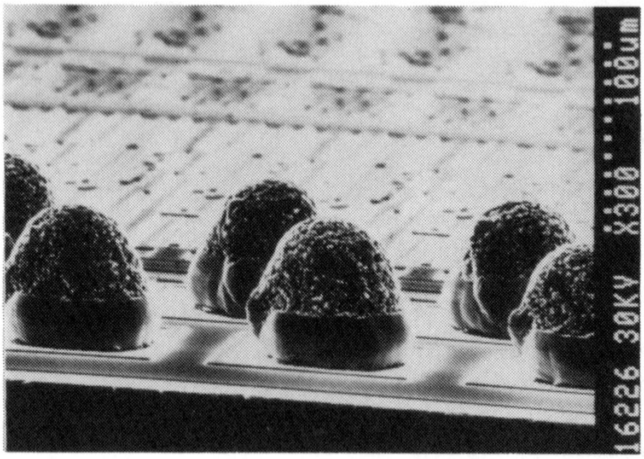

Fig. 7 SEM photograph of Stud-bumps onto which the adhesive has been transferred

COG bonding characteristics

Table 3 specifies the COG bonding characteristics in direct connection to the ITO electrode terminals by Stud-bump type COG mounting technology.

50 μm pitch electrode pads of driver LSI chip can be bonded to ITO electrode terminals on the LCD panel at the present time. In the future, miniaturization of Stud-bump will permit connection at a pitch of 10 μm.

COG contact resistance between ITO electrode terminal and Stud-bump is less than 1.0 Ω. The COG contact resistance was measured by a four point probe method. Insulation resistance between adjacent bumps is more than 10^{12} ohms after sealing by silicone resin. Adhesion strength after curing of the conductive adhesive is approx. 10 kg/cm^2.

The COG contact resistance is the most important parameter for proper movement of the driver LSI chip on the LCD panel. In the Stud-bump type COG mounting technology, the bulk resistance of the Stud-bump, conductive adhesive and ITO electrode terminal and the contact resistance at each phase boundary are included.

Especially so when directly connecting to the ITO electrode terminals, the bulk resistance of the ITO electrode terminal is dominant in the COG contact resistance. That is, ITO electrode terminals' sheet resistance needs to be decreased for lowering of the COG contact resistance. Further, COG contact resistance is proportion to contact area between the conductive adhesive and ITO electrode terminal.

Table 3 COG bonding characteristics

Bonding pitch	50 μm (min.)
Contact resistance	1.0 Ω
Insulation resistance	1.0×10^{12} Ω
Adhesion strength	10 kg/cm^2
Thermal resistance	30 ℃/W

Evaluation of reliability

Several reliability tests of COG bonding portion were carried out. The reliability test are given in Table 4. Test samples were an LSI chip for COG evaluation and an evaluation glass substrate on which ITO test pattern was fabricated.

Evaluation was carried out by the contact resistance between the ITO electrode terminals and Stud-bump and by the insulation resistance between adjacent bumps. Contact and insulation resistance were measured in sealing condition.

In both the high temperature test and the low temperature test, COG contact characteristics were stable without any change. In the high temperature and high humidity test, silicone resin sealing improved COG contact characteristics.

In the bias applying test, the field intensity was restricted to 600 V/mm or less, realizing satisfactory connection without the occurrence of Ag migration as shown in Fig. 8.

In the thermal shock test, COG contact resistance was completely stable even after 500 cycles, as shown in Fig. 9. This is because the conductive adhesive with flexibility relaxes heat stress generated by sudden temperature change in test and maintains stable bonding.

Table 4 Reliability tests of COG bonding

Test	Condition	Duration
High temperature test	125 ℃	2000 h
Low temperature test	- 40 ℃	2000 h
High temperature and high humidity test	85 ℃, 85 %	2000 h
High temperature and humidity bias test	60 ℃, 95 %, d.c.5V	2000 h
Thermal shock test	- 40 ℃ ⇔ 85 ℃	500 cyc.

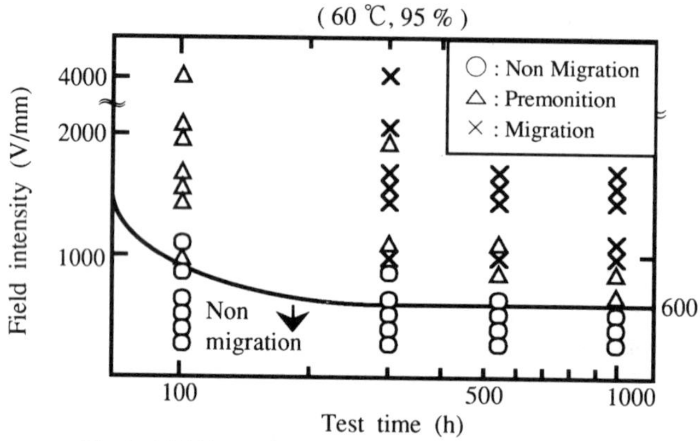

Fig. 8 Field intensity and occurrence of Ag migration

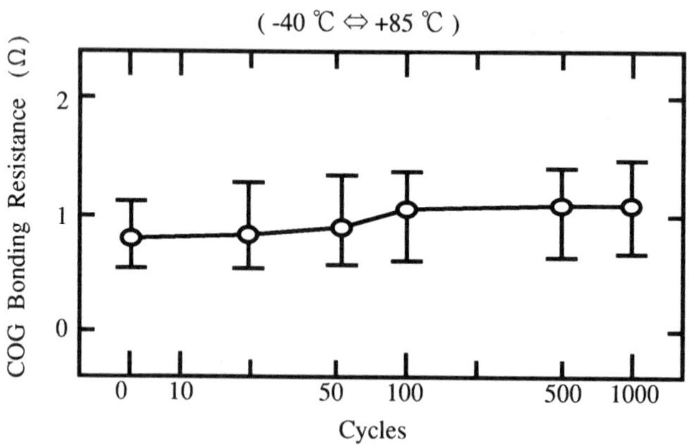

Fig. 9 Results of thermal shock test

COG MOUNTING TECHNIQUE FOR ELECTRONIC VIEWFINDER

Design concept

The color LCD panel for an electronic viewfinder is required to have a high picture quality in terms of resolution, color reproducibility' gradation characteristics and response. Small size and light weight are also indispensable requisites. In addition, considering its operating environment, it must have low power consumption, high vibration resistance and superior temperature characteristics.

Fig. 10 shows the present color LCD panel for electronic viewfinder. The basic components include deflecting plates, a liquid crystal cell, driver LSI chips, and a flexible printed circuit board (FPC). Driver LSI chips are bonded directly onto the TFT array substrate by the COG mounting method, with a bus line formed under each LSI chip to supply signals to the chip.

The COG mounting technique has been adopted for the LCD panel, aiming at substantial increase in the pixel density (243 pixels/mm^2).

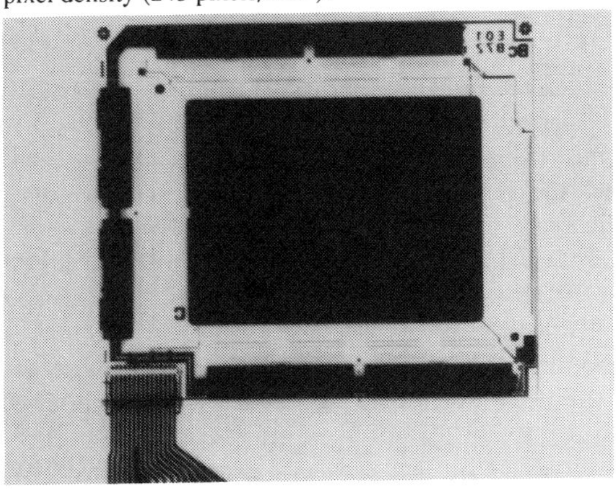

Fig. 10 Color LCD panel for electronic viewfinder

Driver LSI chips and bus lines

In the COG mounting technology, the shape and arrangement of driver LSI chips, and the layout of bus lines for supplying power and signals to the LSI chips are important factors for miniaturization of an LCD panel. The LCD panel we developed has been miniaturized by arranging three rectangular driver LSI chips along three sides of the LCD panel, as shown in Fig. 10, and bus lines under these LSI chips.

Fig. 11 shows an equivalent circuit of the driver block of a color LCD panel. The gate lines of TFT are connected with two X-driver LSI chips for vertical scanning, and the source lines connected with four Y-driver LSI chips for horizontal scanning. The COG mounting technique was used for these connections.

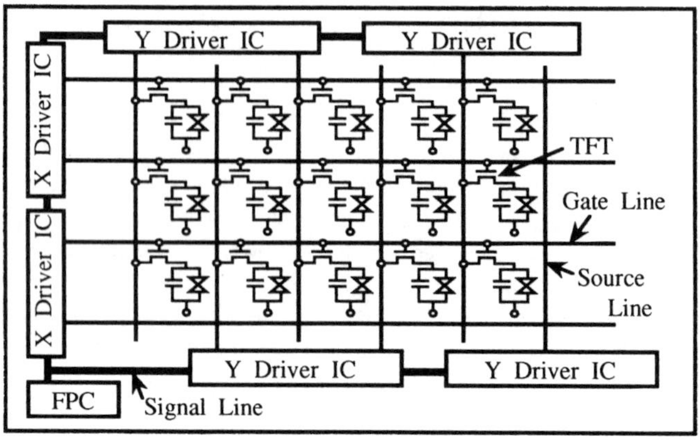

Fig. 11 Equivalent circuit of LCD panel

Basic characteristics

Table 5 shows the various characteristics of a color LCD panel obtained by the present COG mounting technology.

An electronic viewfinder normally presents date, counting number, and letters including "kanji" on the screen. The present LCD panel provides a high picture quality with high legibility of these characters.

Thus, the miniaturized color LCD panel with the same resolution as that of a CRT, manufactured by the COG mounting technique, has made it possible to realize a new, practical electronic viewfinder which can replace the CRT.

The color LCD panel attained by the COG mounting technique is connected, through the FPC on the panel corner, to the external driving chassis in an electronic viewfinder for a video camera.

This LCD panel is already used in our video movie camera "Panasonic (NV-S2 and S9)", and enjoys a high reputation among users.

Table 5 Basic characteristics of LCD panel

Display area	16.3 mm × 21.6 mm (Diagonal 1.1 inch)
Number of pixels	230 (V) × 372 (H) = 85,560
Horizontal resolution	230 TV lines
Contrast ratio	65 : 1
Outside dimensions	31.0 (V) × 36.6 (H) × 3.0(T) mm
Weight	8 g

Application to Next Generation LCD Panels

The COG mounting technique is effective and indispensable in realizing a high-definition LCD panel of the next generation.

We have applied the Stud-bump type COG mounting technique to 20-inch multicolor STN LCD panels[3] for EWS (Engineering Work Station), 3.3-inch ferroelectric LCD panels[4] for projection display, and 2.8-inch TFT LCD panels[5] for projection ED-TV. As a result, the Stud-bump type COG mounting technique has proved to be extremely effective in realizing a miniature, high-definition LCD panel of the next generation.

As an application example in which the effect of the COG mounting technique is the most prominent, Fig. 12 shows an LCD panel for high-definition projection TV.[6] This LCD panel[7] is 2.26-inch in the diagonal size, and contains a-Si TFT array with 960 × 1,422 pixels. Twenty-six driver LSI chips are mounted at a minimum pitch of 60 μm by the COG bonding method.

Fig. 12 LCD panel for high-definition projection TV

CONCLUSION

As described above, the COG mounting technique is the most effective in producing miniature, high-resolution LCD panels for electronic viewfinders and other purposes.

In the future, the most suitable mounting method will be selected for each type of LCD panel. For example, p-Si structure integrated with the driver circuit is selected for 1-inch or smaller LCD modules, and the TAB mounting method for VGA LCD modules with medium resolution.

We will continue our efforts to improve the Stud-bump type COG mounting technique, and to make the best use of the advantages of the COG mounting technique in manufacturing LCD panels.

ACKNOWLEDGMENT

We would like to thank Dr. Nagasawa, the head of Components & Devices Research Center of Matsushita, Mr. Kuchiba, the manager of Liquid Crystal Display SUM, and Mr. Muraida, TFT Plant manager, for their kind and appropriate suggestions and advice for the present study. We also thank all other persons who participated in this study.

REFERENCES

[1] Y.Bessho, et al., "Chip-On-Glass Mounting Technology of LSIs for LCD Module", IMC '90, pp.183-189 (1990)

[2] Y.Yoshida, "PACKAGING TECHNOLOGY FOR FLATPANEL DISPLAYS", SID '86, 5.2 (1986)

[3] H.Yamazoe, et al., "A 20-in.-Diagonal Multicolor Liquid Crystal Display", Japan Display '89, pp.536-539 (1989)

[4] Y.Iwai, et al., "Multi-color, High Resolution Display with Ferroelectric Liquid-Crystal Light-Valves with 4M Pixels", Japan Display '89, pp.180-183 (1989)

[5] I.Kobayashi, et al., "2.8" Defect-Free a-Si TFT-LCD module using Stud-Bump-Bonding COG for Projection ED-TV", EURODISPLAY '90, pp.48-51 (1990)

[6] H.Noda, et al., "High Definition Liquid Crystal Projection TV", Japan Display '89, pp.256-259 (1989)

[7] Y.Takubo, et al., "High Density Reflective Type TFT Array for High Definition Liquid Crystal Projection TV System", Japan Display '89, pp.584-587 (1989)

INSULATION RESISTANCE CHARACTERISTICS AND FAILURE STUDIES OF METAL-FILLED Z-AXIS ADHESIVE INTERCONNECTIONS

D.D. Chang and H.C. Ling
AT&T Bell Laboratories, Princeton, New Jersey

1. INTRODUCTION

Metal-filled z-axis conductive adhesive is an innovative interconnection technology which offers many advantages over soldering. One of the advantages is its capability to interconnect very fine pitch (< 10 mil) devices onto printed circuit board without causing electrical shorts between adjacent leads. The material contains randomly dispersed conductive particles (about 10μm) within a polymer resin. When applying heat and pressure, the resin melts, flows, allowing the conductive particles to contact two opposing leads (or pads) and achieving z direction electrical continuity. The isolation between adjacent leads, in the xy plane, is maintained by the cured polymer resin and a particle concentration below the percolation limit.

In consumer products, z-axis materials have already been used for board-to-board interconnections and interconnection to flat panel displays, where the pitch is larger than 10 mil. The success of coarse pitch applications has increased interests in fine pitch applications, such as flip-chip interconnection. The technical feasibility of flip-chip application has been demonstrated by AT&T.[1] [2] In order to develop this technology into a reliable interconnection technology, we decided to investigate its failure mechanisms. Since metal migration between neighboring leads is always a concern, especially for fine pitch applications, our initial focus was aiming at the conduction and failure characteristics between adjacent leads. Results from this study can also allow us to establish application and testing limitations. This paper describes the test condition, failure developments, and the causes of failures in metal-filled z-axis adhesive interconnections under high electric fields (10^4 to 10^5 V/cm).

2. MATERIALS

The materials tested were thermoset films made by various vendors. The thickness of the films ranges from 1 to 2 mil. The embedded particle sizes were analyzed by SEM; and their compositions were examined by EDX. Table 1 lists the material analysis results. According to manufacturers' specifications, materials A, B, C and D are all designed for fine pitch applications. For the purpose of experimental control, we have included a coarse pitch material, E, from another vendor. Material E has gold coated nickel particles and their sizes are in the range of 16 to 20 μm, and its loading density is very low. The described particle loading concentrations are based on qualitative, visual inspection of assembled test coupons. An example of the particle loading density comparison is shown in Figure 1.

Table 1. Material Analysis

Material	Conductive Particles		
	Type	Size	Loading
A	Ag on Ni	M	High
B	Ag on Glass	M	High
C	Ni on Poly	M	Median
D	Au on Ni	S	Median
E†	Au on Ni	L	Low
L: 16-20µm; M: 8-12µm; S: 5-8µm †: coarse pitch (20 mil) application			

3. TEST VEHICLE AND ASSEMBLY CONDITION

Each test assembly was composed of a sheet of z-axis material film, sandwiched between a quartz chip and a FR4 substrate. The substrate contains six interdigitated conductors of 320 mil length. The spacing between conductors is 4 mil. The conductors consisted of 1 oz copper with 100 µ-inch of nickel as the barrier layer, with a 20 µ-inch electroplated gold as the top layer. The interdigitated conductors were configured as three pairs of anode and cathode, which gives a total electrically stressed area of 6400 mil square (4x5x320). A transparent chip, made from quartz, was cut to a size of 320 mil square. A film material was then placed between a chip and a substrate under 180°C and 100 psi for 2 minutes. The two minutes of cure time was longer than the manufacturers' recommendation of 6 to 20 seconds. This is to ensure the materials were fully cured. Figure 2 shows a test coupon assembly and a close view of interdigitated conductors.

4. EXPERIMENTAL PROCEDURES

An outline of the experimental procedures is shown in Table 2. After the test coupons were assembled, they were subjected to an initial leakage current (I) versus voltage (V) measurement from 0 to 100 V. Each I-V measurement contains a ramped voltage test, from 0 V to 100 V then drops to 0 V, in 40 steps. Leakage current was measured by a picoammeter at each voltage step. The purpose of the I-V characterization was to study the leakage current behavior and to ensure that the concentration of the conductive particles did not exceed the percolation density, which could causes electrical shorting between anode and cathode conductors. [3] All test coupons had to pass this measurement before being subjected to the environmental stress tests. The environmental stress tests were: (1) temperature humidity and bias (THB) test, and (2) temperature and bias (TB) test. The THB test condition was 85°C/85%RH with 100 volts bias across the interdigitated conductors; and the TB test condition was 85°C and 100 volts, with a very low level of relative humidity. The materials between the conductors were under a high electric field; therefore any propensity of dielectric degradation or metal migration would be observed. This allowed us to differentiate the reliability performance of various vendors' materials in a short period of time.

IR was measured in-situ. An increase of three orders of magnitude in the leakage current will stop the test. Post-test I-V characterization was performed before failure mode analysis, which consisted of optical and electron microscopic examination. The coupons were inspected under an optical microscope to see if there was corrosion or metal migration. If metal migration was found, the quartz chip was carefully removed and SEM/EDX was used to identify the metallic elements.

5. RESULTS AND ANALYSES

5.1 I-V Characteristics

Initial I-V characteristics of all samples tested showed very low leakage currents. During the THB testing, some samples showed leakage currents increase substantially. These samples were removed from the test chamber, and followed by a second I-V measurement. A comparison of I-V characteristics before and after THB test is shown in Figure 3(a). Before the THB test, the leakage current increased with the applied voltage, saturating at 10^{-10} amperes at voltages above 30-40 volts. After the THB test, the leakage current is three orders of magnitude higher. A hysteresis loop also developed. The leakage current increased gradually at low voltages, then increased abruptly when the voltage exceeded 45 V. This phenomenon may be due to electric field induced charge emissions across weak links in the polymer dielectrics. We believe that the polymer's dielectric strength has been reduced by the THB stress; while the dielectric strength recovered temporarily when the sample was removed from the test chamber, charge emission initiated again when the voltage exceeded a threshold of 45 V in ramping up. On ramping down, charge emission started to decrease at a lower voltage of about 20 V. Figures 3b and 3c show that the hysteresis loop gradually closed after repeated I-V measurements, indicating the leakage current had reached a steady state condition on cycling.

5.2 Electric Field Effects

Electric field stress plays a major role in the previously described I-V characteristics. For an applied voltage of 100 V across a 4 mil conductor spacing, the average uniform electric field stress is 9.8 kV/cm (100 V/4 mil). However, the actual field stress within the adhesives is greater than 9.8 kV/cm because of the presence of micron-size conductive particles. From electromagnetic theory, the electric field enhancement factor, δ, at the tip of an ellipsoidal conductive protrusion on one of a pair of flat parallel electrodes is described by the modified Larmor equation:[4] [5]

$$\delta = \frac{E_{max}}{E_{avg}} \sim \frac{2\,(\frac{2h}{r}-1)^{1.5}}{(\frac{2h}{r})^{.5}\,\ln(\frac{4h}{r}+2(\frac{h}{r}+(\frac{4h}{r}-2))^{.5}-1)-2(\frac{2h}{r}-1)^{.5}} \qquad (1)$$

where E_{max} is the electric field stress at the tip of the ellipsoid, E_{avg} is the uniform field stress in the absence of the protrusion, h is the height of the protrusion and r is the radius of the tip of the protrusion. This equation is valid only for ellipsoidal protrusions. For a h/r ratio of 8, the enhancement factor is 15.9. For a hemispherical protrusions where r˜=˜h, the stress concentration is equal to 3.[5]

Within the polymer matrix containing a random distribution of particles, complex electric field stress distribution existed. Visual inspection showed that in some areas the particles formed a conductive chain, linking to the electrodes. In other areas they were isolated and formed clusters. Thus, it is reasonable to assume that there was an electric field enhancement between a factor of three to sixteen around the particles and protrusions. Localized field stress intensifications are also expected around dielectric cavities, where a change in the dielectric constant occurs. With an E_{avg} of about 10^4 V/cm, the actual field can exceed 10^5 v/cm around some particles and protrusions. Such electric field stress is approaching the breakdown strength of the polymer matrix, depending on the purity of the polymer resins.

5.3 Observations and Failure Mode Analyses

Table 2 summarizes the test results and microscopic observations.

After 170 hours of THB testing, the test coupons assembly of materials A and B were examined; we found strong anodic corrosion, but no metal migrations. The extent of corrosion is shown in Figure 4. All anode electrodes exhibit brownish black colors. After I-V measurements, THB test of these samples were resumed. We found metallic dendrites growing between 200 to 1000 hours. The dendrites were identified by EDX to be silver. No nickel or gold elements were identified. The corrosion and migration processes are due to the silver from the conductive particles. Under electric field stress, silver atoms from particle attached to the anode easily dissociated into silver ions, and the Ag^+ ions then migrated to cathode and deposited to form the roots of the silver dendritic tree. The silver ions also reacted with $(OH)^-$ and formed the dark brownish silve oxides. The reaction equations are:

$$Ag \rightarrow Ag^+ + e^- \qquad (2)$$

$$2Ag^+ + 2\,(OH)^- \rightarrow Ag_2O\downarrow + H_2O \qquad (3)$$

Table 2. Results and Analyses

Material	A	B	C	D
Size Loading Type	M High Ag/Ni	M High Ag/Glass	M Median Ni/Poly	S Median Au/Ni
Observations:				
• Corrosion	Strong	Strong	Slight	Median
• Migration	No @170 hrs Yes @200-1000 hrs	No @170 hrs Yes @200-1000 hrs	No @90 hrs Minor @193 hrs	No @90 hrs Yes @193 hrs
• e-Emissions	Yes	Yes	Yes	Yes
• Dendrites	Ag	Ag	Cu, Ni	Cu
• Sources	Particles	Particles	Electrodes Particles	Electrodes
Comment	gold dissolution from anode electrode			

Figure 5 illustrates the corrosion and dendritic growth processes observed in one of the test samples. Corrosion was initiated at a few spots on the anode and one spot at the cathode. At the end of silver migration, a silver dendritic tree had formed with the root anchored at the cathode and several branches linking the anode. We believe the high electric field applied in the test ($> 10^5$ V/cm) was the driving failure mechanism. Similar extensive Ag migration had been observed between thick-film silver electrodes fired on ceramics.[6]

The electric field induced metal migrations were also seen on test coupons made from material C and D. However, the migrating element was identified by EDX to be copper. They are shown in Figure 6. The copper came from the electrodes, since that was the only source available. We did not detect any Cu migration in test samples made from material A and B. This is because it takes less energy to dissociate electron from silver than copper, and the mobility of silver ion is greater than copper ion, so silver becomes the dominant dendritic structure. The formation of copper dendrites was concentrated around conductive particles and dielectric cavities, where the electric field stresses were heavily concentrated. The anodic corrosion phenomenon of material C and D was minor , comparing to A and B. The different behavior may be due to the different organic materials constituting the polymer resins.

The electric field presented in material E is probably the lowest due to the large particle size and low particle density. We did not observe IR degradation or find any traces of metal migration and IR degradation in material E. Since it has the largest particle size and the lowest particle concentration, there is minimal electric field enhancement in the polymer matrix. This further substantiated our conjecture that the enhanced electric field effect in particle containing polymer matrix is the main degradation mechanism.

In some of the samples, we also observed dissolution of Au from the anodes. Figure 7(a) shows the SEM of three interdigitated electrodes, with the two exterior anodes losing Au near the edges and revealing the Ni layer below. (Figure 7(b) EDX). We have also observed Ni dendrites growing from embedded Ni particles (Figure 8). Thus, a variety of metals such as Ag, Cu, Au and Ni have been shown to migrate in metal-filled conductive adhesives under high electric fields in the THB condition.

6. CONCLUSIONS

(1) Under an applied field of 100V across 4-mil spaced electrodes in TB and THB tests, we have observed IR degradation of more than three orders of magnitude in vendor-supplied z-axis conductive adhesives claimed to be qualified for fine pitch interconnections.

(2) At various stages of the TB and THB tests, we also identified metallic elements that had either corroded (Ag), migrated to form dendritic structures (Ag, Cu, Ni) or dissolved into the polymer resin (Au).

(3) We proposed an enhanced electric field around the particles and cavities in the polymer matrix as the driving mechanism for the corrosion, migration and dissolution of metallic elements from the metal particles or from the electrodes.

7. ACKNOWLEDGEMENTS

We wish to thank the following people for this study: J.A. Fulton, M.B. Schmidt ,R.E. Sinitski and E.L. Smith. Especially, we would like to thank R.E. Woods for his dedication in SEM and EDX analyses.

References

1. N.R. Basavanhally, D.D. Chang, B.H. Cranston, "Direct Chip Interconnect with Adhesive-Connector Films," Pro. of Electronic Components & Technology Conference, May 18-20, 1992, pp. 487-491.

2. D.D. Chang, J.A. fulton, A.M. Lyons, J.R. Nis, "Design Considerations for the Implementation of Anisotropic Conductive Adhesive Interconnection," Proc. of 92th Nepon West Conference, February 25-27, 1992, pp. 1381-1389.

3. G.R. Ruschau, S. Yoshikawa and R.E. Newnham, "Percolation Constraints in the Use of Conductor-Filled Polymers," IEEE Pro. of 42nd ECTC Conf., Vol. 1, pp. 481-486. 1992.

4. J. Larmor and J.S.B. Larmor, Pro. Royal Soc. Series A, Vol. XC, pp. 312-318, 1914.

5. G. Bahder, G.S. Eager, D.A. Silver and R.G. Lukac, "Criteria for Determining Performance of Cross-linked Polyethylene Insulated Power Cables," IEEE Trans. Pwr. App. 7 systems PAS-95, No. 5, pp. 1552-1566(1976).

6. H.C. Ling and A.M. Jackson, "Correlation of Silver Migration with THB Failures in MLC," IEEE Trans. on Components, Hybrids & Manufacturing Technology, Vol. 12, pp. 130-137(1989).

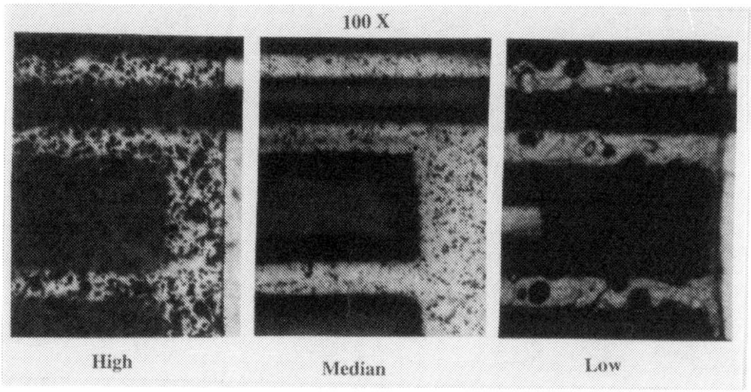

Figure 1. Comparison of conductive particle loading densities

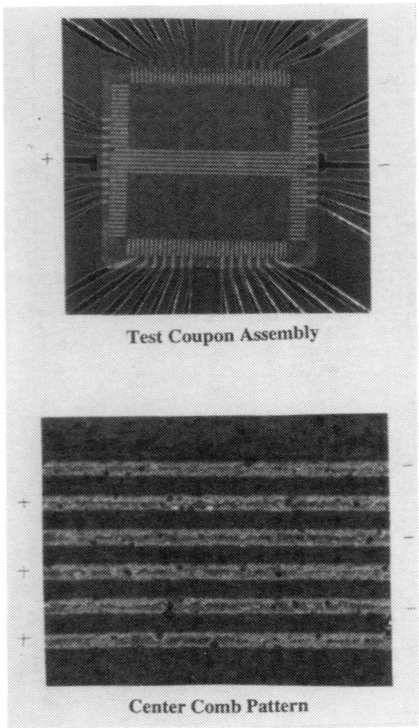

Figure 2. Test coupon assembly and interdigited center conductors

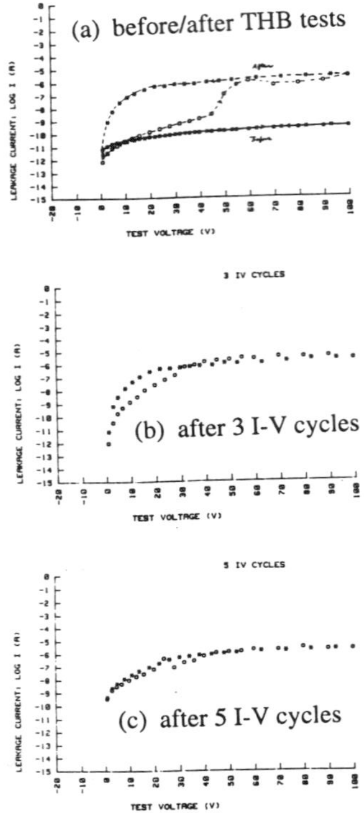

Figure 3. I-V characteristics:

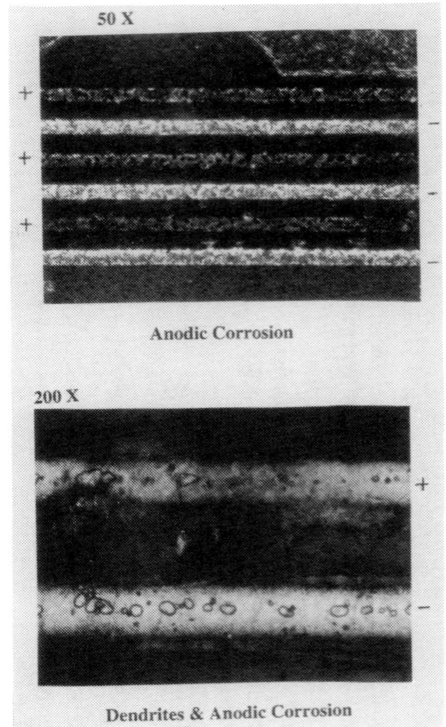

Figure 4. Examples of anodic corrosions and dendrites.

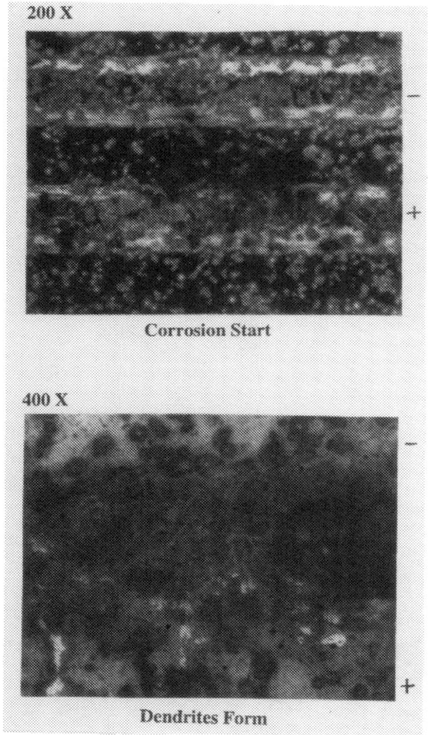

Figure 5. Silver Corrosion and dendritic growth processes.

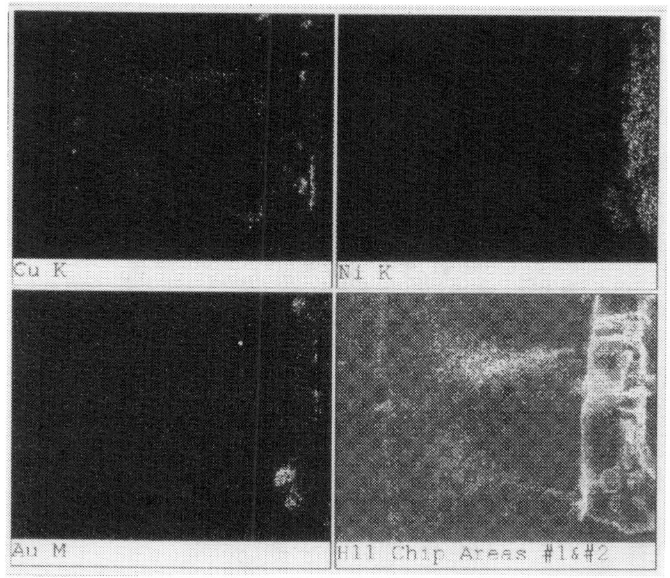

Figure 6. SEM photos of copper dendrites grown between the anode and cathode

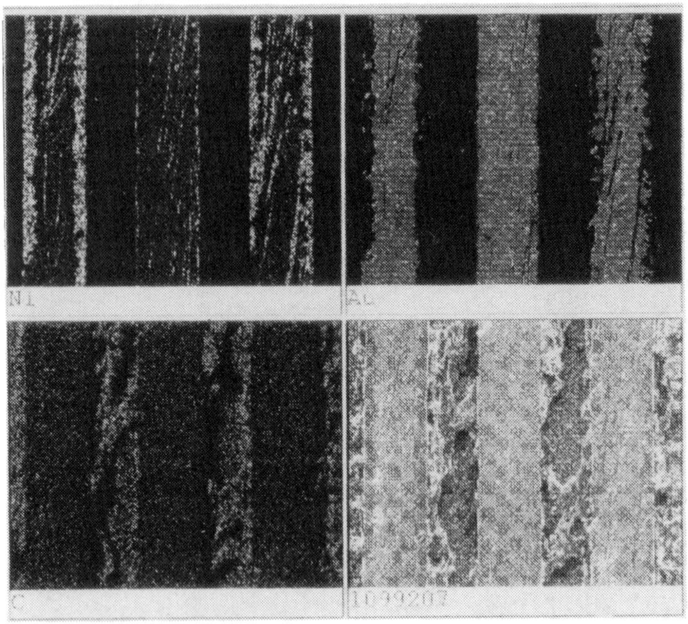

Figure 7. SEM photos of gold dissolution from anodes

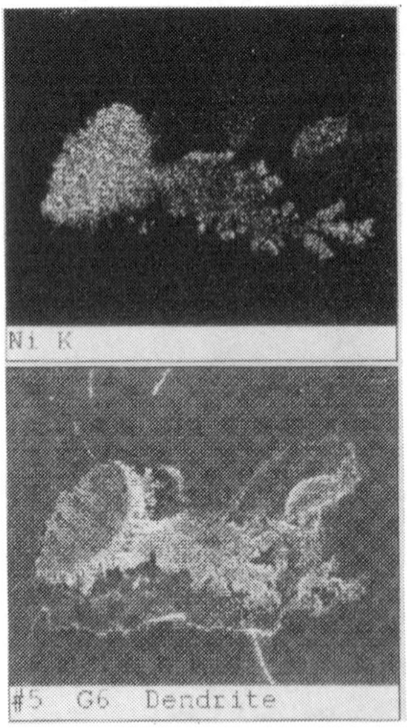

Figure 8. (a) SEM photos of nickel dendrites grown from a nickel coated polymer particle

INDEX